D1036959

The Handbook
of Environmental Chemistry

Volume 3   Anthropogenic Compounds
Part M

O. Hutzinger
Editor-in-Chief

Springer

Berlin
Heidelberg
New York
Barcelona
Hong Kong
London
Milan
Paris
Tokyo

Endocrine Disruptors
Part II

Volume Editor: M. Metzler

With contributions by
M. C. Bosland · T.-Y. Chang · L. E. Gray, Jr.
L. J. Guillette, Jr. · A. Hotchkiss · Y. F. Hu · C. Lambright
G. W. Lucier · M. Metzler · R. R. Newbold
E. F. Orlando · J. S. Ostby · L. Parks · C. J. Portier
I. H. Russo · J. Russo · F. S. vom Saal · J. C. Seely
J. P. Sumpter · S. H. Swan · D. M. Tham · R. W. Tyl
B. A. T. Willems · V. Wilson · C. J. Wolf

Springer

Environmental chemistry is a rather young and interdisciplinary field of science. Its aim is a complete description of the environment and of transformations occurring on a local or global scale. Environmental chemistry also gives an account of the impact of man's activities on the natural environment by describing observed changes.

"The Handbook of Environmental Chemistry" provides the compilation of today's knowledge. Contributions are written by leading experts with practical experience in their fields. The Handbook will grow with the increase in our scientific understanding and should provide a valuable source not only for scientists, but also for environmental managers and decision makers.

As a rule, contributions are specially commissioned. The editors and publishers will, however always be pleased to receive suggestions and supplements information. Papers for *The Handbook of Environmental Chemistry* are accepted in English.

In reference The Handbook of Environmental Chemistry is abbreviated Handb. Environ. Chem. and is cited as a journal.

Springer WWW home page: http://www.springer.de
Visit the HEC home page at http://link.springer.de/series/hec/
or http://link.springer-ny.com/series/hec/

ISSN 1433-6847
ISBN 3-540-42280-3
Springer-Verlag Berlin Heidelberg New York

Library of Congress Cataloging-in-Publication Data
The Natural environment and the biogeochemical cycles / with
contributions by P. Craig ... [et al.].
v. <A–F > : ill. ; 25 cm. – (The Handbook of environmental chemistry :
v. 1) Includes bibliographical references and indexes.
ISBN 0-387-09688-4 (U.S). – ISBN 3-540-55255-3 (pt. F : Berlin). –
ISBN 0-387-55255-3 (pt. F : New York)
1. Biogeochemical cycles. 2. Environmental chemistry.
I. Craig, P. J., 1944– . II. Series.
QD31. H335 vol. 1   [QH344]   628.5 s

Springer-Verlag Berlin Heidelberg New York
a member of BertelsmannSpringer Science+Business Media GmbH
http//www.springer.de
© Springer-Verlag Berlin Heidelberg 2002
Printed in Germany

The use of general descriptive names, registered names, trademarks, etc. in this publication does not imply, even in the absence of a specific statement, that such names are exempt from the relevant protective laws and regulations and therefore free for general use.

Product liability: The publisher cannot guarantee the accuracy of any information about dosage and application contained in this book. In every individual case the user must check such information by consulting the relevant literature.

The instructions given for the practical carrying-out of HPLC steps and preparatory investigations do not absolve the reader from being responsible for safety precautions. Liability is not accepted by the author.

Production Editor: Christiane Messerschmidt, Rheinau
Cover Design: E. Kirchner, Springer-Verlag
Typesetting: Fotosatz-Service Köhler GmbH, Würzburg
Printed on acid-free paper   SPIN: 10833691   52/3020 – 5 4 3 2 1 0

The Handbook of Environmental Chemistry
Now Also Available Electronically

For all customers with a standing order for The Handbook of Environmental Chemistry we offer the electronic form via LINK free of charge. Please contact your librarian who can receive a password for free access to the full articles. By registration at:

http://link.springer.de/series/hec/reg_form.htm

However, if you do not have a standing order, you can browse through the table of contents of the volumes and the abstracts of each article at:

http://link.springer.de/series/hec/
http://link.springer-ny.com/series/hec/

There you will also find information about the

– Editorial Bord
– Aims and Scope
– Instructions for Authors

Preface

Environmental Chemistry is a relatively young science. Interest in this subject, however, is growing very rapidly and, although no agreement has been reached as yet about the exact content and limits of this interdisciplinary discipline, there appears to be increasing interest in seeing environmental topics which are based on chemistry embodied in this subject. One of the first objectives of Environmental Chemistry must be the study of the environment and of natural chemical processes which occur in the environment. A major purpose of this series on Environmental Chemistry, therefore, is to present a reasonably uniform view of various aspects of the chemistry of the environment and chemical reactions occurring in the environment.

The industrial activities of man have given a new dimension to Environmental Chemistry. We have now synthesized and described over five million chemical compounds and chemical industry produces about hundred and fifty million tons of synthetic chemicals annually. We ship billions of tons of oil per year and through mining operations and other geophysical modifications, large quantities of inorganic and organic materials are released from their natural deposits. Cities and metropolitan areas of up to 15 million inhabitants produce large quantities of waste in relatively small and confined areas. Much of the chemical products and waste products of modern society are released into the environment either during production, storage, transport, use or ultimate disposal. These released materials participate in natural cycles and reactions and frequently lead to interference and disturbance of natural systems.

Environmental Chemistry is concerned with reactions in the environment. It is about distribution and equilibria between environmental compartments. It is about reactions, pathways, thermodynamics and kinetics. An important purpose of this Handbook is to aid understanding of the basic distribution and chemical reaction processes which occur in the environment.

Laws regulating toxic substances in various countries are designed to assess and control risk of chemicals to man and his environment. Science can contribute in two areas to this assessment; firstly in the area of toxicology and secondly in the area of chemical exposure. The available concentration ("environmental exposure concentration") depends on the fate of chemical compounds in the environment and thus their distribution and reaction behaviour in the environment. One very important contribution of Environmental Chemistry to the above mentioned toxic substances laws is to develop laboratory test methods, or mathematical correlations and models that predict the environmental fate of new chemical compounds. The third purpose of this Handbook is to help in the basic understanding and development of such test methods and models.

The last explicit purpose of the Handbook is to present, in concise form, the most important properties relating to environmental chemistry and hazard assessment for the most important series of chemical compounds.

At the moment three volumes of the Handbook are planned. Volume 1 deals with the natural environment and the biogeochemical cycles therein, including some background information such as energetics and ecology. Volume 2 is concerned with reactions and processes in the environment and deals with physical factors such as transport and adsorption, and chemical, photochemical and biochemical reactions in the environment, as well as some aspects of pharmacokinetics and metabolism within organisms. Volume 3 deals with anthropogenic compounds, their chemical backgrounds, production methods and information about their use, their environmental behaviour, analytical methodology and some important aspects of their toxic effects. The material for volume 1, 2 and 3 was each more than could easily be fitted into a single volume, and for this reason, as well as for the purpose of rapid publication of available manuscripts, all three volumes were divided in the parts A and B. Part A of all three volumes is now being published and the second part of each of these volumes should appear about six months thereafter. Publisher and editor hope to keep materials of the volumes one to three up to date and to extend coverage in the subject areas by publishing further parts in the future. Plans also exist for volumes dealing with different subject matter such as analysis, chemical technology and toxicology, and readers are encouraged to offer suggestions and advice as to future editions of "The Handbook of Environmental Chemistry".

Most chapters in the Handbook are written to a fairly advanced level and should be of interest to the graduate student and practising scientist. I also hope that the subject matter treated will be of interest to people outside chemistry and to scientists in industry as well as government and regulatory bodies. It would be very satisfying for me to see the books used as a basis for developing graduate courses in Environmental Chemistry.

Due to the breadth of the subject matter, it was not easy to edit this Handbook. Specialists had to be found in quite different areas of science who were willing to contribute a chapter within the prescribed schedule. It is with great satisfaction that I thank all 52 authors from 8 countries for their understanding and for devoting their time to this effort. Special thanks are due to Dr. F. Boschke of Springer for his advice and discussions throughout all stages of preparation of the Handbook. Mrs. A. Heinrich of Springer has significantly contributed to the technical development of the book through her conscientious and efficient work. Finally I like to thank my family, students and colleagues for being so patient with me during several critical phases of preparation for the Handbook, and to some colleagues and the secretaries for technical help.

I consider it a privilege to see my chosen subject grow. My interest in Environmental Chemistry dates back to my early college days in Vienna. I received significant impulses during my postdoctoral period at the University of California and my interest slowly developed during my time with the National Research

Council of Canada, before I could devote my full time of Environmental Chemistry, here in Amsterdam. I hope this Handbook may help deepen the interest of other scientists in this subject.

Amsterdam, May 1980   *O. Hutzinger*

Twentyone years have now passed since the appearance of the first volumes of the Handbook. Although the basic concept has remained the same changes and adjustments were necessary.

Some years ago publishers and editors agreed to expand the Handbook by two new open-end volume series: Air Pollution and Water Pollution. These broad topics could not be fitted easily into the headings of the first three volumes. All five volume series are integrated through the choice of topics and by a system of cross referencing.

The outline of the Handbook is thus as follows:

1. The Natural Environment and the Biochemical Cycles,
2. Reaction and Processes,
3. Anthropogenic Compounds,
4. Air Pollution,
5. Water Pollution.

Rapid developments in Environmental Chemistry and the increasing breadth of the subject matter covered made it necessary to establish volume-editors. Each subject is now supervised by specialists in their respective fields.

A recent development is the accessibility of all new volumes of the Handbook from 1990 onwards, available via the Springer Homepage http://www.springer. de or http://Link.springer.de/series/hec/ or http://Link.springer-ny.com/ series/hec/.

During the last 5 to 10 years there was a growing tendency to include subject matters of societal relevance into a broad view of Environmental Chemistry. Topics include LCA (Life Cycle Analysis), Environmental Management, Sustainable Development and others. Whilst these topics are of great importance for the development and acceptance of Environmental Chemistry Publishers and Editors have decided to keep the Handbook essentially a source of information on "hard sciences".

With books in press and in preparation we have now well over 40 volumes available. Authors, volume-editors and editor-in-chief are rewarded by the broad acceptance of the "Handbook" in the scientific community.

Bayreuth, July 2001   *Otto Hutzinger*

Contents

Foreword XIII

Estrogen and Human Breast Cancer
Y. F. Hu, I. H. Russo, J. Russo 1

The Role of Sex Hormones in Prostate Cancer
M. C. Bosland 27

Beneficial and Adverse Effects of Dietary Estrogens on the Human Endocrine System: Clinical and Epidemiological Data
D. M. Tham 69

Mechanism-Based Carcinogenic Risk Assessment of Estrogens and Estrogen-Like Compounds
B. A. T. Willems, C. J. Portier, G. W. Lucier 109

Alterations in Male Reproductive Development: The Role of Endocrine Disrupting Chemicals
S. H. Swan, F. S. vom Saal 131

Effects of Perinatal Estrogen Exposure on Fertility and Cancer in Mice
R. R. Newbold 171

Genotoxic Potential of Natural and Synthetic Endocrine Active Compounds
M. Metzler 187

Emerging Issues Related to Endocrine Disrupting Chemicals and Environmental Androgens and Antiandrogens
L. E. Gray Jr., C. Lambright, L. Parks, R. W. Tyl, E. F. Orlando, L. J. Guillette Jr., C. J. Wolf, J. C. Seely, T.-Y. Chang, V. Wilson, A. Hotchkiss, J. S. Ostby 209

Developmental and Reproductive Abnormalities Associated with Endocrine Disruptors in Wildlife
E. F. Orlando, L. J. Guillette Jr. 249

Endocrine Disruption in the Aquatic Environment
J. P. Sumpter 271

Subject Index of Part I and Part II 291

Contents of Part I

Mechanisms of Estrogen Receptor-Mediated Agonistic
and Antagonistic Effects
S. O. Mueller, K. S. Korach

In Vitro Methods for Characterizing Chemical Interactions
with Steroid Hormone Receptors
K. W. Gaido, D. P. McDonnell, S. Safe

Antiandrogenic Effects of Environmental Endocrine Disruptors
W. R. Kelce, E. M. Wilson

Chemistry of Natural and Anthropogenic Endocrine Active Compounds
M. Metzler, E. Pfeiffer

Exposure to Endogenous Estrogens During Lifetime
J. Dötsch, H. G. Dörr, L. Wildt

Dietary Estrogens of Plant and Fungal Origin: Occurrence and Exposure
W. E. Ward, L. U. Thompson

Alkylphenols and Bisphenol A as Environmental Estrogens
C. M. Markey, C. L. Michaelson, C. Sonnenschein, A. M. Soto

Hydroxylated Polychlorinated Biphenyls (PCBs) and Organochlorine
Pesticides as Potential Endocrine Disruptors
S. Safe

The Endocrine Disrupting Potential of Phthalates
C. A. Harris, J. P. Sumpter

Foreword

Endocrine disruptors, also called endocrine-active compounds, endocrine modulators, environmental hormones, hormone-related toxicants etc., are compounds that exhibit the potential to interfere with the endocrine system of humans and animals. The endocrine system uses endogenous hormones, i.e. compounds produced by certain glands, to communicate with various tissues and regulate body functions such as growth, development and reproduction. The group of endocrine disruptors comprises a large and still increasing number of natural and anthropogenic agents with diverse chemical structures.

Wide scientific and public interest in endocrine disruptors has evolved about ten years ago, when evidence that chemicals may adversely affect the sexual development of a number of wildlife species was presented at a Workshop convened by Theo Colburn (Proceedings: T. Colburn and C. R. Clement, eds. *Chemically-induced Alterations in Sexual and Functional Development: The Wildlife/Human Connection.* Princeton, NJ: Princeton Scientific Publishing Co. 1992). Subsequent reports, showing that numerous everyday chemicals exhibit hormonal activity, and linking male reproductive problems such as low sperm counts and increased rates of testicular cancer in young men to environmental hormones, increased the concern about endocrine disruptors. As a consequence, endocrine disruptors have become over the past years, and will continue to be over the next years, a "hot" topic at toxicological Meetings, in public media, and also in the political arena.

Although the general interest in endocrine disruptors is relatively recent, scientific interest dates back to the sixties and early seventies when the adverse effects of a synthetic estrogen, diethylstilbestrol (DES), on experimental animals and also on humans were reported for the first time. Based on these observations, a Conference was convened by John A. McLachlan in 1979, at which the basic questions of endocrine disruptors, i.e. "what an estrogen is and how it works, and what effects estrogenic substances might have on human health" (cited from the foreword of the Proceedings: John A. McLachlan, ed. *Estrogens in the Environment.* New York: Elsevier North Holland Inc. 1980) were already raised and addressed for estrogenic compounds, which still constitute the major group of endocrine disruptors. The environmental occurrence and impact of estrogenic agents of natural and man-made origin were also extensively discussed at this Meeting and at a subsequent Conference in 1985 (Proceedings: John A. McLachlan, ed. *Estrogens in the Environment II. Influences on Development.* New York: Elsevier Science Publishing Co., Inc. 1985). These two Conferences and the pioneering work of John McLachlan, Howard Bern and a few other investigators have laid the ground for the present endocrine disruptor field.

Research in this area so far has clearly shown that the answers for the two basic questions of endocrine disruptors, posed above for estrogens, will not be easy and straightforward. The structural requirements and mechanisms of action of endocrine disruptors are complicated by the fact that multiple ways exist to interfere with the endocrine system. For example, endocrine disruptors can (i) mimic or block the binding of endogenous hormones to their receptors, (ii) affect cell signaling pathways in a direct, i.e. non-receptor-mediated manner, or (iii) alter the production or metabolism of endogenous hormones. The effects on human and animal health may be even more difficult to assess due to the complexity of the endocrine system. They appear to depend on the chemical structure and dose of the individual compound, the duration of exposure, and the species, developmental stage (age) and gender of the organism. The exposure to hormonally active compounds does not necessarily lead to adverse effects, as is demonstrated by the putative anti-carcinogenic effects of certain plant estrogens in Asian populations. To sort out the adverse and beneficial effects of endocrine disruptors and the underlying mechanisms will be a challenging scientific and also an important practical task, since exposure to such compounds through food, air, water and many household products is ubiquitous and unavoidable.

The present book provides an overview on important aspects of endocrine disruptors. The handbook comprises 19 chapters and is divided into two parts. Part I addresses the mechanisms and detection of hormone action, and the chemistry of and exposure to the various classes of natural and anthropogenic endocrine disruptors. Part II focuses on the association of sex hormones with diseases in humans, on the effects of endocrine-active compounds in experimental systems, and on the association of endocrine disruptors with environmental effects.

The chapters were authored by scientists who are highly recognized in their areas of research. The editor is greatly indebted to all of them as well as to the staff of Springer for their commitment. It is hoped that the result of this joint effort may prove as an useful and inspiring source of information for everybody interested in the multifaceted issue of endocrine disruptors.

Karlsruhe, September 2001   Manfred Metzler

CHAPTER 1

Estrogen and Human Breast Cancer

Yun Fu Hu, Irma H. Russo, Jose Russo

Breast Cancer Research Laboratory, Fox Chase Cancer Center, 7701 Burholme Avenue, Philadelphia, PA 19111, USA
E-mail: J_Russo@fccc.edu

Breast cancer is the most common neoplastic disease in women worldwide. Development of breast cancer is profoundly influenced by prolonged exposure to estrogens, but the role of estrogen in the development of human breast cancer has remained elusive. There is evidence that in situ metabolism of estrogens by estrone sulfatase and aromatase results in a high level of active estrogens in the breast. Two mechanisms have been considered to be responsible for the potential carcinogenicity of estrogens: receptor-mediated hormonal activity and cytochrome P450-mediated metabolic activation. The receptor-mediated hormonal activity of estrogen has generally been related to stimulation of cellular proliferation, resulting in more opportunities for accumulation of genetic damages leading to carcinogenesis. Since local synthesis of estrogen in the stromal component can increase the estrogen levels and growth rate of breast carcinoma, a paracrine mechanism is likely to account for interactions between aromatase-containing stromal cells and estrogen receptor-containing breast tumor epithelial cells. In addition, expression of the estrogen receptors occurs in cells other than the proliferating cells, suggesting that another paracrine mechanism is operative to mediate the biological response to estrogens. More importantly, estrogen may not need to activate its nuclear receptors to initiate or promote breast carcinogenesis. There is evidence that oxidative catabolism of estrogens mediated by various cytochrome P450 complexes constitutes a pathway of their metabolic activation and generates reactive free radicals and intermediate metabolites that can cause oxidative stress and genomic damage directly. Estrogen-induced genotoxic effects include increased mutation rates and compromised DNA repair system that allows accumulation of genomic lesions essential to estrogen-induced tumorigenesis. However, metabolism of estrogen in normal human breast epithelial cells is largely unclear. More importantly, the carcinogenic potential of estrogens in normal human breast epithelial cells awaits to be elucidated.

Keywords. Estrogen, Estrogen receptor, Estrogen metabolism, Breast cancer, Human

1 Introduction 2

2 Sources of Estrogens in Human Breast Tissue 6

3 Role of Estrogens in Human Breast Proliferation 9

4 Implications of Estrogens in Human Breast Carcinogenesis 12

5 Summary and Future Perspectives 19

6 References 20

The Handbook of Environmental Chemistry Vol. 3, Part M
Endocrine Disruptors, Part II
(ed. by M. Metzler)
© Springer-Verlag Berlin Heidelberg 2002

Abbreviations

ER   estrogen receptor
ERα estrogen receptor type α
ERβ estrogen receptor type β
ERE   estrogen response element
DNA   deoxyribonucleic acid
DHEA   dehydroepiandrosterone
HSD   hydroxysteroid dehydrogenase
Lob 1   lobule type 1
Lob 2   lobule type 2
Lob 3   lobule type 3
Lob 4   lobule type 4
PgR   progesterone receptor
DAB   3,3′-diaminobenzidine
CYP   cytochrome P450

1
Introduction

Breast cancer is the most common neoplastic disease in women worldwide with an incidence of 796,000 new cases and a mortality rate of 314,000 deaths annually [1]. In the United States, the incidence of breast cancer continues to rise and accounts for up to one third of all new cases of women's cancer [2]. In spite of advances in technologies for early detection and intervention of breast cancer, the mortality rate from this disease has remained almost unchanged in the past 5 decades, becoming second only to lung cancer as a cause of cancer deaths [3, 4].

Intensive epidemiological studies have identified a number of genetic risk factors associated with breast cancer, including evidence of *BRCA1* and *BRCA2* susceptibility genes, familiar history of cancer in the breast, ovary or endometrium, and individual history of breast diseases (Table 1) [5]. An increased risk has also been associated with early onset of menstruation, nulliparity or delayed first childbirth, short duration of breast feeding, late menopause, use of hormone replacement therapy and increased bone density (Table 1) [6–10].

A principal culprit common for all these endocrine-related risk factors is the prolonged exposure to female sex hormones [11–13]. The hormonal influences have been mainly attributed to unopposed exposure to elevated levels of estrogens [9, 14, 15], as has been indicated for a variety of female cancers, namely, vaginal, hepatic and cervical carcinomas [16–21]. Exposure to estrogens, particularly during the critical developmental periods (e. g., *in utero*, puberty, pregnancy, menopause), also affects affective behaviors (e.g., depression, aggression, alcohol intake) and increases breast cancer risk [22]. In addition, both environmental and genetic factors are believed to exert their influence by a hormonal mechanism [23–27].

Table 1. Risk factors for human breast cancer

	Biomarkers	Odds ratio
Genetic	Evidence of susceptibility genes *BRCA1* or *BRCA2*	≥4.0
	Evidence of p53 gene (in Li-Fraumeni syndrome)	≥4.0
	Evidence of *PTEN/MAMCI* (in Cowden syndrome)	≥4.0
	Heterozygosity for mutant alleles of *ATM* gene	≥4.0
	Premenopausal breast cancer in mother **and** sister	≥4.0
	Premenopausal breast cancer in mother **or** sister	2.0–4.0
	Postmenopausal breast cancer in first-degree relatives	≤2.0
	Individual history of cancer in one breast	2.0–4.0
	Individual history of ovarian or endometrial cancer	≤2.0
Clinical	Atypical hyperplasia in breast biopsy or aspirate	≥4.0
	Ductal or lobular carcinoma in situ	≥4.0
	Typical hyperplasia in breast biopsy or aspirate	2.0–4.0
	Predominantly nodular densities in mammogram	≤2.0
	Prolonged use of oral contraceptives in women under age 45	≤2.0
	Prolonged estrogen replacement therapy	≤2.0
Biological	Advanced age	2.0–4.0
	Early onset of menstruation (before age 12)	≤2.0
	Delayed first childbirth	≤2.0
	Nulliparity (in women under 40)	≤2.0
	Short duration of breast feeding	≤2.0
	Late onset of menopause (after age 49)	≤2.0
	Postmenopausal obesity	≤2.0
	Tallness in adult life	≤2.0
Social	Smoking	2.0–4.0
	Higher socio-economic status	≤2.0
	Low physical activity	≤2.0
Dietary	Higher alcohol consumption	≤2.0
	Higher fat/energy intake	≤2.0
	Xenobiotics	≤2.0
Environmental	Excess ionizing radiation to chest wall or breasts	≤2.0
	Exposure to chemical carcinogens	≤2.0
	Microbials or infectious agents	≤2.0

It is generally accepted that the biological activities of estrogens are mediated by nuclear estrogen receptors (ER) which, upon activation by cognate ligands, form homodimers with another ER-ligand complex and activate transcription of specific genes containing the estrogen response elements (ERE) (Fig. 1) [28].

According to this classical model, the biological responses to estrogens are mediated by the ER universally identified until recently, which has been termed as ERα after the discovery of a second type of ER (ERβ). The presence of ERα in target tissues or cells is essential to their responsiveness to estrogen action. In fact, the expression levels of ERα in a particular tissue have been used as an index of the degree of estrogen responsiveness [29]. A vast majority of human breast carcinomas are initially positive for ERα, and their growth can be stimulated by estrogens and inhibited by anti-estrogens [30–32]. The ERβ has been

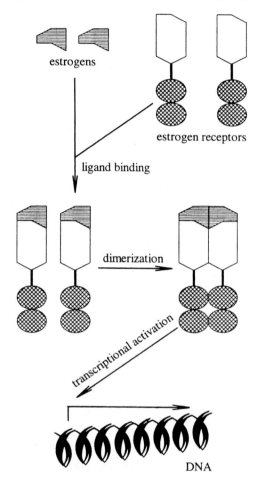

Fig. 1. Classical model of receptor-mediated estrogen action

cloned from the rat [33], mouse [34] and human [35]. ERβ and ERα share high sequence homology, especially in the regions or domains responsible for specific binding to DNA and the ligands [33–35]. ERβ can be activated by estrogen stimulation, and blocked with anti-estrogens [33, 35, 36]. Upon activation, ERβ can form homodimers as well as heterodimers with ERα [36]. The existence of two ER subtypes and their ability to form DNA-binding heterodimers suggests three potential pathways of estrogen signaling: via the ERα or ERβ subtype in tissues exclusively expressing each subtype and via the formation of heterodimers in tissues expressing both ERα and ERβ (Fig. 2) [37–40]. The pathways of the ER-mediated signal transduction have become even more complicated by the recent discovery of other types of ER [41, 42]. In addition, estrogens and anti-estrogens can induce differential activation of ERα and ERβ to control transcription of genes that are under the control of an AP 1 element [43].

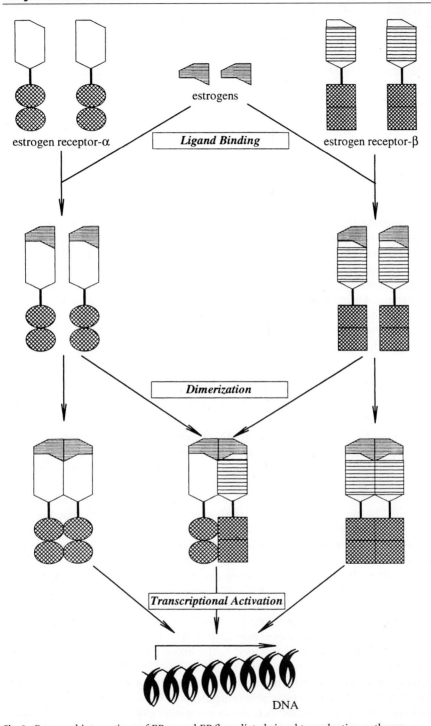

Fig. 2. Proposed interactions of ERα- and ERβ-mediated signal transduction pathways

2
Sources of Estrogens in Human Breast Tissue

17β-Estradiol is biologically the most active estrogen in breast tissue. Circulating estrogens are mainly originated from ovarian steroidogenesis in premenopausal women and peripheral aromatization of ovarian and adrenal androgens in postmenopausal women [8]. The importance of ovarian steroidogenesis in the genesis of breast cancer is highlighted by the fact that occurring naturally or induced early menopause prior to age 40 years significantly reduces the risk of developing breast cancer [8, 44–46]. However, the uptake of 17β-estradiol from the circulation does not appear to contribute significantly to the total content of estrogen in breast tumors, since the majority of estrogen present in the tumor tissues is derived from *de novo* biosynthesis [47–50]. In fact, the concentrations of 17β-estradiol in breast cancer tissues do not differ between premenopausal and postmenopausal women, even though plasma levels of 17β-estradiol decrease by 90 % following menopause [51]. This phenomenon might be explained by the observation that enzymatic transformations of circulating precursors in peripheral tissues contribute 75% of estrogens in premenopausal women and almost 100% in postmenopausal women [52, 53], the data that highlight the importance of *in situ* metabolism of estrogens. Three main enzyme complexes that are involved in the synthesis of biologically active estrogen (i.e. 17β-estradiol) in the breast are:

1) aromatase that converts androstenedione to estrone,
2) estrone sulfatase that hydrolyses the estrogen sulfate to estrone, and
3) 17β-estradiol hydroxysteroid dehydrogenase that preferentially reduces estrone to 17β-estradiol in tumor tissues (Fig. 3) [54, 55].

Aromatase (estrogen synthetase) is the enzyme complex responsible for the final step in estrogen synthesis – the conversion of androstenedione and testosterone to estrone and 17β-estradiol, respectively (Fig. 3). Circulating free and conjugated dehydroepiandrosterone (DHEA) is the major androgen precursor for estrogen synthesis in the peripheral tissues, especially in postmenopausal women. The circulating DHEA is extensively converted to androstenedione and estrone in human breast cancer stromal cells [56], resulting in a tissue plasma concentration gradient of up to an 8-fold higher accumulation of androstenedione in breast cancer tissues [57]. Higher aromatase activities have been observed in areas of the breast bearing cancer than in non-involved quadrants [58]. It has also been shown that local synthesis of estrogen via the aromatase enzyme present predominantly in tumor stromal tissue [59] can increase tumor estrogen levels and growth rate [60–62]. In fact, the aromatase-positive macrophages, the predominant population of leukocytes in some breast carcinomas [63, 64], have been identified as the major source of local estrogen production in breast tissues and breast cancer [65]. In addition, an increase in breast cancer susceptibility has been associated with aberrant aromatase activities [66, 67] as well as genetic polymorphisms in the aromatase gene [68, 69]. It has been suggested that an increase in aromatase expression or activity is related to the malignant phenotype, but not necessarily to the biological behavior

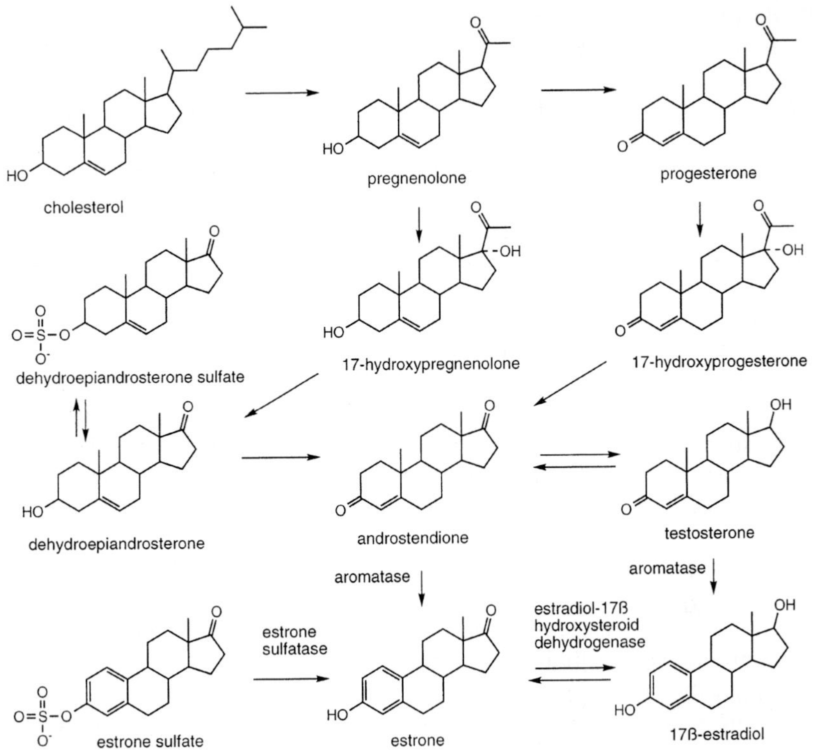

Fig. 3. Steroidogenic pathways in the biosynthesis of estrogens in human breast tissues

or clinical course, of breast cancer [70]. In contrast, an increase in aromatase in the stromal cells of breast adipose tissue may be correlated with the development of, or predisposition to, breast cancer [70]. Ironically, the plasma concentration of DHEA sulfate peaks in the second decade of life and declines markedly during adulthood [71], supporting the notion that concentrations of circulating DHEA are inversely correlated with the risk of breast cancer [72–74]. Nevertheless, aromatization of the androgenic C_{19} steroids by aromatase activity is one of the most important pathways for the biosynthesis of the estrogenic C_{18} steroids.

Even though the major pathway of estrone synthesis is via aromatization of the precursor androgens in the ovary or peripheral tissue, much of the estrone synthesized is converted by estrone sulfotransferase to estrone sulfate, which can be converted back to estrone by estrone sulfatase-mediated hydrolysis [75]. Sulfation is an important process in the metabolism and inactivation of steroids, including estrogens, because the addition of the charged sulfate group protects the hormones from binding to their receptors and serves as reserve material for the biosynthesis of active hormones through the action of endogenous sulfatases [76, 77]. Breast cancer cells lack estrogen sulfotransferase to inactivate estrogens and forced expression of the enzyme in breast carcinoma cells decreases their responsiveness to estrogen-induced growth stimulation [77]. In contrast, breast cancer tissues contain 10–100 times higher sulfatase activity than the aromatase activity and produce estrone mainly through the hydrolysis of estrone sulfate [54, 78]. In fact, it has been proposed that estrone production by sulfatase activity is 10-fold more than that by aromatase activity [79]. A large amount of estrone sulfate and estrone sulfatase activity has been observed in breast tumor tissues, especially in those from postmenopausal women [54]. Thus, estrone sulfate is quantitatively the most important circulating estrogen in women and acts as a large reservoir for the formation of estrone [80, 81].

17β-Hydroxysteroid dehydrogenases (17β-HSD) belong to a family of HSD enzymes that are involved in the interconversion of the oxidized form and the reduced form of steroid hormones. Members of HSD family include 3β-HSD that catalyze the conversion of 3β-hydroxy-5-ene-steroids (e.g., pregnenolone) to corresponding 3-keto-4-ene-steroids (e.g., progesterone), 11β-HSD that catalyze the conversion of glucocorticoids and their inactive metabolites, and 17β-HSD that catalyze the oxido-reduction at carbon 17 of C_{18} and C_{19} steroids [82–84]. There are 7 types of 17β-HSD that have been characterized so far [84–88]. Type I 17β-HSD, which is expressed mainly in the placenta, ovary and breast, catalyzes the reduction of estrone to 17β-estradiol, the most potent estrogen [89–92]. Expression of type I 17β-HSD has also been reported in human breast carcinoma, but its level is variable and not necessarily higher than in non–neoplastic breast tissue [93–96]. In addition, type I 17β-HSD is expressed in both stromal and epithelial components of the breast [96]. However, the level of type I 17β-HSD expression is higher in normal, but not cancerous, breast epithelial cells as compared to their respective stromal counterparts [96]. More importantly, the activity of type I 17β-HSD favors reduction of estrone to 17β-estradiol in primary epithelial and stromal cells derived from cancerous breast, while the oxidative activity of type I 17β-HSD predominates in primary cultures of normal breast epithelial and stromal cells [96]. Similarly, the reductive pathway of type I 17β-HSD is more active in breast tumors [97–100]. Both the expression and the reductive activity of type I 17β-HSD can be up-regulated by cytokines and growth factors [101–104] rich in breast tumors [104–106]. These data suggest an important role of type I 17β-HSD in the local production of biologically active estrogen in human breast.

It should be noted, however, that *de novo* biosynthesis of active estrogen is also influenced by the activities of other enzymes along the steroidogenesis pathways (Fig. 3). For instance, 3β-HSD activity is essential to the formation of androgenic steroids that serve as the substrates for aromatase activity in the breast [107]. It should also be noted that local synthesis of active estrogen could confer growth advantage only if type I 17β-HSD, aromatase and estrogen receptors are coordinately expressed [70]. In this regard, a significant correlation between the expression of type I 17β-HSD and aromatase has been observed in invasive lobular carcinoma, but not in ductal carcinoma [108]. In addition, no consistent correlation has been found between the expression of estrogen receptors and aromatase activity [109, 110]. Since local synthesis of estrogen in the stromal component can increase the estrogen levels and growth rate of

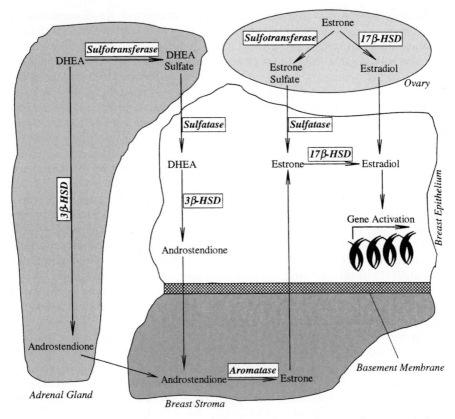

Fig. 4. Stromal and epithelial interactions in the biosynthesis and action of estrogens in human breast tissues

breast carcinoma [60–62], a paracrine mechanism has been proposed to account for stromal-epithelial interactions in the biosynthesis and action of estrogens in human breasts (Fig. 4). Clearly, further studies are warranted to investigate the regulatory mechanisms that are involved in the control of expression of aromatase, type I 17β-HSD and estrogen receptors in the epithelial and stromal components of the breasts.

3
Role of Estrogens in Human Breast Proliferation

Even though the breast is influenced by a myriad of hormones and growth factors [111–118], estrogens are considered to play a major role in promoting the proliferation of both the normal and the neoplastic breast epithelium [111, 112, 115]. Estradiol acts locally in the mammary gland, stimulating DNA synthesis and promoting bud formation, probably through an ER-mediated mechanism [111]. It is also known that the prevailing metabolic condition of an individual animal or human may significantly influence mammary gland responses to hormones. In addition, the mammary gland responds selectively to given hormonal stimuli for either cell proliferation or differentiation, depending upon specific topographic differences in gland development. In either case, the response of the mammary gland to these complex hormonal and metabolic interactions results in developmental changes that permanently modify both the architecture and the biological characteristics of the gland [111, 113]. The fact that the normal epithelium contains receptors for both estrogen and progesterone lends support to the receptor-mediated mechanism as a major player in the hormonal regulation of breast development. The role of these hormones on the proliferative activity of the breast, which is indispensable for its normal growth and development, has been for a long time, and still is, the subject of heated controversies [117–125]. There is little doubt, however, that the proliferative activity of the mammary epithelium in both rodents and humans varies with the degree of differentiation of the mammary parenchyma [111–114, 126–128]. In humans, the highest level of cell proliferation is observed in the undifferentiated lobules type (Lob 1) present in the breast of young nulliparous females [111–114]. The progressive differentiation of Lob 1 into lobules types 2 (Lob 2) and 3 (Lob 3), occurring under the hormonal influences of the menstrual cycle, and the full differentiation into lobules type 4 (Lob 4), as a result of pregnancy, leads to a concomitant reduction in the proliferative activity of the mammary epithelium [111–114, 126–128]. The content of ERα and progesterone receptor (PgR) in the lobular structures of the breast is directly proportional to the rate of cell proliferation, being also maximal in the undifferentiated Lob 1, and decreasing progressively in Lob 2, Lob 3, and Lob 4 (Fig. 5) [113]. Cell proliferation, as determined as the percentage of cycling cells that are positively stained with Ki67 antibody as a brown nuclear reaction characteristic of DAB

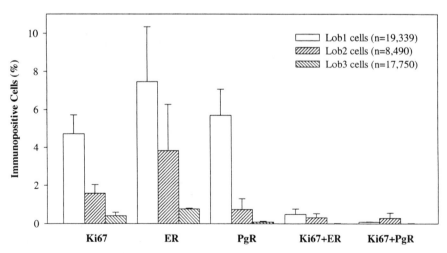

Fig. 5. Percentage of cells positive for estrogen receptor (ER), progesterone receptor (PgR), Ki67, and of cells positive for both ER and Ki67 (ER+Ki67), or PgR and Ki67 (PgR+Ki67) (ordinate). Cells were quantitated in Lob 1, Lob 2, and Lob 3 of the breast (abscissa)

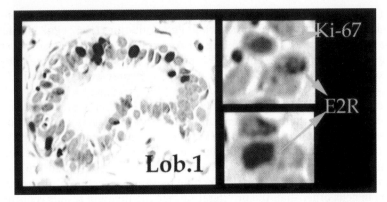

Fig. 6. Lob 1 ductules of the human breast. The single-layered epithelium lining the ductule contains Ki67 positive cells (brown nuclei), and ER positive cells (red-purple nuclei) (×40)

stain, is most frequently found in Lob 1 (Fig. 5) [129]. The percentage of positive cells is reduced by three fold in Lob 2, and by more than ten fold in Lob 3 (Fig. 5) [129]. In all cases, the proliferating cells are almost exclusively found in the epithelium lining ducts and lobules. Only occasionally are positive cells found in the myoepithelium, or in the intralobular and interlobular stroma. The same pattern of reactivity is also observed in tissue sections incubated with the ERα and PgR antibodies. Positive cells are found exclusively in the epithelium. The number of cells positive for ERα or PgR is highest in the Lob 1, and decreases progressively in Lob 2 and Lob 3 (Fig. 5) [129].

It should be noted, however, that it remains unclear from the above studies whether the cells that are positive for steroid receptors are those that are proliferating. The use of the double staining procedure for Ki67 and ERα or PgR has allowed us to quantitatively determine in the same tissue sections the spatial relationship between those cells that are proliferating and those that react with the ERα or PgR antibody. The double stained cells appear purple-red in color due to the alkaline phosphatase-vector red staining (Fig. 6).

The number of cells that express ERα and/or PgR is similar to that of cells positive for Ki67, and the highest percentage of positive cells is also observed in Lob 1 for both steroid hormones. The percentages of ERα- and PgR-positive cells in Lob 1 are 7.5% and 5.7%, respectively, which do not differ significantly (Fig. 5). The percentages of ERα- and PgR-positive cells in Lob 2 are reduced to 3.8% and 0.7%, respectively, and become negligible in Lob 3 (Fig. 5).

Of interest is the observation that even though there are similarities in the relative percentages of Ki67-, ERα- and PgR-positive cells, and in the progressive reduction in the percentage of positive cells as the lobular differentiation progresses, those cells positive for Ki67 are not the same as those positive for ERα or PgR. Very few cells, less than 0.5% in Lob 1, and even fewer in Lob 2 and Lob 3, appear positive for both Ki67 and ERα (Ki67+ER) (Fig. 5). This double reactivity is identified by the darker staining of the nuclei, which appear dark purple-brown. Whereas the percentage of cells double labeled with Ki67 and

ERα (Ki67+ER) decreases gradually from Lob 1 to Lob 3, the percentage of cells exhibiting double labeling with Ki67 and PgR antibodies (Ki67+PgR) is high in Lob 2 but low in Lob 1 and Lob 3 (Fig. 5). As a result, Ki67+PgR-positive cells is lower than the percentage of Ki67+ER-positive cells in Lob 1, but the percentages of Ki67+PgR- and Ki67+ER-positive cells become quite similar in Lob 2 and Lob 3 (Fig. 5).

Our data indicate that the contents of ERα and PgR in the normal breast tissue, as detected immunocytochemically, vary with the degree of lobular development, but are linearly related to the rate of cell proliferation of the same structures. The utilization of a double labeling immunocytochemical technique to stain the same tissue section for steroid hormone receptors and Ki67 proliferating antigen has allowed us to conclude that the expression of the receptors occurs in cells other than the proliferating cells, confirming results reported by others [122]. The findings that proliferating cells are different from those that are ERα- and PgR-positive support data that indicate that estrogen controls cell proliferation by an indirect mechanism. This phenomenon has been demonstrated using supernatant of estrogen-treated ERα-positive cells that stimulates the growth of ERα-negative cell lines in culture. The same phenomenon has been shown *in vivo* in nude mice bearing ER-negative breast tumor xenografts [130, 131]. ERα-positive cells treated with antiestrogens secrete transforming growth factor-β that inhibits the proliferation of ERα-negative cells [132]. The findings that proliferating cells in the human breast are different from those that contain steroid hormone receptors explain many of the *in vitro* data [133–135]. Of interest are the observations that while the ERα-positive MCF-7 cells respond to estrogen treatment with increased cell proliferation, and that the enhanced expression of the ERα by transfection also increases the proliferative response to estrogen [133–136], ERα-negative cells, such as MDA-MB 468 and others, when transfected with ERα, exhibit inhibition of cell growth under the same type of treatment [134–137]. Although the negative effect of estrogen on those ERα-negative cells transfected with the ERα has been interpreted as an interference of the transcription factor used to maintain estrogen independent growth [136], there is no definitive explanation for their lack of survival. However, it can be explained by the finding that proliferating and ERα-positive cells are two separate populations. Further support is the finding that when Lob 1 of normal breast tissue are placed in culture, they lose the ERα-positive cells, indicating that only proliferating cells that are also ERα-negative can survive and constitute the stem cells [137, 138].

4
Implications of Estrogens in Human Breast Carcinogenesis

Although 67% of breast cancers are manifested during the postmenopausal period, a vast majority, 95%, are initially hormone-dependent [14]. This indicates that estrogens play a crucial role in their development and evolution. It has been established that *in situ* metabolism of estrogens through aromatase-mediated pathway is correlated with the risk of developing breast cancer [66–69]. A recent finding that expression of estrone sulfatase is inversely correlated with relapse-free survival of human breast cancer patients [78] reiterates the importance of estrone sulfatase-mediated local production of estrogen in the development and progression of human breast cancer. However, it is still unclear whether estrogens are carcinogenic to the human breast. Most of the current understanding of carcinogenicity of estrogens is based on studies in experimental animal systems and clinical observations of a greater risk of endometrial hyperplasia and neoplasia associated with estrogen supplementation or polycystic ovarian syndrome [19–21].

There are two mechanisms that have been considered to be responsible for the carcinogenicity of estrogens: receptor-mediated hormonal activity, which has generally been related to stimulation of cellular proliferation, resulting in more opportunities for accumulation of genetic damages leading to carcinogenesis [139, 140], and cytochrome P450-mediated metabolic activation, which elicits direct genotoxic effects by increasing mutation rates (Fig. 7) [140–142]. There is also evidence that estrogen compromises the DNA repair system and allows accumulation of lesions in the genome essential to estrogen-induced tumorigenesis [143].

The receptor-mediated activity of estrogen is generally related to induction of expression of the genes involved in the control of cell cycle progression and growth of human breast epithelium [144]. The biological response to estrogen depends upon the local concentrations of the active hormone and its receptors. The level of ER expression is higher in breast cancer patients than in control subjects and is related to breast cancer risk in postmenopausal women [145]. It has been suggested that overexpression of ER in normal human breast epithelium may augment estrogen responsiveness and hence the risk of breast cancer [145]. The proliferative activity and the percentage of ERα-positive cells are

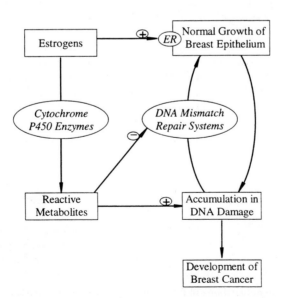

Fig. 7. Potential mechanisms of estrogen-induced carcinogenesis in human breast tissues

highest in Lob 1 in comparison with the various lobular structures composing the normal breast. These findings provide a mechanistic explanation for the higher susceptibility of these structures to be transformed by chemical carcinogens *in vitro* [146, 147], supporting as well the observations that Lob 1 are the site of origin of ductal carcinomas [148].

The presence of ERα-positive and ERα-negative cells with different proliferative activity in the normal human breast may help to elucidate the genesis of ERα-positive and ERα-negative breast cancers [149, 150]. It has been suggested that either ERα-negative breast cancers result from the loss of the ability of the cells to synthesize ERα during clinical evolution of ERα-positive cancers, or that ERα-positive and ERα-negative cancers are different entities [150, 151]. Based on our observations, it is postulated that Lob 1 contain at least three cell types, ERα-positive cells that do not proliferate, ERα-negative cells that are capable of proliferating, and a small proportion of ERα-positive cells that can proliferate as well (Fig. 8) [129]. Therefore, estrogen might stimulate ERα-positive cells to produce a growth factor that in turn stimulates neighboring ERα-

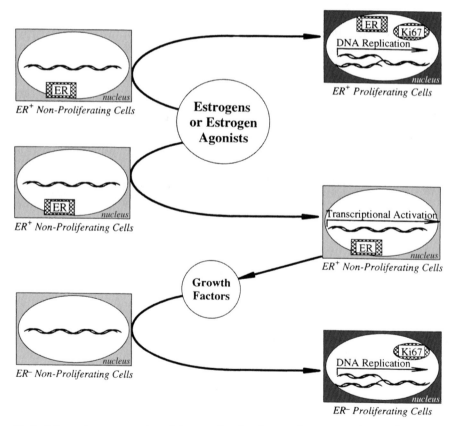

Fig. 8. Schematic representation of the postulated pathways of estrogen actions on breast epithelial cells

(A)

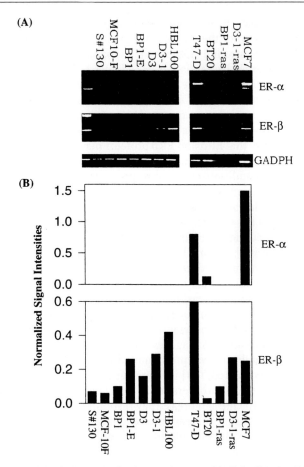

Fig. 9 A, B. Expression of ERα or ERβ in human breast epithelial cells. Signal intensities of the ERα or ERβ products for each cell line were normalized using GAPDH products to produce arbitrary units of relative abundance. The average value for each cell line from three independent RT-PCR reactions is plotted in Fig. 9 B, while a representative picture of the gel for ERα or ERβ reactions is shown in Fig. 9 A (Reproduced with permission from ref. [154])

negative cells capable of proliferating (Fig. 8) [129]. In the same fashion, the small proportion of cells that are ERα-positive and can proliferate could be the source of ERα-positive tumors. The possibility exists, as well, that the ERα-negative cells convert to ERα-positive cells (Fig. 8) [129].

The newly discovered ERβ opens another possibility that those cells traditionally considered negative for ERα might be positive for ERβ [33–35, 152–158]. It has recently been found that ERβ is expressed during the immortalization and transformation of ER-negative human breast epithelial cells (Fig. 9) [154], supporting the hypothesis of conversion from a negative to a positive receptor cell. The functional role of ERβ-mediated estrogen signaling pathways in the pathogenesis of malignant diseases is essentially unknown. In the rats, ERβ-mediated mechanisms have been implicated in the up-regulation of PgR expression in the dysplastic acini of the dorsolateral prostate in response to treatment of testosterone and 17β-estradiol [159]. In the human, ERβ has been detected in both normal and cancerous breast tissues or cell lines [153], and is the predominant ER type in normal breast tissue [158]. Expression of ERβ in breast tumors is inversely correlated with the PgR status [157] and variant transcripts of ERβ have been observed in some breast tumors [155, 156]. ERβ and ERα are co-expressed in some breast tumors and a few breast cell lines [153–155], suggesting an interesting possibility that ERα and ERβ proteins may interact with each other and discriminate between target sequences leading to differential responsiveness to estrogens (Fig. 2). In addition, estrogen responses mediated by ERα and ERβ may vary with different composition of their co-activators that transmit the effect of ER-ligand complex to the transcription complex at the promotor of target genes [160]. Recently, it has been shown that an increase in the expression of ERα with a concomitant reduction in ERβ expression occurs during tumorigenesis of the breast [161] and ovary [162], but breast tumors expressing both ERα and ERβ are lymph node-positive and tend to be of higher histopathological grade [158]. These data suggest a change in the interplay of ERα- and ERβ-mediated signal transduction pathways during breast tumorigenesis.

Even though it is now generally believed that alterations in the ER-mediated signal transduction pathways contribute to breast cancer progression toward hormonal independence and more aggressive phenotypes, there is also mounting evidence that a membrane receptor coupled to alternative second messenger signaling mechanisms [163, 164] is operational, and may stimulate the cascade of events leading to cell proliferation. This knowledge suggests that ERα-negative cells found in the human breast may respond to estrogens through this or other pathways. The biological responses elicited by estrogens are mediated, at least in part, by the production of autocrine and paracrine growth factors from the epithelium and the stroma in the breast [165, 166]. In addition, evidence has accumulated over the last decade supporting the existence of ER variants, mainly a truncated ER and an exon-deleted ER [167–171]. It has been suggested that expression of ER variants may contribute to breast cancer progression toward hormone independence [171]. Although more studies need to be done in this direction, it is clear that the findings that in the normal breast the proliferating and steroid hormone receptor-positive cells are different open new possibilities for clarifying the mechanisms through which estrogens might act on the proliferating cells to initiate the cascade of events leading to cancer.

More importantly, estrogen may not need to activate its nuclear receptors to initiate or promote breast carcinogenesis. There is evidence that oxidative catabolism of estrogens mediated by various cytochrome P450 (CYP) complexes constitutes a pathway of their metabolic activation and generates reactive free radicals and intermediate metabolites, reactive intermediates that can cause oxidative stress and genomic damage directly [141, 142].

17β-Estradiol and estrone, which are continuously interconverted by 17β-estradiol hydroxysteroid dehydrogenase (or 17β-oxidoreductase), are the two major endogenous estrogens (Fig. 10). They are generally metabolized via two major pathways: hydroxylation at C16α position and at the C2 or C4 positions

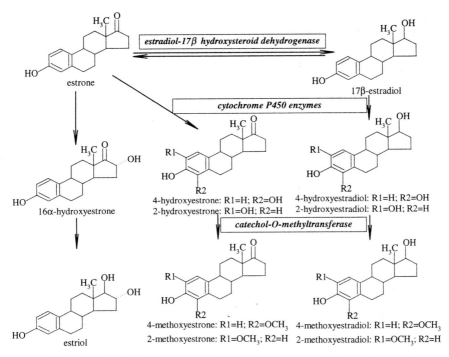

Fig. 10. Biosynthesis and steady-state control of catechol estrogens in human breast tissues

(Fig. 10) [172–174]. The carbon position of the estrogen molecules to be hydroxylated differs among various tissues and each reaction is probably catalyzed by various CYP isoforms [175]. For example, in MCF-7 human breast cancer cells, which produce catechol estrogens in culture [176, 177], CYP1A1 catalyzes hydroxylation of 17β-estradiol at C2, C15α and C16α, CYP1A2 predominantly at C2 [178], and a member of the CYP1B1 subfamily is responsible for the C4 hydroxylation of 17β-estradiol [179–181]. CYP3A4 and CYP3A5 have also been shown to play a role in the 16α-hydroxylation of estrogens in human [182, 183].

The hydroxylated estrogens are catechol estrogens that will easily be autooxidated to semiquinones and subsequently quinones, both of which are electrophiles capable of covalently binding to nucleophilic groups on DNA via a Michael addition and, thus, serve as the ultimate carcinogenic reactive intermediates in the peroxidative activation of catechol estrogens (Fig. 11) [184]. In addition, a redox cycle consisting of the reversible formation of the semiquinones and quinones of catechol estrogens catalyzed by microsomal P450 and cytochrome P450-reductase can locally generate superoxide and hydroxyl radicals to produce additional DNA damage (Fig. 11) [141, 142]. Furthermore, catechol estrogens have been shown to interact synergistically with nitric oxide present in human breast generating a potent oxidant that induces DNA strand breakage [185].

Steady-state concentrations of catechol estrogens are determined by the cytochrome P450-mediated hydroxylations of estrogens and monomethylation of

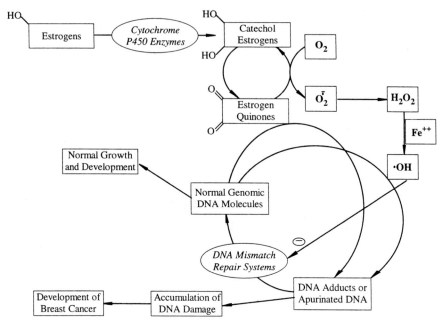

Fig. 11. Carcinogenic effects associated with the metabolisms of catechol estrogens in human breast tissues

catechols catalyzed by blood-borne catechol O-methyltransferase (Fig. 10) [173, 178, 186, 187]. Increased formation of catechol estrogens as a result of elevated hydroxylations of 17β-estradiol at C4 [181] and C16α [188–190] positions occurs in human breast cancer patients and in women at a higher risk of developing this disease. There is also evidence that lactoperoxidase, present in milk, saliva, tears and mammary glands, catalyzes the metabolism of 17β-estradiol to its phenoxyl radical intermediates, with subsequent formation of superoxide and hydrogen peroxide that might be involved in estrogen-mediated oxidative stress [191]. A substantial increase in base lesions observed in the DNA of invasive ductal carcinoma of the breast [192, 193] has been postulated to result from the oxidative stress associated with metabolism of 17β-estradiol [191].

Clearly, the ability of the mammary gland to metabolize 17β-estradiol and/or to accumulate "genotoxic" metabolites could profoundly influence the neoplastic transformation of the epithelium [194]. Oxidative biotransformation of estradiol occurs in human mammary explant cultures [194, 195). Treatment of normal mouse mammary epithelial cells with the mutagenic polycyclic hydrocarbon 7,12-dimethylbenzo[a]anthracene results in production of 16α-hydroxyestrone as the predominant metabolite of estrogens, which increases unscheduled DNA synthesis, cellular proliferation and anchorage-independent growth, indicative of preneoplastic transformation [196]. In experimental animals, catechols have been implicated as mediators of estrogen-induced carcinogenesis [197]. Elevated metabolic conversion of 17β-estradiol to catechol estrogens has been documented in a number of organs susceptible to estrogen-induced carcinogenesis, including hamster kidney [174, 178, 198], mouse uterus [199, 200] and rat pituitary [177, 198]. Modulation of catechol estrogen concentrations influences susceptibility to estrogen-inducible carcinogenesis. For instance, catecholamines, which are substrates and competitive inhibitors of the catechol O-methyltransferase, are present at much higher levels in the organs susceptible to estrogen-induced carcinogenesis [201]. Inhibition of the catechol O-methyltransferase-catalyzed O-methylation of 2- and 4-hydroxy-17β-estradiol by quercetin, a flavonoid, increases accumulation of catechol estrogens [202] and augments the induction of estradiol-induced carcinogenesis [203]. In addition, a genetic polymorphism of catechol O-methyltransferase [i.e., low-activity allele] has been positively associated with an increased breast cancer risk in premenopausal women, especially in those overweight [204]. The carcinogenic potential of estrogens in normal human breast epithelial cells is currently under active investigation.

5
Summary and Future Perspectives

Prolonged exposure to estrogen has long been identified as a risk factor for human breast cancer, but the role of estrogen in the development of human breast cancer has been difficult to ascertain. One of the difficulties is to relate breast cancer development to circulating levels of estrogens. The levels of free estrogens are significantly higher in the primary breast cancer tissues than those in the circulation [47], highlighting the importance of *in situ* metabolism of estrogens by estrone sulfatase and aromatase [66, 205]. There are two mechanisms that have been considered to be responsible for the carcinogenicity of estrogens: a receptor-mediated hormonal activity and cytochrome P450-mediated metabolic activation. The receptor-mediated hormonal activity of estrogen has generally been related to stimulation of cellular proliferation, resulting in more opportunities for accumulation of genetic damages leading to carcinogenesis [139, 140]. Since local synthesis of estrogen in the stromal component can increase the estrogen levels and growth rate of breast carcinoma [60–62], a paracrine mechanism is likely to account for interactions between aromatase-containing stromal cells and ER-containing breast tumor epithelial cells [60]. Further studies are warranted to investigate the regulatory mechanisms that are involved in the control of expression of aromatase, type I 17β-HSD and ER. In addition, expression of the ER occurs in cells other than the proliferating cells [122, 129], suggesting that another paracrine mechanism is operative to mediate the biological response to estrogens. More importantly, estrogen may not need to activate its nuclear receptors to initiate or promote breast carcinogenesis. There is evidence that oxidative catabolism of estrogens mediated by various CYP complexes constitutes a pathway of their metabolic activation and generates reactive free radicals and other reactive intermediates that can cause oxidative stress and genomic damage directly [141, 142]. Estrogen-induced genotoxic effects include increased mutation rates [140–142] and compromised DNA repair system that allows accumulation of genomic lesions essential to estrogen-induced tumorigenesis [143]. Oxidative biotransformation of estrogen does occur in human mammary explant culture [194, 195]. Increased formation of catechol estrogens as a result of elevated hydroxylations of 17β-estradiol at C4 [181] and C16α [188–190] positions has been observed in human breast cancer patients and in women at a higher risk of developing this disease. There is also evidence that formation of superoxide and hydrogen peroxide as a result of the metabolism of estrogen might also be involved in estrogen-mediated oxidative stress [191]. In fact, a substantial increase in base lesions observed in the DNA of invasive ductal carcinoma of the breast [192, 193] has been postulated to result from the oxidative stress associated with metabolism of 17β-estradiol [191]. However, metabolism of estrogen in normal human breast epithelial cells is largely unclear. More importantly, the carcinogenic potential of estrogens in normal human breast epithelial cells awaits to be elucidated.

6
References

1. Parkin DM, Psiani P, Ferlay J (1999) CA Cancer J Clin 49: 33
2. Landis SH, Murray T, Bolden S, Wingo PA (1999) CA Cancer J Clin 49: 8
3. Harris JR, Lippman ME, Veronesi U, Willet W (1992) N Engl J Med 327: 319
4. Wingo PA, Ries LA, Rosenberg HM, Miller DS, Edwards BK (1998) Cancer 82: 1197
5. Stoll (1995) Reducing Breast Cancer Risk in Women. Kluwer Academic, Amsterdam, pp 3
6. Pike MC, Spicer DV, Dahmoush L, Press MF (1993) Epidemiol Rev 15: 17
7. Kelsey JL, Gammon MD, John EM (1993) Epidemiol Rev 15: 36
8. Bernstein L, Ross RK (1993) Epidemiol Rev 15: 48
9. Toniolo PG (1997) Environ Health Perspect 105 (Suppl 3): 587
10. Colditz GA (1998) J Natl Cancer Inst 90: 814
11. Henderson BE, Ross RK, Pike MC, Casagrande JT (1982) Cancer Res 42: 3232
12. Henderson BE, Ross RK, Pike MC (1993) Science 259: 633
13. Spencer-Feigelson H, Ross RK, Yu MC, Coetzee GA, Reichardt JKV, Henderson BE (1996) J Cell Biochem 25S: 15
14. Henderson BE, Ross R, Bernstein L (1988) Cancer Res 48: 246
15. Thomas HV, Reeves GK, Key TJ (1997) Cancer Cause Control 8: 922
16. Greenwald P, Barolom JJ, Nasca PC (1971) N Engl J Med 285: 390
17. Nissen ED, Kent DR (1975) Obst Gynecol 46: 460
18. Herbst AL (1981) Cancer 48: 484
19. Shaw RW (1987) Br J Obst Gynecol 94: 724
20. Chivers C, Mant D, Pike MC (1987) Br Med J 295: 1446
21. Beral V, Hannaford P, Kay C (1988) Lancet II: 1331
22. Hilakivi-Clarke L (1997) Breast Cancer Res Treat 46: 143
23. Cuzick J, Baum M (1985) Lancet 2: 28
24. Davis DL, Telang NT, Osborne MP, Bradlow HL (1997) Environ Health Perspect 105 (Suppl 3): 571
25. Ferguson AT, Davidson NE (1997) Crit Rev Oncogenesis 8: 29
26. Sonnenschein C, Soto AM (1998) J Steroid Biochem Mol Biol 65: 143
27. Zava DT, Dollbaum CM, Blen M (1998) Proc Soc Exp Biol Med 217: 369
28. Tsai MJ, O'Malley BW (1994) Annu Rev Biochem 63: 451
29. Katzenellenbogen BS (1980) Annu Rev Physiol 42: 17
30. Topper J, Freedman C (1980) Physiol Rev 60: 1049

31. Dickson RB, Lippman ME (1988) Cellular, molecular biology. In: Lippman ME, Dickson RB (eds), Breast Cancer. Kluwer Academic, Boston, pp 119
32. Jordan C (1997) J Natl Cancer Inst 89: 747
33. Kuiper GGJM, Enmark E, Pelto-Huikko M, Nilsson S, Gustaffson JA (1996) Proc Natl Acad Sci USA 93: 5925
34. Tremblay GB, Tremblay A, Copeland NG, Gilbert DJ, Jenkins NA, Labrie F, Giguere V (1997) Mol Endocrinol 11: 353
35. Mosselman S, Polma J, Dijkema R (1996) FEBS Lett 392: 49
36. Kuiper GGJM, Carlsson B, Grandien K, Enmark E, Haggblad J, Nilsson S, Gustafsson JA (1997) Endocrinology 138: 863
37. Cowley SM, Hoare S, Mosselman S, Parker MG (1997) J Biol Chem 272: 19858
38. Kuiper GGJM, Gustafsson JA (1997) FEBS Lett 410: 87
39. Pace P, Taylor J, Suntharalingam S, Coombes RC, Ali S (1997) J Biol Chem 272: 25832
40. Ogawa S, Inoue S, Watanabe T, Hiroi H, Orimo A, Hosoi T, Ouchi Y, Muramatsu M (1998) Biochem Biophys Res Commun 243: 122
41. Rao BR (1998) J Steroid Biochem Mol Biol 65: 3
42. Bhat RA, Harnish DC, Stevis PE, Lyttle CR, Komm BS (1998) J Steroid Biochem Mol Biol 67: 233
43. Paech K, Webb P, Kuiper GGJM, Nilsson S, Gustafsson JA, Kushner PJ, Scanlan TS (1997) Science 277: 1508
44. Feinleib M (1968) J Natl Cancer Inst 41: 315
45. Trichopouloos D, MacMahon B, Cole P (1972) J Natl Cancer Inst 48: 605
46. Nissen-Meyer R (1991) Ann Oncol 2: 343
47. McNeill JM, Reed MJ, Beranek PA, Bonney RC, Ghilchik MW, Robinson DJ, James VH (1986) Int J Cancer 38: 193
48. Reed MJ, Owen AM, Lai LC, Coldham NG, Ghilchik MW, Shaikh NA, James VH (1989) Int J Cancer 44: 233
49. Blankenstein MA, Maitimu-Smeele I, Donker, GH Daroszewski J, Milewicz A, Thijssen JH (1992) J Steroid Biochem Mol Biol 41: 891
50. Duncan LJ, Reed MJ (1995) J Steroid Biochem Mol Biol 55: 565
51. van Landeghem AAJ, Poortman J, Nabuurs M, Thijssen JHH (1985) Cancer Res 45: 2900
52. Labrie F (1991) Mol Cell Endocrinol 78: C113
53. Labrie F, Simard J, Luu-The V, Pelletier G, Belghtni K, Belanger A (1994) Bailliere's Clin Endocrinol Metab 8: 451
54. Pasqualini JR, Chetrite G, Nguyen BL, Maloche C, Talbi M, Feinstein MC, Blacker C, Botella J, Paris J (1995) J Steroid Biochem Mol Biol 53: 407
55. Reed MJ, Purohit A, Duncan LJ, Singh A, Roberts CJ, Williams GJ, Potter BV (1995) J Steroid Biochem Mol Biol 53: 413
56. Killinger DW, Strutt BJ, Roncari DA, Khalil MW (1995) J Steroid Biochem Mol Biol 52: 195
57. Vermeulen A, Deslypere JP, Parideans R, Lecleereq C, Roy F, Henson JC (1986) Eur J Cancer Clin Oncol 22: 515
58. Miller WR, Mullen P, Sourdaine P, Watson C, Dixon JM, Telford J (1997) J Steroid Biochem Mol Biol 61: 193
59. Simpson ER, Mahendroo MS, Means GD, Kilgore MW, Graham-Lorence S, Amarneh B, Ito Y, Fisher CR, Michael MD (1994) Endocrine Rev 15: 342
60. Santen RJ, Santner SJ, Pauley RJ, Tait L, Kaseta J, Demers LM, Hamilton C, Yue W, Wang JP (1997) J Steroid Biochem Mol Biol 61: 267
61. Brodie A, Lu Q, Nakamura J (1997) J Steroid Biochem Mol Biol 61: 281
62. Koh JI, Kubota T, Sasano H, Hashimoto M, Hosoda Y, Kitajima M (1998) Anticancer Res 18: 2375
63. Steele R, Brown M, Eremin O (1985) Br J Cancer 51: 135
64. Mantovani A, Bottazzi B, Colotta F, Sozzani S, Ruco L (1993) Immunol Today 13: 265
65. Mor G, Yue W, Santen RJ, Gutierrez L, Eliza M, Berstein LM, Harada N, Wang J, Lysiak J, Diano S, Naftolin F (1998) J Steroid Biochem Mol Biol 67: 403

66. Miller WR, O'Neill J (1987) Steroids 50: 537
67. Bulun SE, Price TM, Aitken J (1993) J Clin Endocrinol Metab 77: 1622
68. Dowsett M (1997) J Steroid Biochem Mol Biol 61: 261
69. Kristensen VN, Andersen TI, Lindblom A, Erikstein B, Magnus P, Borresen-Dale AL (1998) Pharmacogenetics 8: 43
70. Sasano H, Ozaki M (1997) J Steroid Biochem Mol Biol 61: 293
71. Orentreich N, Brind JL, Rizer RL (1984) J Clin Endocrinol Metab 59: 551
72. Gordon CB, Shantz LM, Talalay P (1987) Adv Enzyme Regul 26: 355
73. Gordon GB, Bush TL, Helzlsouer KJ, Miller SR, Constock GW (1990) Cancer Res 50: 3859
74. Secrete G, Toniolo P, Berrino F, Recchione C, Cavalleri A, Pisani P, Totis A, Fariselli G, di Pietro A (1991) Cancer Res 51: 2572
75. Reed MJ, Purohit A (1993) Cancer 45: 51
76. Falany JL, Falany CN (1996) Cancer Res 56: 1551
77. Falany JL, Falany CN (1997) Oncol Res 9: 589
78. Utsumi T, Yoshimura N, Takeuchi S, Ando J, Maruta M, Maeda K, Harada N (1999) Cancer Res 59: 377
79. Santner SJ, Feil PD, Santen RJ (1984) J Clin Endocrinol Metab 59: 29
80. Loriaux DL, Ruder HJ, Lipsett MB (1971) Steroids 18: 463
81. Roberts KD, Rochefort JG, Bleau G, Chapdelaine A (1980) Steroids 35: 179
82. Simard J, Sanchez F, Durocher F, Rheaume E, Turgeon C, Labrie Y, Luu-The V, Mebarki F, Morel Y, de Launoit Y, Labrie F (1995) J Steroid Biochem Mol Biol 55: 489
83. Yang K, Khail MW, Strutt BJ, Killinger DW (1997) J Steroid Biochem Mol Biol 60: 247
84. Tremblay MR, Poirier D (1998) J Steroid Biochem Mol Biol 66: 179
85. Anderson S (1995) J Endocrinol 146: 197
86. Adamski J, Normand T, Leenders F, Monte A, Begue A, Stehelin D, Jungblut PW, De Launoit Y (1995) Biochem J 311: 437
87. Biswas MG, Russell DW (1997) J Biol Chem 272: 15959
88. Nokelainen PA, Pectoketo H, Vihko RK, Vihko PT (1998) Proc Am Endocrinol Soc 1: 279
89. Luu-The V, Labrie C, Zhao HR, Couet J, Lachance Y, Simard J, Leblanc G, Cote J, Berube D, Gagne R, Labrie F (1989) Mol Endocrinol 3: 1301
90. Martel C, Rheaume E, Takahashi M, Trudel C, Couet J, Luu-The V, Simard J, Labrie F (1992) J Steroid Biochem Mol Biol 41: 597
91. Blomquist CH (1995) J Steroid Biochem Mol Biol 55: 515
92. Luu-The V, Zhang Y, Poirier D, Labrie F (1995) J Steroid Biochem Mol Biol 55: 581
93. Pollow K, Boquoi E, Baumann J, Schmidt-Gollwitzer M, Pollow B (1977) Mol Cell Endocrinol 6: 333
94. Poutanen M, Isomaa V, Lehto VP, Vihko R (1992) Int J Cancer 50: 386
95. Isomaa VV, Ghersevich SA, Maentausta OK, Peotoketo EH, Poutanen MH, Vihko RK (1993) Ann Med 25: 91
96. Speirs V, Green AR, Atkin SL (1998) J Steroid Biochem Mol Biol 67: 267
97. Adams EF, Coldham NG, James VHT (1988) J Endocrinol 118: 149
98. Reed MJ, Sigh A, Ghilchik MW, Coldham NG, Purohit A (1991) J Steroid Biochem Mol Biol 39: 791
99. Poutanen M, Isomaa V, Peltoketo H, Vihko R (1995) J Steroid Biochem Mol Biol 55: 525
100. Castegnetta LA, Granata OM, Taibi G, Casto ML, Comito L, Oliveri G, Di Falco M, Carruba G (1996) J Endocrinol 150: S73
101. Speirs V, Adams EF, Rafferty B, White MC (1993) J Steroid Biochem Mol Biol 46: 11
102. Duncan LJ, Coldham NG, Reed MJ (1994) J Steroid Biochem Mol Biol 49: 63
103. Turgeon C, Gingras S, Carriere M-C, Blais Y, Labrie F, Simard J (1998) J Steroid Biochem Mol Biol 65: 151
104. Ducan LJ, Reed MJ (1995) J Steroid Biochem Mol Biol 55: 565
105. Speirs V, Green AR, White MC (1996) Int J Cancer 66: 551
106. Green AR, Green VL, White MC, Speirs V (1997) Int J Cancer 72: 937
107. Simard J, Durocher F, Mebarki F, Turgeon C, Sanchez P, Labrie Y, Couet J, Trudel C, Rheaume E, Morel Y, Luu-The V, Labrie F (1996) J Endocrinol 50: S189

108. Sasano H, Frost AR, Saitoh R, Harada N, Poutanen M, Vihko R, Bulun SE, Silverberg SG, Nagura H (1996) J Clin Endocrinol Metab 81: 4042
109. Miller WR, Anderson TJ, Lack WJL (1990) J Steroid Biochem Mol Biol 37: 1055
110. Esteban JM, Warsi Z, Haniu M, Hall P, Shively JE, Chen S (1992) Am J Pathol 940: 337
111. Russo J, Russo IH (1997) Role of hormones in human breast development: The menopausal breast. In: Wren BG (ed), Progress in the Management of Menopause. Parthenon, New York, pp 184
112. Russo IH, Russo J (1998) J Mammary Gland Biol Neoplasia 3: 49–61
113. Russo J, Russo IH (1997) Endocr Related Cancer 4: 7
114. Calaf G, Alvarado ME, Bonney GE, Amfoh KK, Russo J (1995) Int J Oncol 7: 1285
115. Lippman ME, Dickson RB, Gelmann EP, Rosen N, Knabbe C, Bates S, Bronzert D, Huff K, Kasid A (1987) J Cell Biochem 35: 1
116. Meyer JS (1977) Hum Path 8: 67
117. Masters JRW, Drife JO, Scarisbrick JJ (1977) J Natl Cancer Inst 58: 1263
118. Ferguson DJP, Anderson TJ (1981) Br J Cancer 44: 177
119. Longacre TA, Bartow SA (1986) Am J Surg Pathol 10: 382
120. Going JJ, Anderson TJ, Battersby S (1988) Am J Pathol 130: 193
121. Potten CS, Watson RJ, Williams GT (1988) Br J Cancer 58: 163
122. Clark RB, Howell A, Potter CS, Anderson E (1997) Cancer Res 57: 4987
123. Laidlaw IJ, Clark RB, Howell A, Owen AWMC, Potten CS, Anderson E (1995) Endocrinology 136: 164
124. Clarke RB, Howell A, Anderson E (1997) Breast Cancer Res Treat 45: 121
125. Goodman HM (ed) (1994) Basic Medical Endocrinology. Raven Press, New York, p 288
126. Russo J, Russo IH (1980) Cancer Res 40: 2677
127. Russo J, Russo IH (1987) Lab Invest 57: 112
128. Russo J, Rivera R, Russo IH (1992) Breast Cancer Res Treat 23: 211
129. Russo J, Ao X, Grill C, Russo IH (1999) Breast Cancer Res Treat 53: 217
130. Clarke R, Dickson RB, Lippman ME (1992) Crit Rev Oncol Hematol 12: 1
131. Lippman ME, Dickson RB, Bates S, Knabbe C, Huff K, Swain S, McManaway M, Bronzert D, Kasid A, Gelmann EP (1986) Breast Cancer Res Treat 7: 59
132. Knabbe C, Lippman ME, Wakefield LM, Flanders KC, Kasid A, Derynck R, Dickson RB (1987) Cell 48: 417
133. Foster JS, Wimalasena J (1996) Mol Endocrinol 10: 488
134. Wang W, Smith R, Burghardt R, Safe SH (1997) Mol Cell Endocrinol 133: 49
135. Levenson AS, Jordan VC (1994) J Steroid Biochem Mol Biol 51: 229
136. Zajchowski DA, Sager R, Webster L (1993) Cancer Res 53: 5004
137. Pilat MJ, Christman JK, Brooks SC (1996) Breast Cancer Res Treat 37: 253
138. Calaf G, Tahin Q, Alvarado ME, Estrada S, Cox T, Russo J (1993) Breast Cancer Res Treat 29: 169
139. Nandi S, Guzman RC, Yang J (1995) Proc Natl Acad Sci USA 92: 3650
140. Adlercreutz H, Gorbach SL, Goldin BR, Woods MN, Hamalainen E (1994) J Natl Cancer Inst 86: 1644
141. Liehr JG, Ulubelen AA, Strobel HW (1986) J Biol Chem 261: 16865
142. Roy D, Liehr JG (1988) J Biol Chem 263: 3646
143. Yan Z-J, Roy D (1995) Biochem Mol Biol Int 37: 175
144. Prall OWJ, Rogan EM, Sutherland RL (1998) J Steroid Biochem Mol Biol 65: 169
145. Khan SA, Rogers MA, Khurana KK, Meguid MM, Numann PJ (1998) J Natl Cancer Inst 90: 37
146. Russo J, Reina D, Frederick J, Russo IH (1988) Cancer Res 48: 2837
147. Russo J, Calaf G, Russo IH (1993) CRC Crit Rev Oncogen 4: 403
148. Russo J, Gusterson BA, Rogers A, Russo IH, Wellings SR, van Zwieten MJ (1990) Lab Invest 62: 244
149. Harlan LC, Coates RJ, Block G (1993) Epidemiology 4: 25
150. Habel LA, Stamford JL (1993) Epidemiol Rev 15: 209
151. Moolgavkar SH, Day NE, Stevens RG (1980) J Natl Cancer Inst 65: 559

152. Vladusic EA, Hornby AE, Guerra-Vladusic FK, Lupu R (1997) Proc Am Assoc Cancer Res 38: 297
153. Dotzlaw H, Leygue E, Watson PH, Murphy LC (1997) J Clin Endocrinol Metab 82: 2371
154. Hu YF, Lau KM, Ho SM, Russo J (1998) Int J Oncol 12: 1225
155. Vladusic EA, Hornby AE, Guerra-Vladusic FK, Lupu R (1998) Cancer Res 58: 210
156. Lu B, Leygue E, Dotslaw H, Murphy LJ, Murphy LC, Watson PH (1998) Mol Cell Endocrinol 138: 199
157. Dotzlaw H, Leygue E, Watson PH, Murphy LC (1999) Cancer Res 59: 529
158. Speirs V, Parkes AT, Kerin MJ, Walton DS, Carleton PJ, Fox JN, Atkin SL (1999) Cancer Res 59: 525
159. Lau KM, Leav I, Ho SM (1998) Endocrinology 139: 424
160. Watanabe T, Inoue S, Ogawa S, Ishii Y, Hiroi H, Ikeda K, Orimo A, Muramatsu M (1997) Biochem Biophys Res Commun 236: 140
161. Leygue E, Dotzlaw H, Watson PH, Murphy LC (1998) Cancer Res 58: 3197
162. Brandenberger AW, Tee MK, Jaffe RB (1998) J Clin Endocrinol Metab 83: 1025
163. Aronica SM, Kraus WL, Katzenellenbogen BS (1994) Proc Natl Acad Sci USA 91: 8517
164. Pappos TC, Gametahu B, Watson CS (1994) FASEB J 9: 404
165. Dickson RB, Johnson MD, Bano M, Shi E, Kurebayashi J, Ziff B, Martinez-Lacaci I, Amundadottir LT, Lippman ME (1992) J Steroid Biochem Mol Biol 43: 69
166. Rosen JM, Humphreys R, Krnacik S, Juo P, Raught B (1994) Prog Clin Biol Res 387: 95
167. Dotz1aw H, Alkahalaf M, Murphy LC (1992) Mol Endocrinol 6: 773
168. Fuqua SA, Allred DC, Auchus RJ (1993) J Cell Biochem Suppl 17G: 194
169. Fuqua SA, Wolf DM (1995) Breast Cancer Res Treat 35: 233
170. Murphy L, Leygue E, Dotzlaw H, Douglas D, Courts A, Watson P (1997) Ann Med 29: 221
171. Murphy LC, Dotzlaw H, Leygue E, Courts A, Watson P (1998) J Steroid Biochem Mol Biol 65: 175
172. Adlercreutz H, Gorbach SL, Goldin BR, Woods MN, Hamalainen (1994) J Natl Cancer Inst 86: 1644
173. Ball P, Knuppen R (1980) Acta Endocrinol (Copenh) 232 (Suppl): 1
174. Zhu BT, Bui QD, Weisz J, Liehr JG (1994) Endocrinology 135: 1772
175. Spink DC, Eugster H-P, Lincolin DW II, Schuetz JD, Schuetz EG, Johnson JA, Kaminsky LS, Gierthy JF (1992) Arch Biochem Biophys 37: 235
176. Brueggemeier RW, Katlic NE, Palmer CW Jr, Stevens JM (1989) Mol Cell Endocrinol 64: 161
177. Bui QD, Weisz J (1988) Pharmacology 36: 356
178. Ashburn SP, Han X, Liehr JG (1993) Mol Pharmacol 43: 534
179. Niwa T, Bradlow HL, Fishman J, Swaneck GE (1989) J Steroid Biochem 33: 311
180. Spink DC, Hayes CL, Young NR, Christou M, Sutter TR, Jefcoat CR, Gierthy JF (1994) J Steroid Biochem Mol Biol 51: 251
181. Liehr JG, Ricci MJ (1996) Proc Natl Acad Sci USA 93: 3294
182. Shou M, Korzekawa K, Brooks EN, Krausz KW, Gonzalez FJ, Gelboin HV (1997) Carcinogenesis 18: 207
183. Huang Z, Guengerich FP, Kaminsky LS (1998) Carcinogenesis 19: 867
184. Dwivedy I, Devanesan P, Cremonesi P, Rogan E, Cavalieri E (1992) Chem Res Toxicol 5: 828
185. Yoshic Y, Ohshima H (1998) Free Rad Biol Med 24: 341
186. Knuppen R, Ball P, Emons G (1986) J Steroid Biochem 24: 193
187. Creveling CR, Inoue K (1994) Polycyclic Arom Comp 6: 253
188. Schneider J, Kinne D, Fracchia A, Pierce V, Anderson KE, Bradlow HL, Fishman J (1982) Proc Natl Acad Sci USA 79: 3047
189. Bradlow HL, Hershcopf J, Martucci C, Fishman J (1986) Ann NY Acad Sci 464: 138
190. Osborne MP, Bradlow HL, Wong GYC, Telang NT (1993) J Natl Cancer Inst 85: 1917
191. Sipe FU Jr, Jordan SJ, Hanna PM, Mason RP (1994) Carcinogenesis 15: 2637
192. Malins DC, Haimanot R (1990) Cancer Res 51: 5430
193. Malins DC, Holmes EH, Polissar NL, Gunselman SJ (1993) Cancer 71: 3036

194. Telang NT, Axelrod DM, Bradlow HL, Osborne MP (1990) Ann NY Acad Sci 586: 70
195. Fishman J, Osborne MP, Telang NT (1995) Ann NY Acad Sci 768: 91
196. Telang NT, Suto A, Wong GY, Osborne MP, Bradlow HL (1992) J Natl Cancer Inst 84: 634
197. Stalford AC, Maggs JL, Gilchrist TL, Park BK (1994) Mol Pharmacol 45: 1259
198. Weisz J, Bui QD, Roy D, Liehr JG (1992) Endocrinology 131: 655
199. Bunyagidj C, McLachlan JA (1988) J Steroid Biochem 31: 795
200. Patin BC, Chakraborty C, Dey SK (1990) Mol Cell Endocrinol 69: 25
201. Zhu BT, Liehr JG (1993) Arch Biochem Biophys 304: 248
202. Zhu BT, Liehr JG (1996) J Biol Chem 271: 1357
203. Zhu BT, Liehr JG (1994) Toxicol Appl Pharmacol 125: 149
204. Thompson PA, Shields PG, Freudenheim JL, Stone A, Vena JE, Marshall JR, Graham S, Laughlin R, Nemoto T, Kadlubar FF, Ambrosone CB (1998) Cancer Res 58: 2107
205. Pasqualini JR, Schatz B, Varin C, Nguyen BL (1992) J Steroid Biochem Mol Biol 41: 323

CHAPTER 2

The Role of Sex Hormones in Prostate Cancer

Maarten C. Bosland

Departments of Environmental Medicine and Urology, New York University School of
Medicine, 57 Old Forge Road, Tuxedo, NY 10987, USA
E-mail: maarten.bosland@med.nyu.edu

Prostate cancer is the most frequently diagnosed malignancy and second leading cause of
cancer death in men in Western countries. Little is understood about its causes, but steroid
hormones, particularly androgens, are suspected to play a major role. Human populations at
high risk for prostate cancer may have slightly higher androgen production, circulating androgens, 5α-reductase activity, or androgen receptor transactivation activity than low-risk
populations. Elevated circulating estrogens have been found in some high risk populations,
such as African American men, suggesting estrogen involvement. Studies in rats have clearly
demonstrated that testosterone is a weak complete carcinogen and a strong tumor promotor
for the prostate, but the exact mechanisms of these activities are not clear. Treatment of rats
with a combination of testosterone and 17β-estradiol can induce prostate cancer at high incidence, while androgens alone cause a low prostate cancer incidence, indicating involvement
of estrogens as well. This effect may involve genotoxic activity of estrogens, as well as receptor-mediated changes in prostatic sex steroid metabolism and receptors. Aromatase in the
prostate may provide a local source of estrogens. Perinatal estrogen exposure is carcinogenic
for rodent male accessory sex glands. Hyperplastic/metaplastic changes have been reported
in human prostatic tissues following prenatal DES exposure, indicating that this exposure
also targets the human prostate. The mechanisms of prenatal estrogen effects may involve
hormonal imprinting. A multifactorial hypothesis of the role of steroid hormones in prostate
carcinogenesis may involve androgens as strong tumor promotors acting via androgen receptor-mediated mechanisms to enhance the carcinogenic activity of strong endogenous
genotoxic carcinogens, such as reactive estrogen metabolites and reactive oxygen species, and
possibly weak environmental carcinogens; these processes are modulated by environmental
factors such as diet and by genetic determinants such as hereditary susceptibility and polymorphic genes that encode for receptors and enzymes involved in the metabolism and action
of steroid hormones.

Keywords. Androgens, Carcinogenesis, Epidemiology, Estrogens, Prostate, Prostate cancer,
Steroid hormones

The Handbook of Environmental Chemistry Vol. 3, Part M
Endocrine Disruptors, Part II
(ed. by M. Metzler)
© Springer-Verlag Berlin Heidelberg 2002

1	Introduction	29

2	**The Epidemiology of Prostate Cancer**	30
2.1	General Risk Factors	30
2.2	Risk Factors Possibly Associated with Steroid Hormonal Factors	31
2.2.1	Diet and Nutrition	31
2.2.2	Vasectomy	32
2.2.3	Sexual Factors	33
2.2.4	Physical Activity and Anthropometric Correlates of Risk	33
2.3	Epidemiological Studies of Endogenous Hormones and Hormone Metabolism	34
2.3.1	Steroid Hormonal Factors in Populations with Different Risk for Prostate Cancer	34
2.3.2	Comparisons of Steroid Hormonal Factors in Nested Case-Control Studies	40
2.3.3	Critical Interpretation of the Studies	44
2.4	Epidemiology: Summary and Conclusions	46

3	**Hormonal Induction of Prostate Cancer in Laboratory Animals**	47
3.1	Induction of Prostate Cancer with Sex Hormones	48
3.1.1	Androgens	48
3.1.2	Estrogen and Androgen Combinations	48
3.1.3	Perinatal Estrogen Exposure	49
3.2	Induction of Prostate Cancer with Chemical Carcinogens and Sex Hormones	50
3.2.1	Chemical Carcinogens Combined with Hormonal Stimulation of Cell Proliferation	50
3.2.2	Testosterone as Tumor Promotor of Prostate Carcinogenesis	50
3.2.3	Effects of Androgens on Prostate Cancer Induction by Cadmium and Ionizing Radiation	51
3.3	Conclusions	51

4	**Mechanisms of Hormonal Prostate Carcinogenesis**	52
4.1	Androgens and Prostate Carcinogenesis	53
4.1.1	Stimulation of Cell Proliferation and Carcinogenic and Tumor-Promoting Effects of Androgens	53
4.1.2	Androgen Metabolism and Androgen Receptor Sensitivity	54
4.2	Estrogens and Prostate Carcinogenesis	55
4.2.1	Estrogen Receptor-Mediated Mechanisms	56
4.2.2	Non-Receptor Mechanisms	57
4.2.3	Perinatal Estrogen Exposure: Hormonal Imprinting	58

5	**Summary and Conclusion**	59

6	**References**	61

Abbreviations

BOP   *N*-nitroso-bis-(oxopropyl)amine
BPH   benign prostatic hypertrophy
CI   confidence interval
DES   diethylstilbestrol
DHEA   dehydroepiandrosterone
DHT 5α-dihydrotestosterone
DMAB   3,2'-dimethyl-4-aminobiphenyl
4-HPR   *N*-(4-hydroxyphenyl)retinamide
MNU   *N*-methyl-*N*-nitrosourea
PIN   prostatic intraepithelial neoplasia
PSA   prostate specific antigen
RAR   retinoic acid receptor
RXR   retinoic acid X receptor
SHBG   sex hormone binding globulin

1
Introduction

Carcinoma of the prostate is the most frequently occurring malignancy and the second leading cause of cancer death in men in the US and many Western countries [1]. Notwithstanding the importance of this malignancy, little is understood about its causes. Sex steroids, particularly androgenic hormones, are suspected to play a major role in human prostate carcinogenesis. However, the precise mechanisms by which androgens affect this process and the possible involvement of estrogenic hormones are not clear. A causal relation between androgens and prostate cancer development is biologically very plausible because most human prostate cancers are androgen-sensitive and respond to hormonal therapy by remission which is later followed by relapse to a hormone-refractory state. The purpose of this overview is to summarize the literature about the role of sex steroids in prostate carcinogenesis. Although the objective of this overview is to explore the involvement of not only endogenous sex steroids but also a possible role of environmental hormonal factors, very little is know about exogenous chemicals with sex hormone activity in relation to prostate cancer.

The prostate is rarely a site of tumor development in carcinogenesis bioassays in rodents [2, 3], or in aging male laboratory rodents, with the exception of ventral prostatic neoplasms in some rat strains [4–9]. Prostate cancer is also very uncommon in male farm and companion animals, with the exception of the dog that is the only species besides man that develops this malignancy. As will be discussed in this review, sex steroid hormones can induce and substantially enhance prostate cancer development in rodents, and this phenomenon can serve as a model to study the involvement and mechanisms of sex steroids in prostate cancer etiology.

In this review, the epidemiological evidence for a role of sex steroids in prostate carcinogenesis is summarized first, followed by an overview of experimental data and a discussion of the possible mechanisms whereby sex steroids, androgens as well as estrogens, may be involved in prostate cancer causation. Although there are several hypotheses about the role of sex steroids in prostate cancer etiology, the available data are often contradictory and incomplete, and an in-depth mechanistic understanding of how sex steroid hormones are involved in prostate carcinogenesis is poor at best.

2
The Epidemiology of Prostate Cancer

Prostate cancer incidence and mortality rates have increased in the US over the few decades preceding the frequent use of prostate specific antigen (PSA) for early detection [10]. Incidence rates have increased substantially since the mid-1980s because of use of PSA screening for early detection [1], but rates have recently begun to decline [11]. In 1999, 179,300 new cases of prostate cancer are expected and 37,000 deaths from this malignancy [12]. The epidemiology of prostatic cancer has been reviewed in depth elsewhere [10, 13–17]. The most important potential risk factors are summarized below, with special attention to those possibly related to hormonal factors.

2.1
General Risk Factors

Many studies have demonstrated that prostate cancer is more frequent in men with a family history of prostate cancer by twofold to over tenfold, depending on the number of affected blood relatives as discussed elsewhere [10, 13, 18–22]. Although a strong risk factor per se, inherited risk for prostate cancer only explains a small proportion of prostate cancer cases, less than 10% [10, 18, 20]. A range of genetic alterations have been identified in human prostate carcinomas [23]; however, few of these have thus far been linked to the heritability of prostate cancer. Susceptibility loci have been identified on chromosome 1 at 1q24–25 (termed HPC1) and at 1q42.2–43 and on the X chromosome at Xq27–28, which may be involved in hereditary prostate cancer [23–30]. Breast and prostate cancer cluster in some families, which is perhaps related to BRCA1 and BRCA2 mutations in some cases [18, 31]. None of these loci have thus far been associated with hormonal factors.

A history of venereal disease is a consistent risk factor across studies, except in Japan, but detailed studies are lacking [13, 32–34]. In a few studies elevated relative risks were found for a history of prostatitis [13, 34, 35]. An association between prostate cancer risk and the prior occurrence of benign prostatic hypertrophy (BPH) is biologically unlikely because these diseases develop in different parts of the prostate and their epidemiology is dissimilar (see [13]).

Although in a few studies a modest relationship between smoking and risk for prostate cancer has been revealed [34, 36, 37], in the vast majority of studies no such relationship was found [37–40]. In addition, smoking appears to have no effect on circulating levels of testosterone and other hormones that may be involved in prostate carcinogenesis [41, 42]. In most studies of alcohol use as a potential risk factor for prostate cancer, no evidence for an association was found (see [12, 43]). A notable exception is a study by Hayes et al. [44], who found a positive association in a US case-control study, but only for heavy use of alcohol. This may be related to the fact that prostate cancer risk is elevated in alcoholics with liver disease (see [13, 43]), probably because of an impaired clearance of estrogens described in men with liver cirrhosis [45, 46].

Increased risk has been observed for a variety of occupations in studies of occupational factors and prostate cancer [13, 16, 47, 48]. There is no evidence for an association of cadmium exposure or exposures in the rubber industry and prostate cancer risk (see [13, 49, 50]). There is weak to inconclusive evidence for a positive association between farming and prostate cancer risk [13, 16, 47, 48, 51]. Workers in the nuclear industry and armed services personnel may be at increased risk for prostate cancer [13, 17, 48, 52]. However the evidence for exposure to ionizing radiation as a risk factor is equivocal [13, 17, 48, 52–54].

2.2
Risk Factors Possibly Associated with Steroid Hormonal Factors

A variety of epidemiologic studies have suggested risk factors that may be related with hormonal mechanisms, including dietary factors, vasectomy, various sexual factors, physical activity, and obesity. These are summarized in the following paragraphs.

2.2.1
Diet and Nutrition

Associations between dietary factors and prostate cancer risk have been summarized elsewhere [10, 13, 17, 55–57]. Most studies indicate that a high intake of fat, particularly total fat and saturated fat, is a risk factor for prostate cancer, but the strength of the associations is modest [57] and may be greater in African Americans than European Americans [58]. As much as 25% of prostate cancer in the US may be attributable to a high saturated fat intake [59]. However, Whittemore et al. [22] estimated that dietary fat intake may account for only 10–15% of the difference in prostate cancer occurrence between European Americans and African Americans or Asians. A mechanism of an enhancing effect of fat on prostate carcinogenesis is not understood, but several hypotheses, including hormonal mediation, have been discussed elsewhere [13, 17, 57, 60]. In addition, a high intake of protein and energy and a low intake of dietary fiber and complex carbohydrates have been found associated with increased risk for prostate cancer in some studies [10, 13, 17, 56]. The associations with prostate cancer risk reported for individual nutrients or foods are not very strong, but migration from low-risk areas such as Japan to high risk countries, such as the US, increases risk considerably (see [10, 13]). These changes in risk are thought to be due to differences in environment including life-style, particularly dietary habits (see [10, 13]). It is conceivable that the combined effects on prostate carcinogenesis of several dietary factors are more important than the separate effects of any individual dietary factor [61]. This notion is consistent with the observed lack of any effect of dietary fat as a single factor on the induction of prostate cancer in animal models whereas epidemiological studies consistently show a positive association between prostate cancer risk and dietary fat intake (see [13, 17, 61]).

Most studies of effects of dietary changes and the consumption of vegetarian or health food diets on hormonal status ([62–66], summarized in [13]) have not separately addressed effects of dietary fat. However, they clearly indicate that diet can influence circulating hormone levels by changing androgen production rates or the metabolism and clearance of androgens and estrogens. Recently, Dorgan et al. [67] reported that the combination of a high fat-low fiber diet given to healthy men increased total testosterone and testosterone bound to SHBG (sex hormone binding globulin) in the plasma and urinary testosterone excretion as compared to a low fat-high fiber diet, but lowered urinary excretion of estrone, 17β-estradiol, and the 2-hydroxy metabolites of these estrogens. Although these studies did not address the separate effects of single dietary factors such as total fat intake, they indicate that diet can affect sex steroid hormonal status.

There is no consistency among epidemiologic studies of prostate cancer risk and intake of dietary vitamin A and β-carotene (see [10, 13, 17]), and it is possible that retinoids or carotenes enhance rather than inhibit prostatic carcinogenesis under certain circumstances or in certain populations [68]. A major retinol metabolite in mammalian species, 9-cis-retinoic acid, strongly inhibited the induction of prostate cancer in a rat model [69], but N-(4-hydroxyphenyl)-retinamide (4-HPR), a synthetic retinoid, did not have any effect [70]. In vitro, however, both 9-cis-retinoic acid and 4-HPR inhibit the growth and induce apoptosis of the androgen-sensitive human prostate cancer LNCaP cell line, and so does all-$trans$-retinoic acid which only binds to RAR receptors [71–73]. 9-cis-Retinoic acid is a pan-agonist for retinoic acid receptors binding both RAR and RXR receptors, but 4-HPR may act via a non-receptor mechanism [73]. Retinoic acid receptors and androgen receptors both belong to the steroid receptor superfamily [74], which raises the possibility that retinoids may be able to bind to and activate mutated androgen receptors, or that the retinoic acid RAR and/or RXR receptors may activate transcription of androgen-regulated genes, as suggested by studies on regulation by sex steroids and retinoic acid of glutathione-S-transferase in hamster smooth muscle tumor cells [75] and on androgen receptor gene expression in human breast cancer cells [76].

2.2.2
Vasectomy

Vasectomy has been identified as a possible risk factor for prostate cancer in seven case-control studies [34, 77–83] and two cohort studies [84, 85], but no elevation of risk was found in six other case control studies [86–91] and two retrospective cohort studies [92–94]. Although a meta-analysis of 14 studies indicated that there is no causal relation between vasectomy and prostate cancer [95], further studies will be required to establish definitively whether vasectomy is a true risk factor for prostate cancer [96–98]. Three mechanisms by which vasectomy could enhance risk have been proposed: elevation of circulating androgens, immunologic mechanisms involving anti-sperm antibodies, and reduction of seminal fluid production [34, 77, 78, 84, 89, 97, 99]. Most studies of pituitary-gonadal function did not find any effect of vasectomy [100–104], but some studies found changes in hormone levels [89, 99, 105–109]. There are reports in vasectomized men of slightly elevated circulating testosterone levels [34, 99, 107, 110], elevated levels of 5α-dihydrotestosterone (DHT), the active metabolite of testosterone in the prostate, a decrease in SHBG [89], and an increase in the ratio of testosterone to SHBG [34]. These reports suggest an elevation of circulating free testosterone following vasectomy which may be a critical factor associated with risk for prostate cancer, but specific mechanisms whereby vasectomy could influence the hypothalamo-pituitary-gonadal axis are not known.

2.2.3
Sexual Factors

Several case-control studies have addressed the possibility that sexual factors play a role in prostate cancer etiology (see [13, 32–34, 111–113]). In general, the results of these studies suggest that prostate cancer risk may be associated with (1) an early onset of sexual activity, (2) high sexual drive, particularly at young age, (3) a low frequency of intercourse, especially at older age, or (4) a high frequency of intercourse up to 50 years of age and a low frequency thereafter (see [13]). A positive association has been reported between the level of sexual activity and circulating total testosterone levels in men of 60–79 [114]. These findings suggest that a hormonal mechanism may underlie the associations between prostate cancer risk and sexual activity observed in the case-control studies.

2.2.4
Physical Activity and Anthropometric Correlates of Risk

The level of physical activity may be a possible risk factor for prostate cancer, but the evidence for such an association is inconclusive at present [17, 61, 115]. Exercise may influence androgen concentrations [116, 117], and thus it is possible that the type and extent of physical activity influence circulating androgen concentrations and, thereby, possibly prostate cancer risk. The evidence that obesity or an increased body mass index is a prostate cancer risk factor is contradictory at present (see [17, 61, 118]), but an increase in prostate cancer risk with increasing upper arm circumference and upper arm muscle area, but not fat area, has been observed [119]. This positive association between prostate cancer risk and muscle mass, but not fat mass, suggests exposure to endogenous or exogenous androgenic hormones or other anabolic factors [119, 120]. There are reports that body mass index is inversely correlated with plasma testosterone and SHBG levels and positively correlated with 17β-estradiol levels [118, 121, 122]; see also [13, 42].

2.3
Epidemiological Studies of Endogenous Hormones and Hormone Metabolism

As mentioned earlier, a causal relation between androgens and prostate cancer development is biologically plausible. Prostate cancer develops in an androgen-dependent epithelium and these cancers are androgen-sensitive and respond to hormonal therapy by temporary remission followed by relapse to a hormone-refractory state. There are also case reports of prostate cancer in men that used androgenic steroids either as anabolic agents or for medical purposes [123–128]. The endocrine status of prostate cancer patients has been compared with that of control subjects, but these studies have not provided a consistent pattern [129–132] and are probably not very meaningful, because the presence of the cancer may by itself alter hormonal status, and they are not likely to be informative about the endocrine status prior to the onset of the disease [13, 17, 123]. More meaningful are nested case-control studies in ongoing cohorts and studies comparing healthy males in populations that are at high risk for prostate cancer with populations at lower risk. The two major hypotheses in these studies were that an increased testicular production of testosterone [133] or an increased conversion of testosterone to DHT (increased 5α-reductase activity) [134] are associated with increased risk for prostate cancer [135]. Functional genetic polymorphisms in the 5α-reductase gene or in genes involved in testosterone biosynthesis (the CYP17 gene) or DHT catabolism (the 3β-hydroxy-steroid dehydrogenase gene) could be responsible for increased testosterone production or increased DHT levels (increased conversion of testosterone and/or decreased DHT catabolism) [135]. Polymorphisms with functional significance have also been discovered in the androgen receptor gene [136–138]. These various polymorphisms have been postulated to be critical determinants of prostate cancer risk at the population or individual level by determining intraprostatic DHT concentrations and androgen receptor sensitivity [18, 134].

2.3.1
Steroid Hormonal Factors in Populations with Different Risk for Prostate Cancer

A number of studies have compared circulating levels of steroid hormones and other hormonal factors in very high risk African Americans with those in high risk European Americans, lower risk Asian Americans, and very low risk Asians living in Asia or African black men. The details of these studies are summarized in the following paragraphs.

Ahluwalia et al. [139] studied African Americans and black Nigerian men that were matched controls in a case control study of prostate cancer. Plasma levels of testosterone and estrone were significantly higher in the Americans than in the Nigerian men, whereas levels of DHT and 17β-estradiol were not different. Similar differences were found for the prostate cancer cases. Hill et al. [62–64] compared the hormonal status of small groups of middle-aged African American, European American, and black (rural) South African men consuming their customary diets. In a separate study, African American, European American, and black South African boys and young African American and black South African men were compared [140]. In the older men, plasma levels of the testosterone precursor dehydroepiandrosterone (DHEA) were lower in the two groups of black men than in the white men, and estrone levels were higher. Plasma levels of the testosterone precursor androstenedione and 17β-estradiol were higher in the African men than in the two American groups, but there were no differences among these groups in testosterone levels. In the study with the boys and young men, similar findings were obtained for testosterone and DHEA. However, androstenedione levels were lower (and not higher) in the African than in the American subjects, and 17β-estradiol was lower in young black boys than in white boys but higher in older black boys and young black men than in white boys and men. These interesting data suggest a complex interaction between ethnic background and environmental differences that change during sexual maturation. In these studies by Hill and co-workers among South African black men, the young men were different from the older men for androstenedione and DHEA. This observation suggests that it may be important for the interpretation of hormonal profiles to separately consider younger and older men.

Ross et al. [133] compared healthy young African-American men and young US Caucasian males. Total circulating testosterone and free testosterone were ~20% higher in the black men than in the white group. Serum estrone concentrations were also higher (by 16%) in the blacks than in the whites. There were no differences between the groups in circulating 17β-estradiol and SHBG. The 19–21% difference in circulating testosterone may be sufficiently large to explain the twofold difference in prostate cancer risk between white and black men in the US. This study suggests an association between prostate cancer risk and high concentrations of circulating androgens and, perhaps, estrogens. Henderson et al. [141] compared hormone levels in pregnant African-American and white women in their first trimester. Serum testosterone levels were 47% higher in black women than in white women, and 17β-estradiol levels were 37% higher. No differences were observed in SHBG and human chorionic gonadotrophin. These findings may suggest that US black males are exposed to higher androgen concentrations than white men even before birth.

The US black and white men from the study by Ross et al. [133] were compared with Japanese men of the same age [134]. The serum testosterone levels of Japanese men were intermediate between those of the US whites and blacks, but their SHBG levels were lower. This suggests a higher free testosterone in the Japanese than in the US men, but free testosterone was not measured. The two US groups had higher circulating levels of the conjugated androgen metabolites androsterone glucuronide and $3\alpha,17\beta$-androstanediol glucuronide than the Japanese men. This suggests that in comparison with the high risk US groups, the low risk Japanese population has a lower rate of testosterone metabolism, most likely a lower activity of the enzyme 5α-reductase that converts testosterone to DHT and the testosterone precursor androstenedione to androsterone. However, the higher levels of androsterone glucuronide in US men could also indicate a higher testosterone production rate than in Japanese men. Lookingbill et al. [142] reported similar observations, comparing young healthy US Caucasians and Chinese males in Hong Kong. The Caucasian men had 67%

higher serum levels of androsterone glucuronide and 76% higher levels of $3\alpha,17\beta$-androstanediol glucuronide than the Chinese men. Circulating levels of testosterone, free testosterone, or DHT were not different, but Caucasian men had higher serum levels of the androgen precursors DHEA sulfate and androstenedione. These data are also suggestive of a higher 5α-reductase activity in high risk Caucasians than in low risk Chinese men, and they suggest an increased production of androgen precursors in the Caucasians.

In contrast to the observations of Ross et al. [133, 134] and Lookingbill et al. [142], De Jong and co-workers [143] found 71% higher circulating total testosterone levels in 50–79 year-old Caucasian Dutch men (high risk) than in Japanese men (low risk). DHT levels were not different, but the ratio of DHT to testosterone was slightly lower in Dutch than Japanese men, perhaps indicating lower 5α-reductase activity, but no androgen metabolites were measured. Serum levels of 17β-estradiol were higher in the Dutch than the Japanese men. SHBG levels were not different, but the ratio of testosterone to SHBG concentrations was higher in Dutch than Japanese men, suggesting higher amounts of free testosterone in Dutch men, but this was not measured. Ellis and Nyberg [144] compared serum testosterone levels in non-Hispanic white US Army Vietnam veterans with those in African American, Hispanic, Asian/Pacific Islander, and Native American veterans. The serum testosterone levels in the African American men were significantly higher than those in the non-Hispanic white men, but the differences among the other groups were not significant. The serum testosterone difference between black and white men was larger in men 31–35 years of age (6.6%) than for men 35–40 (3.7%) or 40–50 of age (0.5%).

Wu et al. [145] compared circulating hormone levels in a population-based study of healthy African-American, Caucasian, Chinese-American, and Japanese-American men 35–89 years of age, 8.2% of whom were 60 years old or younger. Serum levels of total testosterone were 9–11% (significant) higher in Asian-Americans than in Caucasians, and they were intermediate and not different from the two other groups in African-Americans. The serum levels of bioavailable testosterone (not bound to SHBG) and free testosterone (not bound to either SHBG or albumin) was 11–12% (significant) higher in Chinese-Americans than Caucasian men. SHBG levels were not different among the groups. In comparison with Caucasian men, DHT levels were 7% higher (significant) in African-Americans and Japanese-Americans, but similar in Chinese-Americans. The ratio of DHT to testosterone was 10% lower (significant) in Chinese-Americans than in Caucasians, but not significantly different in African- and Japanese-Americans than Caucasians. These data do not support the notion of a relation between increased 5α-reductase activity or testosterone production and prostate cancer risk, but no direct indicators of 5α-reductase activity such as androsterone glucuronide and $3\alpha,17\beta$-androstanediol glucuronide were measured.

Santner et al. [146] conducted the only study of direct measurement of androgen production and metabolism by 5α-reductase in populations with different risk for prostate cancer. A radioisotope method using intravenous injection of tritiated testosterone was used to measure the conversion of testosterone to

DHT in young healthy US Caucasians, Chinese-Americans, and Chinese men living in Beijing, China. No differences in the conversion of testosterone to DHT were found among these three groups. Circulating testosterone and SHBG levels were lower in the Beijing Chinese than in the two US groups, and the differences with the US Chinese subjects were significant, but there were no differences in free testosterone. There was an insignificant trend towards lower metabolic conversion rates of testosterone comparing US Caucasians with the Chinese groups and US Chinese with Beijing Chinese. Testosterone production rates were lower in Beijing Chinese than in the two US groups, and the difference with American-Chinese was significant. The ratios of urinary 5β- to 5α-reduced steroids, which is an indicator of overall 5α-reductase activity, were not different in US Caucasian male students compared to Chinese students living in Hong Kong. Urinary excretion of androsterone, etiocholanolone, and total ketosteroids was lower in the Chinese than in the US students. Together, these data indicate that 5α-reductase activity is not different in Asian and Caucasian men and is not affected by the environment in which Asian men live, but they suggest that the living environment influences testosterone production in Asian men.

The *SRD5A2* gene, which encodes for human type II 5α-reductase enzyme and is expressed in the prostate [147, 148], contains polymorphic *TA* dinucleotide repeats in its transcribed 3′ untranslated region [149]. Reichardt et al. [150] demonstrated that alleles containing longer *TA* dinucleotide repeats are unique to African Americans, which may be related to their extremely high risk for prostate cancer. However, the functional significance of these polymorphisms is not yet known.

Makridakis et al. [151] identified another polymorphism in the *SRD5A2* gene, the presence of a valine to leucine mutation at codon 89. In a homozygous state, this mutation confers lower 5α-reductase activity as measured in Asian men with this genotype compared to heterozygous men and men without the mutation [151]. The frequency of the 89 leucine-leucine genotype (lower 5α-reductase activity) was 3–4% in African American and white men, 15% in Latinos, and 22% in Asian Americans. The higher frequency of the 89 leucine-leucine genotype in Latino American men and particularly Asian Americans may be related to their low risk for prostate cancer. These findings were confirmed by Lunn et al. [153].

Makridakis et al. [156] also identified yet another polymorphism in the *SRD5A2* gene, a mis-sense alanine to threonine mutation at codon 49, associated with a substantial increase in activity of the enzyme. The frequency of the mutation was very low, 1.0 and 2.3%, in healthy high risk African Americans and lower risk Hispanic men, respectively, and it seems unlikely that this mutation is responsible for the large ethnic/racial variations in prostate cancer risk.

The CYP17 gene, which encodes for the cytochrome P450C17α enzyme that has both 17α-hydroxylase and 17,20-lyase activity in the biosynthesis of androgens in the adrenal and testis, is polymorphic with two common alleles, the wild type CYP17A1 allele and the CYP17A2 allele containing a single base-pair mutation [157]. This mutation creates an additional Sp1 site in the promoter region, which suggests increased expression potential [157]. The functional significance of this polymorphism in men is not known, but pre- and postmenopausal women with the A2 allele had higher circulating estradiol and progesterone levels than women homozygous for the A1 allele, and circulating levels of DHEA and androstenedione, but not testosterone, were increased in post-menopausal women. Lunn et al. [153] reported frequencies of the A1/A1 and A1/A2 genotype of between 40 and 44% and frequencies of the A2/A2 genotype between 16 and 17% in both African American and European American men, resulting in an A2 allele frequency of 0.36–0.38. In Asians (Taiwanese), however, the frequency of the A1/A1 genotype was 24%, that of the A1/A2 genotype 49%, and that of the A2/A2 genotype 27%, resulting in an A2 allele frequency of 0.52. The frequency differences between the Asians and the American groups were statistically significant, and are possibly related with the low prostate cancer risk in Asian men.

The human *HSD3B2* gene, which is located on chromosome 1p13 and encodes for type II 3β-hydroxysteroid dehydrogenase (an enzyme that catabolizes DHT and is expressed in the adrenals and testes), contains several complex dinucleotide polymorphisms [152]. Devgan et al. [154] reported that the frequency of *HSD3B2* alleles differs among ethnic groups. African American men are unique in one minor allele, and the most common allele is more frequent in European Americans than in either African Americans or Asian men. The second most common allele is more frequent in African Americans than in either Asians and European men. However, the functional significance of these *HSD3B2* gene polymorphisms is unknown.

The human androgen receptor gene, located on the X chromosome, contains polymorphisms which are CAG and GGC (or GGN) microsatellite repeats of different length in exon 1 encoding for the N-terminus of the protein associated with transactivation activity [138]. The CAG repeat length is involved in determining transactivation activity of the androgen receptor; 40 or more repeats are associated with human androgen insensitivity syndromes and reduction of repeat length results in increased transactivation activity in vitro [136–138]. However, the functional significance of the GGC repeat length is not clear. Irvine et al. [155] reported that most African American men have CAG repeat lengths of less than 22, whereas fewer European and Asian Americans have such shorter alleles. Very short alleles of less than 17 repeats were almost exclusively found in African Americans. The most common GGC allele occurred in most Asian American men, fewer European Americans, and only one in five African Americans. The frequency of short GGC repeats of less than 16 was highest in African Americans, lower in Asian Americans, and lowest in European Americans. GGC repeats longer than 16 were rare in the Asian American men, but more frequent in African Americans and European Americans. Thus, shorter CAG repeat alleles, which may be associated with greater androgen receptor transactivation, were the most frequent in the highest risk population (African Americans) and lowest in the lowest risk group (Asian Americans), and the frequency was intermediate in intermediate risk European American men. The high frequency of short GGC repeats in African Americans is perhaps also related with their extremely high risk for prostate cancer.

In conclusion, the above summary indicates few clear or convincing patterns about associations between circulating hormone concentrations and prostate cancer risk at the population level. Two studies examined the levels of 5α-reduced androgen metabolites (androsterone glucuronide and 3α,17β-androstanediol glucuronide), which are considered indicators of 5α-reductase activity [158]. In both studies the levels of these metabolites were lower in low risk Asian populations than in high risk European Americans [134, 142]. These observations suggest lower 5α-reductase activity in the Asians and, therefore, reduced formation of DHT and androgenic stimulation of the prostate. This notion is supported by the higher frequency in Asian men than European Americans or African Americans of a 5α-reductase (SRD5A2) gene polymorphism that appears to be associated with lower 5α-reductase activity [135, 151]. However, when overall conversion of testosterone into DHT was directly measured, no differences were found between Asian and European American men [146]. Furthermore, levels of 5α-reduced androgen metabolites were not higher in very high risk African Americans than in intermediate risk European Americans, and circulating levels of DHT and the ratio of DHT to testosterone were similar in ethnic populations that differ in prostate cancer risk [142, 143, 145]. Although circulating levels of testosterone and free testosterone were slightly higher in African Americans than in European Americans in all studies addressing this issue, this was statistically significant in only one study. In addition, lower as well as higher testosterone concentrations have been reported to occur in lower risk Asian or African men as compared with higher risk European or African Americans, but testosterone levels were lower in Japanese and Chinese men living in Asia than in any ethnic group living in the US. In conclusion, there is at present no substantive evidence in favor of the hypothesis that elevated 5α-reductase activity is causally related to prostate cancer risk at the population level, and only limited evidence for the hypothesis that elevated (free) testosterone levels are associated with prostate cancer risk.

The only two other patterns appearing from the above summarized studies are that levels of estrogens are higher and those of DHEA lower in black Africans and African Americans than in men of European descent (there are virtually no data on Asians in this regard) [62–64, 133, 139, 140, 143]. Higher estrogen levels were only found in black men younger than 50 years. Although the biological significance of these findings is not clear, they may be related to the high susceptibility of black men to prostate cancer when they live in an US environment. Whether this is associated with dietary or other life-style factors or exposures to environmental pollutants is not clear. There are, to the knowledge of the author, no epidemiologic studies that explored a possible role of environmental chemicals with hormonal activity in the etiology of prostate cancer. It is noteworthy that the above summarized endocrine differences between very high risk African Americans and high risk European American were not consistent in younger and older men, and they were not similar to the differences observed between the high risk US populations and the low risk African black men [62–64, 139]. These inconsistencies may suggest that the factors and endocrine mechanisms involved in determining the difference in risk between African black men and African-Americans are different from those that determine the difference in risk between African Americans and European Americans [13].

Finally, androgen receptor transactivation activity may be an important factor. A CAG repeat length polymorphism in the androgen receptor gene was found to be associated with prostate cancer risk in two studies, and short CAG repeat alleles are probably associated with greater androgen receptor activity. Short CAG repeat alleles were most frequent in African Americans (at very high risk for prostate cancer), least frequent in Asian Americans (at low risk), and intermediate in European Americans (at intermediate risk).

2.3.2
Comparisons of Steroid Hormonal Factors in Nested Case-Control Studies

In a number of studies circulating levels of steroid hormones and other hormonal factors have been evaluated in nested case-control studies derived from ongoing cohort studies. The details of each study are summarized in the following paragraphs.

Nomura et al. [159] compared prostate cancer cases with matched controls from a cohort of Hawaiian Japanese men followed for approximately 14 years. There were no significant differences between cases and controls or associations with risk for circulating testosterone, DHT, estrone, 17β-estradiol, and SHBG measured once at the start of the cohort study. There was a relation between risk and an increasing ratio of testosterone to DHT, which was borderline significant $(0.05 < p < 0.1)$. The results from this nested case-control study may suggest an inverse relation between (peripheral) 5α-reductase activity and prostate cancer risk.

Barrett-Connor et al. [160] followed a cohort of white upper-middle class Californian men for a period of 14 years. There was no significant relation between risk for prostate cancer and base-line serum concentrations of testosterone, estrone, and SHBG. However, relative risk increased linearly with increasing serum level of androstenedione, a testosterone precursor. Relative risk also increased with increasing serum level of 17β-estradiol, but this was not significant.

Hsing and colleagues [161, 162] conducted a population-based nested case-control study in a cohort of mostly European-American men. Blood samples were obtained in 1974, but exact follow-up time was not specified. Men of 70 years and older and younger than 70 were studied separately. There were no significant differences between cases and controls or associations with risk for base-line testosterone, DHT, DHEA, DHEA sulfate, estrone, or 17β-estradiol. The ratio of testosterone to DHT was higher in cases than in controls, regardless of age, and for men younger than 70 years, prostate cancer risk was increased with an increasing testosterone/DHT ratio; these associations were only borderline significant. These findings may suggest an inverse relation between 5α-reductase activity and prostate cancer risk.

Nomura et al. [163] reported on a 20-year follow-up of their cohort study of Hawaiian Japanese men. In this follow-up, no significant differences were found between cases and controls or associations with risk for base-line testosterone, free testosterone, DHT, ratio of testosterone to DHT, androsterone glucuronide, $3\alpha,17\beta$-androstanediol glucuronide, and androstenedione. These findings mostly confirmed those of their earlier report [159].

Gann et al. [164] conducted a prospective nested case-control study using probably largely white cases of prostate cancer and controls from the US Physician's Health Study with a follow-up of approximately six years. There were no significant differences between cases and controls for testosterone, SHBG, DHT, ratio of testosterone to DHT, $3\alpha,17\beta$-androstanediol glucuronide, or 17β-estradiol. Because several highly significant associations were found between plasma levels of the various steroid hormones and SHBG studied, odd ratios were calculated after simultaneous adjustment for all these hormonal factors. This approach yielded a significant positive association with risk for testosterone and the ratio of testosterone to DHT, and inverse associations with risk for SHBG and 17β-estradiol. A positive association with risk for $3\alpha,17\beta$-androstanediol glucuronide was borderline significant. There was no association with risk for DHT. These observations contradict earlier mentioned findings suggesting a relation between (peripheral) 5α-reductase activity and prostate cancer risk.

Guess et al. [165] conducted a population-based case-control study from a cohort of European-American men in the Kaiser Permanente Medical Care Program, with a median follow-up of 14 years. There were no significant differences between cases and controls or associations with risk for base-line serum testosterone, free testosterone, DHT, androsterone glucuronide, or $3\alpha,17\beta$-androstanediol glucuronide.

Vatten et al. [166] reported results of a population-based nested case-control study from a cohort of Norwegian men with a mean follow-up of 10 years. There were no significant differences between cases and controls or associations with risk for base-line serum testosterone, DHT, ratio of testosterone to DHT, or $3\alpha,17\beta$-androstanediol glucuronide.

Dorgan et al. [167] reported results from population-based nested case-control study from a cohort of Finnish men from the Alpha-Tocopherol, Beta-Carotene Cancer Prevention Study of cigarette smokers with a follow-up of 5– 8 years. There were no significant differences between cases and controls or associations with risk for base-line serum testosterone, free testosterone, SHBG, DHT, DHEA sulfate, $3\alpha,17\beta$-androstanediol glucuronide, androstenedione, estrone, or 17β-estradiol. There was a non-significant trend towards a higher ratio of testosterone to DHT in cases than controls and a non-significant positive association with risk for this ratio. These finding perhaps suggest an inverse relation between 5α-reductase activity and prostate cancer risk.

Heikkilä et al. [174] conducted a population-based nested case-control study in a Finnish cohort study where serum was collected and stored from a cohort of 16,481 men which was followed for up to 24 years. Over this period 166 prostate cancer cases were identified and 300 matched controls were obtained from the cohort. There were no differences between cases and control and there was association with prostate cancer risk for serum levels of testosterone, SHBG, and androstenedione. However, there was a borderline significant (p= 0.06) trend for increasing risk with increasing levels of testosterone, but only when cases identified in the first eight years of follow-up were excluded. This finding supports the notion of a relationship between elevated androgen levels and prostate cancer risk.

Kantoff et al. [168] studied the association between prostate cancer risk and the in *TA* dinucleotide repeat polymorphisms in the human *SRD5A2* gene, which encodes for type II 5α-reductase enzyme, in a nested case-control study using the Physician's Health Study cohort. As indicated earlier, the functional significance of these polymorphisms is not known. These investigators observed that the frequency of the three most common genotypes was not associated with risk. Men that were homozygous for long repeats were at lower risk for prostate cancer than men with the predominant shorter repeat length genotype, with a borderline significantly decreased odds ratio. These findings are in sharp contrast with the earlier mentioned observation that long alleles are unique to African Americans who are at very high risk for prostate cancer [150].

Another polymorphism in the *SRD5A2* gene was identified by Makridakis et al. [156] as indicated earlier, which is a mis-sense alanine to threonine mutation at codon 49, probably associated with an increase in 5α-reductase activity. Although the frequency of the mutation was low in a nested case-control study using the Hawaii-Los Angeles Multiethnic Cohort Study of Diet and Cancer, it appeared to be responsible for 8–9% of cases in African Americans and Hispanic men [156]. The age-adjusted relative risk of prostate cancer for possessing a mutated allele was 3.28 (significant at the $p=0.05$ level) in African American men and 2.50 (not significant) in Hispanics. For advanced prostate cancer, the relative risks for possessing a mutated allele were more significant: 7.22 (95% confidence interval (CI): 2.17–27.91) in African American men and 3.60 (95% CI: 1.09–12.27) in Hispanics. The results of this study support the notion that increased 5α-reductase activity may be related to prostate cancer risk. However, it seems unlikely that the alanine to threonine mutation at codon 49 in the *SRD5A2* gene is involved in a substantial proportion of prostate cancer cases.

The relation between prostate cancer risk and the occurrence of the aforementioned valine to leucine mutation at codon 89 in the *SRD5A2* gene, a polymorphism that is associated with reduced 5α-reductase activity, was examined by Febbo et al. [175] and Lunn et al. [153]. Febbo et al. [175], who conducted a nested case-control study using the Physician's Health Study cohort, found the valine-leucine and leucine-leucine genotypes in 50% of cases and 51% of controls and these were not associated with elevated prostate cancer risk as compared with the valine-valine genotype. Lunn et al. [153] confirmed these findings in a case-control study that employed prostate cancer patients from urology clinics in North Carolina. Controls, not matched to cases, were drawn from BPH and impotence patients from the same clinics. Most men were European American and 5–11% were African American. The valine-leucine and leucine-leucine genotypes were found in 56% of cases and 49% of controls and were not associated with prostate cancer risk as compared with the valine-valine genotype. These observations are consistent with the results of Febbo et al. [175] who did not find a relation between plasma concentrations of 3α,17β-androstanediol glucuronide and the three different *SRD5A2* gene codon 89 genotypes.

Lunn et al. [153], in the same case-control study, also examined the association between prostate cancer risk and the earlier mentioned single base-pair mutation polymorphism in CYP17 gene [157] which encodes for cytochrome

P450C17α 17α-hydroxylase and 17,20-lyase enzyme activity. The CYP17A2 allele that contains a single base-pair mutation was found in 69% of cases and 57% of controls and it was significantly associated with prostate cancer risk (odds ratio of 1.7). The association was limited to men younger than 65 years with no increased odds ratio for men 65 years or older. Opposite findings of this association between prostate cancer risk and the presence of the CYP17A2 allele were reported from a Swedish case-control study by Wadelius et al. [176]. The frequency of the A1/A2 or A2/A2 genotype was 61% in prostate cancer cases (n = 178) and 71% in population controls (n = 160). The odds ratio of having the A1/A1 genotype as compared to the A1/A2 or A2/A2 genotype was 1.61 (significant at the 95% level). This latter finding is consistent with a preliminary report of higher circulating testosterone levels found in men that are homozygous for the A1 allele as compared with men with an CYP17A2 allele [176].

In six case-control studies the association was examined between the aforementioned CAG and GGC (or GGN) repeat polymorphisms in the human androgen receptor gene [155, 169–171, 177, 178]. In these studies, CAG repeat lengths of shorter than 22, 20, or less than 20 were non-significantly associated with slightly increased risk for prostate cancer [155, 169–171]. In one study [169], the trend of increased risk with decreasing repeat length was significant, but not in two other studies looking at this [170, 171]. In summary, these findings suggest that prostate cancer risk may be slightly increased with shorter CAG repeat alleles and that it may be related with a greater androgen receptor transactivation activity, which is associated with shorter CAG repeat alleles. Three of these studies also addressed GGN or GGC repeats in the androgen receptor gene, the functional significance of which is unknown at present, but contradictory findings were obtained [155, 171, 178]. In all three studies the combined effects of CAG and GGN repeat length were found to be greater than those of either polymorphism separately [155, 171, 178]. However, the functional significance of this combined effect is not clear.

From the studies discussed above it must again be concluded that, with few exceptions, no clear or convincing patterns are apparent about associations between circulating hormone concentrations and prostate cancer risk at the individual level. In most studies an association was found between increased risk and increased ratios of testosterone to DHT, but this was statistically significant in only three of six studies. Although these findings suggest a relation between reduced 5α-reductase activity and prostate cancer risk, no associations were found between risk and the levels of 5α-reduced androgen metabolites that are indicators of 5α-reductase activity. Significant associations were found between prostate cancer risk and elevated levels of testosterone and androstenedione and decreased levels of SHBG and 17β-estradiol, but they each occurred in only a single study [160, 164] and were not found in other studies. It is possible that relevant associations may have been missed in most studies, because, except in the study by Gann et al. [164], the individual hormone data were not adjusted for concentrations of other hormones, although there are many inter-correlations between these. Eaton et al. [179] conducted a meta-analysis study using most, but not all, studies included in this overview and some unpublished data not available to us. They found no significant differences for the ratios of mean hormone levels between cases and controls, except for slightly elevated levels of $3\alpha,17\beta$-androstanediol glucuronide. This meta-analysis is in agreement with our analysis, with the only consistent finding being slightly elevated ratios between cases and controls of $3\alpha,17\beta$-androstanediol glucuronide in five of five studies. However, Eaton et al. [179] did not take into account the risk estimates produced by these studies, which seriously limits their conclusions.

The results of three nested case-control studies on prostate cancer risk and polymorphisms in the human type II 5α-reductase enzyme gene (*SRD5A2*) do not support the hypothesis of an association between risk and increased 5α-reductase activity [153, 168, 175]. There are only contradictory data on a relation between prostate cancer risk and a polymorphism with unknown functional significance in the CYP17 gene encoding for the cytochrome P450C17α 17α-hydroxylase and 17,20-lyase activity which is involved in androgen biosynthesis [153, 177].

In five similar studies of polymorphisms in trinucleotide repeats in the androgen receptor gene, a weak association was found between risk and shorter CAG repeat alleles, which may be related with greater androgen receptor transactivation activity, but this association was significant in only one study [155, 169–171, 177]. However, short CAG repeats were also correlated with early onset of prostate cancer [172]. An association between risk and polymorphisms in androgen receptor GGC or GGN repeat lengths is not clear because results of three studies of this were inconsistent, and the functional significance of these polymorphisms is not known [155, 171, 178]. There is possibly an interaction between CAG and GGC/GGN repeat length in relation to prostate cancer risk, but results of the studies examining this were inconsistent [155, 171, 178].

2.3.3
Critical Interpretation of the Studies

The results of the studies summarized above do not provide unequivocal or strong evidence for any particular association between prostate cancer risk and hormone exposure (circulating levels of hormones) or polymorphisms in genes that encode for proteins involved in steroid hormone action or metabolism. The only associations with prostate cancer risk that have been observed consistently in at least three studies are slightly higher circulating testosterone and estrogen levels in high risk African American men as compared with lower risk European American men and a functional polymorphism in the androgen receptor gene; however, these associations are weak at best and mostly not statistically significant. There are no epidemiologic data about prostate cancer risk and exposure to exogenous hormones, including environmental agents with hormonal activity.

Several important points should be taken into consideration in interpreting the above-mentioned observations. First, there are many potential problems with studies that measure circulating hormone levels such as the usually large inter- and intra-assay variability in the immunoassays used [121, 173], the fact that typically only single blood samples are studied, and the problem of within-subject variations over time and circadian rhythms. Another difficulty relates to the interrelationships between various hormones [143, 163] which are taken into account in very few studies during data analysis [164]. Second, young Asian men are probably at least partially Westernized in their life-style, and they cannot simply be compared with older Asians. In addition, young men studied today with respect to their hormonal status probably have a prostate cancer risk that is different from the risk currently recorded in older men of the same population, as suggested by the rising prostate cancer rates in Japan. Third, as indicated earlier, the factors that cause the differences in prostate cancer risk between black, white, and Asian men in the US may be different from those that determine differences in prostate cancer risk between Asian or African populations and populations in the US or West European countries. Fourth, circulating hormone levels provide very little information about concentrations at the molecular targets of these hormones in the prostate gland or about steroid hormone metabolic processes within the prostate. For example, less than 10% of circulating DHT is produced by the prostate and a substantial amount of serum $3\alpha,17\beta$-androstanediol glucuronide is derived from non-prostatic sources. Therefore, these two steroids are not very good indicators of prostatic 5α-reductase activity [156–158]. Also, aromatase activity has been reported to occur in the human prostate and the LNCaP prostate cancer cell line [180, 181], although there are contradictory findings [182, 183]. In addition, estrogen levels in the human prostate exceed those found in the circulation [184], suggesting that local formation of estrogens may occur in the prostate and may contribute to disregulation of prostate growth. Finally, although there are reports of functional polymorphisms in genes encoding for proteins involved in steroid hormone action or metabolism, their influence on steroid hormone metabolizing enzyme activity or steroid hormone receptor activity *within the prostate* is not known. It is likely that these polymorphisms have only limited and probably cell type-specific influences on these regulatory processes, given the highly complex and often tissue-specific mechanisms of regulation of gene expression.

On the basis of the above-summarized studies, one may hypothesize that prostate cancer risk is positively associated with one or more of the following hormonal factors:

1. Serum levels of (bioavailable) testosterone as indicated by studies comparing healthy low and high risk men [133, 134, 139, 143, 146]
2. Serum levels of androgen precursors (DHEA) as indicated by a prospective study [160] and a study comparing healthy Chinese and US Caucasian men (androstenedione) [142]
3. Peripheral and possibly prostatic activity of 5α-reductase [134, 142, 151, 164]
4. Serum levels of estrogens as indicated by studies comparing healthy low and high risk men [62–64, 133, 134, 143]
5. Increased androgen receptor transactivation activity [155, 169, 170].

However, for each of these hypotheses there are, as indicated earlier, contradictory data, and the observed associations were at best weak.

Two of the four hypotheses implicate higher bioavailable circulating androgen levels in high risk men in comparison with low risk populations [173], suggesting increased androgen production [135]. However, this seems not to be true for all high risk groups. Meikle and co-workers [185, 186] reported that serum levels of testosterone and DHT were significantly lower, not higher, in blood relatives of prostate cancer patients (brothers and sons of prostate cancer patients who have a three- to fourfold excess risk) than in unrelated control subjects. Zumoff et al. [187] observed that circulating levels of testosterone, but not DHT, in prostate cancer patients were markedly lower in those younger than 65 years than those of 65 years and older, whereas control subjects had testosterone levels that were similar to those of prostate cancer patients of 65 years and older. In several of the earlier summarized epidemiologic studies, some hormonal findings were markedly different when comparing younger (18 to 25 – 40 years) and older healthy men (over 40 years of age), or comparing younger (less than 62 to 70 years) and older (older than 65 to 70 years) prostate cancer patients with their age-matched controls. Circulating testosterone levels paradoxically decrease with aging, while prostate cancer risk increases [143 – 145, 188]. At the same time, SHBG levels increase with age and estrogen levels remain constant or increase [143, 145, 188]. Thus, bioavailable estrogens and particularly testosterone decrease with increasing age and increasing risk for prostate cancer. This may explain the lower prostatic concentrations of DHT with aging reported by Krieg et al. [184], but is in contrast to increasing prostatic estrogen levels with aging observed by this group.

These studies suggest that the role of androgens and estrogens in prostatic carcinogenesis may differ in younger men (or perhaps men with early onset prostate cancer) and older men (or men with late onset cancer), and may be different in men that are at high risk because of familial predisposition and those at high risk associated with their ethnic background or living environment. It is also possible that risk-increasing effects of elevated circulating levels of androgens and possibly estrogens may be effectuated early in life [133, 134, 140, 142] or even before birth [141], rather than in one or two decades preceding the diagnosis of prostate cancer.

2.4
Epidemiology: Summary and Conclusions

There are only few clearly established and strong risk factors for prostate cancer which are consistently observed in epidemiologic studies: (1) African-American descent, (2) a Western life style, in particular Western dietary habits, and (3) a family history of prostate cancer. Less consistently found and weaker risk factors are a history of venereal disease, and employment in farming, the armed services, and the nuclear industry.

Elevation of bioavailable and bioactive androgens in the circulation and in the prostatic target tissue may be an important and biologically very plausible risk factor. As will be detailed later, the results of several animal model studies strongly support this contention, but more research is needed to confirm and further define this association in humans, and to establish its underlying biological mechanisms (increased androgen production or 5α-reductase activity and decreased DHT catabolism) (see also [173]). Since circulating testosterone levels may be lower in men with a family history of prostate cancer than in other men, hormonal involvement in familial aggregation of prostate cancer risk seems paradoxical and the data suggest that the involvement of androgens in hereditary prostate cancer is different from sporadic prostate cancer. Similarly paradoxical is the well-established inverse relationship between decreasing circulating androgen levels and increasing prostate cancer risk with aging. Although body mass index or obesity does not appear to be a risk factor, there are some indications that muscle mass is positively correlated with risk, perhaps reflecting exposure to endogenous androgens or anabolic steroids. Another risk factor may be increased androgen receptor activity related to genetic polymorphisms in the androgen receptor gene. Heavy alcohol use accompanied with liver disease may increase risk and be related with decreased clearance of estrogens and elevated circulating estrogen levels. Estrogen levels were also elevated in healthy black men living in the US or Africa as compared with European American men, and this is perhaps associated with the very high risk for prostate cancer of black men living in the US. However, no association between risk and circulating estrogen levels was found in nested case-control studies in predominantly European American cohorts. Thus, the epidemiologic evidence for involvement of androgenic and estrogenic steroid hormones in human prostate carcinogenesis remains inconclusive (see also [173]).

The strongest single risk factor appears to be a Western life style, particularly Western dietary habits, including a high fat intake. It is conceivable that dietary risk factors exert their enhancing effects mediated by a hormonal mechanism that involves androgens. However, it is unlikely that life style is the sole factor that explains the differences in prostate cancer risk between Asian and American populations [10, 135]. The single most important combination of risk factors is to be of sub-Saharan African descent and to reside in the US – African Americans have, as a group, twice the risk of white Americans. The reasons for the black-white disparity in prostate cancer rates in the US are not understood. While genetic factors, such as genetically determined elevated 5α-reductase activity, are possibly involved, environmental exposures (in the broadest sense of the term) are probably responsible for a large fraction of this disparity [10, 13, 17, 189]. A relation with similar racial disparities in exposure to potential carcinogens and high-risk dietary habits has been proposed [13, 22, 58]. A hormonal mechanism may be involved as well, possibly acting *in utero*, because young African-American men and pregnant African-American women have been reported to have higher circulating levels of androgens and estrogens than European Americans [141].

3
Hormonal Induction of Prostate Cancer in Laboratory Animals

With the exception of dogs and humans, spontaneously occurring prostate cancer is rare in most species [5–7, 190]. It not understood why prostate cancer is so common in men whereas it is very rare in almost all other species. As indicated earlier, there are compelling reasons to implicate hormones, particularly androgenic and estrogenic steroids, in human prostate carcinogenesis. As will be demonstrated in the following sections, the same steroid hormones are also very powerful factors in the induction of prostate cancer in rodent species in which spontaneous prostate cancer is rare [17, 56, 191]. However, it is important first to point out that the various lobes of the rat prostate differ in their propensity to develop prostate carcinomas, either spontaneously or induced by carcinogens or hormones [17, 190, 191]. The rodent prostate, unlike the human or canine prostate, consists of distinct paired lobes – the ventral, dorsal, lateral, and anterior lobes; the dorsal and lateral lobes are often referred to as the dorsolateral prostate, and the anterior lobe is commonly called the coagulating gland. These lobes have merged into one gland in the human and canine prostate, in which different zones have been defined [192]. However, a homologue of the rodent ventral lobe is not present in the human gland [193].

3.1
Induction of Prostate Cancer with Sex Hormones

3.1.1
Androgens

Chronic administration of testosterone induces a low to moderate (5–56%) incidence of prostate cancer in several rat strains [190, 194–199], but not in all strains [200]. The induced tumors were adenocarcinomas in all studies but one, in which also some squamous cell carcinomas were observed [197], and these carcinomas appeared to develop from the dorsolateral prostate and/or coagulating gland, but not the ventral prostate lobe [190, 194–198]. The prostate carcinoma incidence in most of these studies was low (5–20%) [190, 194, 197, 198]. Only the studies reported by Pollard and co-workers using the Lobund Wistar strain sometimes had higher carcinoma incidences, but the incidences varied considerably (0–60%) [195, 196, 201–204]. In the only other study with the Lobund Wistar strain, a 7% incidence was found [198]. The actual dose of testosterone considerably fluctuated over time in many of these studies from five to ten times control values down to control values [198, 200], but even when the level of circulating testosterone was kept steadily elevated by two- to threefold, prostate carcinomas were induced [199]. These data indicate that testosterone acts as a complete carcinogen for the rat prostate.

3.1.2
Estrogen and Androgen Combinations

Noble first demonstrated that testosterone is carcinogenic for the rat prostate, and he established that sequential treatment with testosterone and estrogens was even more effective than testosterone per se. Noble used the Noble (or NBL) rat strain which he developed [194]. Long-term treatment of NBL rats with a combination of testosterone and 17β-estradiol leads to a high incidence of adenocarcinomas, that develop from the periurethral ducts of the dorsolateral and anterior prostate [205–207]. The development of these tumors is preceded by the appearance of epithelial dysplasia in these ducts and in the acini of the dorsolateral prostate [206–208]. Carcinomas developing from the acinar dysplasia in the periphery of the prostate gland have not been reported, and the malignant potential of these lesions, which are morphologically similar to human prostatic intraepithelial neoplasia (PIN), is not certain [206, 207]. Treatment of NBL rats with diethylstilbestrol (DES) combined with testosterone resulted in wide spread dysplasia in the ventral prostate but less dysplasia in the dorsolateral prostate [208]. Long-term treatment of NBL rats with DES and testosterone induced a low carcinoma incidence in the dorsolateral prostate and some early-stage carcinomas in the ventral lobe [207]. In summary, the addition of estrogen to testosterone treatment of NBL rats markedly increases the incidence of prostate cancer from 35–40% to 100%.

3.1.3
Perinatal Estrogen Exposure

Carcinogenic effects on the male accessory sex glands have been reported after perinatal exposure of mice, rats, and hamsters to DES (see [17, 209–211]). McLachlan and co-workers [209, 212] found that 25% of the male offspring of CD-1 mice that had been treated with DES on days 9–16 of gestation had nodular enlargements of the coagulating gland, ampullary glands, and colliculus seminalis at an age of 9–10 months. One animal had a lesion in the area of the coagulating gland and colliculus seminalis that resembled early cancer [212]. Of eight prenatally DES-exposed male mice that survived for 20–26 months, one had an adenocarcinoma of the coagulating gland, three had hyperplasia of the coagulating gland, two had hyperplasia of the ventral prostate, one had a carcinoma of the seminal vesicle, and two had squamous metaplasia of the seminal vesicle, but no such lesions occurred in control animals [209, 210]. Prenatal DES exposure of mice also induced testicular tumors (particularly of the rete testis) and non-neoplastic lesions in the testes and epididymis [213].

Treatment of Han:NMRI mice with DES or 17β-estradiol on the first 3 days of life resulted in a 90–100% incidence of epithelial dysplasia of the periurethral glands, and the periurethral proximal parts of the dorsolateral prostate, coagulating glands, and seminal vesicles after 12–18 months [214, 215]. Subsequent treatment with DHT and 17β-estradiol from 9 to 12 months of age increased the severity of the dysplasia when the prostates were examined at 12 months, suggesting permanent estrogen hypersensitivity of these tissues. Arai et al. [210] treated Wistar rats with DES for the first 30 days of life. One group was neonatally castrated and the second group remained intact. Two of 11 castrated, DES-exposed rats developed squamous cell carcinomas in the area of the dorsolateral prostate, coagulating gland, and ejaculatory ducts, and all these animals had papillary hyperplasia and squamous metaplasia of the coagulating gland and ejaculatory duct. Squamous metaplasia was also found in some of eight non-castrated, DES-exposed rats, but no hyperplasia or neoplasia developed. Vorherr et al. [216] obtained similar results in rats exposed prenatally and/or neonatally to DES.

In summary, the results of these studies demonstrate that prenatal and neonatal estrogen exposure of rodents can be carcinogenic for the prostate. There are also data to suggest that these treatments may imprint permanent alterations in the hormonal sensitivity of the prostate which may play a role in the carcinogenic effect of perinatal estrogen exposure.

3.2
Induction of Prostate Cancer with Chemical Carcinogens and Sex Hormones

3.2.1
Chemical Carcinogens Combined with Hormonal Stimulation of Cell Proliferation

Induction of prostatic adenocarcinomas by chemical carcinogens is rare. Only two organic chemical carcinogens, i.e., *N*-nitroso-bis-(oxopropyl)amine (BOP) and 3,2′-dimethyl-4-aminobiphenyl (DMAB), cause prostate adenocarcinomas upon systemic administration, without any additional concomitant or subsequent treatment [217, 218]. Direct application of chemical carcinogens to prostate tissue in experimental animals produces sarcomas or squamous cell carcinomas (see [5, 219]).

Hormonal stimulation of cell proliferation in the prostate at the time of carcinogen administration has been demonstrated to increase the sensitivity of the target cells for tumor induction [199, 220–223]. Dorsolateral prostate adenocarcinomas have been produced at 5–25% incidence when prostatic cell proliferation was stimulated in combination with treatment with indirect-acting carcinogens, such as DMAB and 9,12-dimethylbenz[*a*]anthracene, and direct-acting chemical carcinogens, such as *N*-methyl-*N*-nitrosourea (MNU); except for DMAB, these carcinogens do not induce these tumors when administered alone [199, 220–223]. However, not in all studies has this enhancing effect of stimulation of prostatic cell proliferation on prostate carcinoma induction been found [197, 200, 224, 225]. Nevertheless, stimulation of cell proliferation can be considered to be co-carcinogenic for prostate cancer induction in rats by many chemical carcinogens.

3.2.2
Testosterone as Tumor Promotor of Prostate Carcinogenesis

Long-term administration of testosterone to rats markedly enhances prostatic carcinogenesis following initial treatment with chemical carcinogens that target the prostate because of tissue-specific metabolism (DMAB and BOP) and/or concurrent hormonal stimulation of prostatic cell proliferation [69, 70, 191, 196–200, 202, 203, 226]. The presence and magnitude of this enhancement appears to depend on several factors [190, 191, 197]. For example, after a single injection of BOP or MNU given to F344 rats without concurrent stimulation of prostatic cell proliferation, long-term testosterone treatment did not enhance prostatic carcinogenesis [200]. High incidences (66–83%) of adenocarcinomas of the dorsolateral and/or anterior prostate were induced by chronic treatment with testosterone following a single administration of MNU or BOP given during stimulation of prostatic cell proliferation in Wistar rats, or during and after ten repeated biweekly injections of DMAB in F344 rats [69, 70, 190, 191, 197, 199, 200, 226]. This effect is somewhat strain-dependent, because when similar treatments were given to Lobund Wistar rats, rather variable incidences of between 50% and 97% were reported by Pollard and co-workers [196, 203, 207], and only a 24% incidence was found by Hoover et al. [198].

The enhancing effect of testosterone on prostate carcinogenesis is remarkably confined to the dorsolateral and anterior prostate, and no tumors occur in the ventral prostate. Moreover, chronic testosterone treatment produces a shift of the site of DMAB- and BOP-induced carcinoma occurrence from exclusively the ventral lobe to predominantly the dorsolateral and anterior lobes [197, 200, 226]. The dose-response relationship between testosterone dose and prostate carcinoma yield is very steep: elevation of circulating testosterone levels by less than 1.5-fold is sufficient for a near-maximal enhancement of the tumor response, and a two- to threefold elevation is sufficient for a maximal response; these concentrations are within the normal range of circulating testosterone levels in the rat [199]. Thus, testosterone is a powerful tumor promotor for the rat prostate.

3.2.3
Effects of Androgens on Prostate Cancer Induction by Cadmium and Ionizing Radiation

Cadmium can be carcinogenic for the rat ventral prostate as demonstrated by Waalkes and co-workers [227, 228]. The selective sensitivity of the ventral prostate lobe for the carcinogenic action of cadmium is most likely due to the lack of cadmium-binding proteins in this lobe [229]. A single injection of cadmium chloride produced early stage carcinomas in the ventral lobe, but only when cadmium-induced testicular toxicity was avoided, either by keeping the cadmium dose low or by antagonizing the testicular toxicity of cadmium by simultaneous administration of zinc. These observations indicate that cadmium induces proliferative lesions in the rat ventral prostate only when testicular function, conceivably testosterone production, is intact. In addition, these data suggest that androgens also act as tumor promotors in this system, but this hypothesis has not been tested. Other mechanisms may also be involved, because, for example, testosterone considerably increases disposition and retention of cadmium in the rat ventral prostate [230].

Local radiation (X-ray) exposure of the pelvis has been shown to induce prostate carcinomas in mice [231] and rats [232]. Prostate carcinomas (33% incidence) developed only in rats that were castrated and received androgen-replacement prior to irradiation, whereas intact and castrated rats did not develop prostate cancer following irradiation. These observations suggest that testosterone treatment was required for tumor development, perhaps as tumor promotor [232].

3.3
Conclusions

Stimulation of prostatic epithelial cell proliferation by androgens during exposure to chemical carcinogens increases the susceptibility of the rat prostate to cancer induction as a co-carcinogen. Testosterone is a weak complete carcinogen, but is a very strong tumor promotor for the rat prostate at near-physiological plasma concentrations [199]. The very strong tumor-promoting activity of androgens is perhaps responsible for their weak complete carcinogenic activity on the rat prostate. A slight elevation of circulating testosterone can lead to a marked increase in prostate cancer in rat models. This observation is very important in view of the aforementioned weak association between human prostate cancer risk and slightly elevated circulating androgen levels found in some epidemiological studies [173]. Thus, the experimental animal data provide strong support for the concept that minimal increases in circulating androgens may have substantial enhancing effects on prostate cancer risk. The addition of 17β-estradiol to chronic androgen treatment strongly enhances the carcinogenic activity of testosterone for the rat dorsolateral prostate, particularly the proximal periurethral ducts of the dorsolateral and anterior prostate. The 17β-estradiol plus testosterone treatment also induces acinar lesions that are similar to human PIN. These observations strongly suggest a critical role for estrogens in prostate carcinogenesis. Perinatal estrogen exposure can also be carcinogenic for the male rodent accessory sex glands. The proximal periurethral ducts of the dorsolateral and anterior prostate and seminal vesicle and the intraprostatic urethral epithelium appear to be the male rodent genital tract tissues most sensitive to the carcinogenic effects of perinatal estrogen exposure. Interesting in this regard is the report by Driscoll and Taylor [233] of hypertrophy and squamous metaplasia of the prostatic utricle and prostatic ducts in 55 –71% of 31 male infants that had been exposed to DES in utero and had died perinatally from unrelated causes. Such squamous metaplastic changes have also been reported to occur in human fetal prostatic tissue transplanted into nude mice that were subsequently treated with DES [234]. These human observations suggest that the DES findings in rodents may have human relevance. There are no other data on carcinogenic effects of environmental hormonal agents for the prostate in animal models.

4
Mechanisms of Hormonal Prostate Carcinogenesis

As stipulated earlier, there are compelling reasons to assume that androgens play a critical role in prostate carcinogenesis, and there is experimental evidence to suggest that estrogens are involved as well [55]. Because of the hormonal nature of these steroids, receptor mediation has been proposed as the major mechanism by which androgens and estrogens act in the causation of prostate cancer [235]. For estrogens, however, non-receptor-mediated genotoxic effects are conceivable, in addition to receptor-mediated processes [55]. For androgens, mechanisms other than those mediated by androgen receptors seem unlikely, except for the generation of estrogens via aromatization. These potential mechanisms are discussed in the following sections.

Prostatic mesenchyme is known to be a mediator of androgen action in the developing and adult rodent prostate and possibly the human prostate [236, 237]. Therefore, interactions between epithelial and stromal cells in the normal prostate are undoubtedly critical and may be essential in prostate carcinogenesis. However, there are hardly any studies that have directly addressed the role of stromal-epithelial interaction in human or rodent prostate carcinogenesis [238]. Krieg et al. [184] measured steroid hormone concentrations in stromal and epithelial compartments of normal human prostates from subjects varying from 20 to 80 years of age. Epithelial DHT concentrations decreased considerably with aging, but they remained stable in stromal cells and testosterone concentrations appeared unaffected by age in either cell type. Thus, the activity of 5α-reductase in the epithelium may decrease with aging, but perhaps remains intact in the stroma. However, concentrations of 17β-estradiol and estrone in the stroma, but not the epithelium, increased markedly with aging. These findings suggest that the prostatic stroma is an important site for both androgen and estrogen action and metabolism such as aromatase activity, which seems to increase with aging because estrogens accumulate with aging and androgen levels remain stable, in contrast to concentrations of estrogens and androgens in the circulation or in epithelial cells which decrease with aging. Thus, it is conceivable that, with increasing age and increasing risk for prostate cancer, the prostatic stroma continues to be an important androgen signal mediator to the epithelium and is an increasingly important local producer of estrogens.

4.1
Androgens and Prostate Carcinogenesis

Results from the earlier summarized rodent experiments clearly indicate carcinogenic and strong tumor promoting properties of androgens, and the results of a limited number of epidemiological studies provide some support for the concept that androgens may have such effects in humans. The mechanisms of the carcinogenic and tumor-promoting effects of androgens on the rodent prostate are not known with certainty. The very steep relationship between testosterone dose and prostate carcinoma response in rat models suggests involvement of an androgen receptor-mediated mechanism [199]. However, other mechanisms may also be involved. For example, Ripple et al. [239] observed increased oxidative stress in androgen-sensitive LNCaP human prostate cancer cells exposed to DHT, but it is possible that these effects were androgen receptor-mediated.

4.1.1
Stimulation of Cell Proliferation and Carcinogenic and Tumor-Promoting Effects of Androgens

Androgen receptor-mediated stimulation of prostatic cell proliferation by androgens has been implicated in human prostate carcinogenesis [135, 235]. However, there is no direct evidence that elevation of circulating testosterone leads to increased cell proliferation in the human prostate. Androgen administration to castrated rodents causes elevation of prostatic cell proliferation similar to that observed in cell cycle synchronization experiments with cells in vitro [240]. The increase in prostatic cell proliferation caused by androgen administration to castrated rodents is only transient, and after a few days cell turnover returns to its normal very low level [240]. Thus, continued androgen treatment of rodents does not result in constantly elevated rates of cell proliferation in the male accessory sex glands, but probably only supports differentiation. DHT may even suppress prostatic cell proliferation in intact rats [206]. Thus, continuous stimulation of cell proliferation is unlikely be the major mechanism of the enhancing effects of androgens on prostate carcinogenesis in rodents and possibly humans. There are conceivably also other, non-hormonal, factors that affect prostatic cell proliferation, such as inflammatory insults (prostatitis) [241, 242] and possibly sexual activity. Strong support for a cell proliferation hypothesis is derived from rodent experiments that indicate that increased prostatic cell proliferation at the time of exposure to carcinogens can enhance the sensitivity of the tissue to the carcinogenic effects of these agents [199, 220–223]. Stimulation of cell proliferation during carcinogen exposure increases the likelihood that promutagenic DNA damage, such as carcinogen-DNA adducts, will be fixed as permanent somatic mutations. Increased cell proliferation may thus enhance the carcinogenic effects of low level exposure to environmental and endogenous carcinogens in humans.

Importantly, the rate of cell proliferation at the time of carcinogen exposure may be only one of several androgen-related factors that determine sensitivity of the prostate to cancer induction by carcinogens through androgen-receptor mediated mechanisms. For example, Sukumar et al. [243] have hypothesized that prostate cells that have undergone critical genetic alterations (activating mutations in oncogenes or inactivating alterations in tumor suppressor genes) may be selectively sensitive to stimulation of cell proliferation by androgens. These cells could thus have a selective growth advantage over normal cells, which do not respond to chronic testosterone treatment with sustained proliferation, but with cellular differentiation [200]. However, this hypothesis has not been critically tested. It is also possible that androgens, in addition to other factors, influence the effectiveness of indirect-acting carcinogens that are metabolized in the prostate itself.

4.1.2
Androgen Metabolism and Androgen Receptor Sensitivity

Any hypothesis implicating androgens in prostate carcinogenesis has to consider androgen receptor function and androgen metabolism. Ross et al. [135] have developed the concept that genetically determined differences in androgen receptor activity as well as the activities of steroid biosynthetic enzymes, 5α-reductase, and enzymes that metabolize DHT are major determinants of risk both at the population and individual levels (see also [18]). Functional polymorphisms in the genes that encode for these enzymes and the androgen receptor have been hypothesized to be at the core of this notion [18, 135].

The earlier summarized and evaluated results of endocrinological studies and evidence for a role of these polymorphisms in human prostate carcinogenesis lead to the following conclusions: to date there is little evidence that functional polymorphisms in the 5α-reductase gene and differences in 5α-reductase activity are important determinants of prostate cancer risk. However there is some evidence [155, 169–171, 177] to suggest that risk is associated with a functional polymorphism in the androgen receptor gene resulting in short lengths of CAG repeats in the transactivation domain of the protein which are linked with increased transactivation activity in vitro [136–138]. Several other polymorphisms with as yet unknown functional significance have been identified in genes encoding for androgen metabolizing enzymes (cytochrome P45017α (CYP17) involved in androgen biosynthesis, and type II 3β-hydroxysteroid dehydrogenase catabolizing DHT) and the androgen receptor (GGC/GGN repeat length). These have been found unevenly distributed among populations that differ in prostate cancer risk [153–155] or to be associated with risk in case-control studies [154, 155, 171, 176, 178].

The observation of slightly higher circulating testosterone levels in high risk African American men compared to lower risk European men suggests that their rates of androgen biosynthesis may be higher. Lower as well as higher serum testosterone concentrations have been found in low risk Asian or African men as compared with higher risk European or African Americans. However, testosterone levels were lower in Asians living in Asia than in American populations, regardless of their ethnicity, in the only two studies that included Asian populations. In addition, testosterone production rates measured directly were lower in Chinese in China than in either Chinese Americans or European Americans [146]. These observations are consistent with the hypothesis that environmental factors such as diet determine prostate cancer risk at the population level by influencing androgen production.

Evaluation of the role of androgens in prostate carcinogenesis is complicated by the fact that not only the epithelium but also the stroma is an important site for androgen action and metabolism in the prostate. Clearly, studies of circulating androgenic (or other) hormone levels or genomic polymorphisms in relevant genes do not necessarily provide relevant information about processes at the level of the prostatic epithelial cell and its important immediate environment, the prostatic stroma.

4.2
Estrogens and Prostate Carcinogenesis

As summarized earlier, the results of epidemiological studies provide limited evidence for an association between prostate cancer risk and circulating levels of estrogens, which appear to be higher in African American men (under 50 years of age) than in European American men. This observation suggests that estrogens may be involved in prostate carcinogenesis, because men of African descent living in an American environment have the highest risk for prostate cancer world-wide. However, most direct evidence in support for a role of estrogens in prostate carcinogenesis is derived from studies with treatment of NBL rats with testosterone and 17β-estradiol [206, 207, 244]. The mechanisms involved in the prostatic effects in this model are a mixture of estrogen receptor-mediated and non-receptor processes. In addition, there is evidence to suggest that the mechanisms involved in hormonal induction of rat prostate carcinomas, which originate from the periurethral prostatic ducts, are different from those involved in the induction dysplastic lesions, the dorsolateral prostate acini in this model.

There is evidence for the presence of the CYP19 enzyme aromatase in the human prostate which could provide a local source of estrogens from conversion of testosterone [177–181], but reports are contradictory [182, 183]. The local production of estrogens in the prostate is possibly a stromal process, and stromal aromatase activity may increase with aging [184]. Data on the presence of aromatase in the rodent prostate are also somewhat contradictory, because aromatase activity has been reported in the rat ventral prostate and a transplantable rat prostate carcinoma [245] but it was not detectable in mouse prostate [246]. These discrepancies may be due to species or methodological differences.

4.2.1
Estrogen Receptor-Mediated Mechanisms

The prostate contains estrogen receptors, and both the estrogen receptor-α and -β are present in the rat prostate [247]. Thus, involvement of direct receptor-mediated effects of estrogens in prostate carcinogenesis is plausible, but rodent studies using anti-estrogen treatments (such as tamoxifen and ICI-182,780) have yielded contradictory results. The prostate tumor-promoting effects of testosterone may involve estrogen generated by aromatization. However, simultaneous administration of testosterone and tamoxifen did not alter the prostate carcinogenesis enhancing effect of the androgen in experiments in rats injected with prostatic carcinogens prior to the hormone treatment [248; McCormick and Bosland, unpublished data]. On the other hand, ICI-182,780 blocked the induction of epithelial dysplasia in the prostatic periphery in NBL rats treated with testosterone and 17β-estradiol [249]; the effects of this antiestrogen on induction of periurethral prostate carcinomas are not known.

Dorsolateral prostatic tissue with epithelial dysplasia from NBL rats treated with testosterone and 17β-estradiol for 16 weeks accumulates 17β-estradiol and 5α-androstane-3β,17β-diol, a weak estrogenic agonist; this does not occur in the ventral lobe, which also does not develop dysplasia [206, 208]. In rats treated with testosterone and DES for 16 weeks, dysplasia developed more distinctly in the ventral than in the dorsolateral prostate, coinciding with a preferential accumulation of 17β-estradiol and 5α-androstane-3β,17β-diol in the ventral prostate [208]. These observations suggest that increased levels of estrogenic species in prostatic target tissue may be causally related with the development of hormone-induced dysplasia and possibly carcinomas in the NBL rat model [208].

In dorsolateral prostatic peripheral tissue with epithelial dysplasia from rats treated with testosterone and 17β-estradiol for 16 weeks, elevated levels have been found of nuclear but not cytosolic type II (intermediate-affinity) estrogen binding sites, but not type I (high-affinity) binding sites [206, 248]. The type II estrogen receptor is believed to be a cell proliferation marker and a key factor in normal and aberrant growth regulation in female estrogen target tissues [206, 248]. In addition, mitotic indices in testosterone plus estrogen-treated

NBL rat dorsolateral prostate were increased over control values; this increased mitotic activity was largely confined to the dysplastic lesions [206, 248]. These data indicate that continued stimulation of cell proliferation may be involved in the development of hormone-induced NBL rat prostate dysplasia [206, 248].

Estrogen treatment in rodents stimulates prolactin secretion, raising the possibility that some or all estrogen effects on the rodent prostate may be mediated through prolactin. Transplantation of a prolactin-producing pituitary tumor into rats treated with the carcinogen DMAB enhanced the formation of pre-neoplastic lesions, but not carcinomas, in the ventral prostate; treatment with bromocryptine, a prolactin secretion suppressing agent, counteracted this effect [250]. Bromocryptine also decreased the formation of ventral prostatic atypical hyperplasia and carcinomas in rats treated with only DMAB [250], and it blocked the development of epithelial dysplasia in the dorsolateral prostatic periphery of NBL rats treated with testosterone and 17β-estradiol, but effects on periurethral carcinoma development were not studied [251]. Thus, there is evidence to suggest that prolactin may modulate the induction of preneoplastic lesions in the rat prostate, but the relevance of these findings for prostate cancer development is not clear.

In conclusion, there are several lines of evidence suggesting that estrogen receptor-mediated mechanisms contribute to the induction of prostate cancer in rats by hormonal treatments, but conclusive evidence is lacking and there are no human data in this regard.

4.2.2
Non-Receptor Mechanisms

Estrogens are capable of producing DNA damage in the target tissues for estrogen carcinogenicity, independent of their interaction with the estrogen receptor [252]. Gladek and Liehr [253] have found a direct DES-DNA adduct in the kidney of male hamsters treated with DES, as well as indirect estrogen-generated DNA adducts (perhaps of endogenous origin and of undetermined structure) detectable by ^{32}P-postlabeling [254]. Both observations are thought to be related with the formation of catechol estrogens that undergo redox cycling during which reactive intermediates and reactive oxygen species are generated and lipid peroxidation can be initiated [252]. In the NBL rat, treatment for 16 weeks with testosterone plus 17β-estradiol enhanced the formation of a chromato-graphically unique endogenous adduct selectively in the periurethral region of the rat dorsolateral prostate, which is the site of the carcinogenic effect of this treatment [255, Bosland, unpublished data]. Ho and Roy [256] found increased single strand DNA breaks and accumulation of fluorescent lipid peroxidation products in the dorsolateral prostate of NBL rats after this treatment, but they did not separately analyze the periurethral and peripheral areas of the prostate. In addition, elevated levels of 8-hydroxydeoxyguanosine and lipid hydroperoxides have been found at the periurethral tissue but not in the peripheral area of these glands [Bosland, unpublished data]. Lower, but still elevated, levels of the endogenous DNA adduct detectable by ^{32}P-postlabeling, 8-hydroxydeoxyguan-osine, and lipid hydroperoxides were also found in the periuretheral prostate of rats treated only with testosterone, perhaps due to formation of estrogens by aromatization [Bosland, unpublished data]. The enhancement of endogenous DNA adduct formation, oxidative DNA damage, and lipid peroxidation selectively at the site of tumor formation and preceding it strongly suggest that these effects are causally involved in the carcinogenic effect of the hormone treatment. It is conceivable that exogenously administered estrogens or formation of estrogen from testosterone by aromatization, as well as a genotoxic mechanism are critical to the carcinogenic effect of this hormone combination for the rat prostate, rather than receptor-mediated mechanisms.

This hypothesis implicates that catechol estrogen formation occurs at the relevant site within the prostate [252]. Lane et al. [257] demonstrated that microsomes isolated from the dorsolateral prostate of NBL rats treated with testosterone plus 17β-estradiol do not generate the catechol estrogens 2- and 4-hydroxy-estradiol and 2- and 4-hydroxy-estrone. However, because periurethral prostate tissue was not analyzed and the relevance of microsomal assays for the in vivo situation is unclear, these data do not refute the possibility of catechol estrogen formation in the periurethral prostate. In addition, it is conceivable that the mechanisms of induction of dysplasia in the dorsolateral glandular prostate are estrogen receptor-mediated, possibly not involving estrogen-generated genotoxic processes, and that the mechanism involved in the induction of periurethral prostatic carcinomas is estrogen-generated genotoxicity, but possibly no estrogen receptor mediation. Thus, testosterone may act as a tumor promotor and estrogens as genotoxic "tumor initiators" in the testosterone-plus 17β-estradiol-treated NBL rat model of (periurethral) prostate carcinogenesis; the androgen may also act as enhancer of induction of dysplasia (periphery) in this model, which requires conjunct action of estrogen via estrogen receptors. These hypotheses remain to be critically tested.

The human relevance of these findings in the NBL rat model remains unclear at present. However, oxidative DNA damage and lipid peroxidation reflective of reactive oxygen damage have been observed in the human prostate [258], and increased oxidative stress has been found in prostate cancer patients as compared with controls [259]. Whether these observations are related with estrogen exposure or associated with other risk factors such as a high fat diet [242] is not known, but they suggest that endogenous oxidative stress may be important in human prostate carcinogenesis, and they are consistent with the hypothesis of estrogen-generated oxidative DNA damage as an important factor.

4.2.3
Perinatal Estrogen Exposure: Hormonal Imprinting

Perinatal estrogen exposure of mice resulted in epithelial dysplasia of the periurethral proximal parts of the dorsolateral and anterior prostate and seminal vesicles [214, 215], as well as in carcinomas in these areas [212], as summarized earlier. In addition, mice that were neonatally estrogenized hyper-responded to secondary estrogen treatment [214]. These very same tissue areas contain estrogen receptors which indicates their estrogen sensitivity [214]. However, the activity of 17β-estradiol hydroxysteroid oxidoreductase, believed to be a marker of estrogen sensitivity, and incorporation of tritiated thymidine in epithelial compartments of these tissues were not changed in response to secondary treatments with estrogen in neonatally estrogenized mice [214, 215]. In response to secondary androgen treatment, tritiated thymidine incorporation was markedly increased selectively only in the stromal cells of the anterior and ventral prostate, indicating a lasting effect of neonatal estrogen exposure on the androgen-responsiveness of the stromal component of the mouse prostate [214]. These observations suggest that perinatal estrogen exposure of mice imprints lasting alterations in estrogen as well as androgen responsiveness of the male accessory sex glands.

The precise mechanism of these complex imprinting effects is not clear. Perinatal estrogen treatment may act indirectly on the male accessory sex glands by imprinting permanent alterations in the secretion of pituitary hormones and testicular androgen, or directly by imprinting altered expression of androgen, estrogen, and prolactin receptors or changes in steroid metabolism in the accessory sex gland [260–263]. For example, in neonatally estrogenized mice, the plasma levels of luteinizing hormone and follicle stimulating hormone were elevated [261], while circulating testosterone levels were decreased [260, 263] or unchanged [261]. Prostatic DHT formation via 5α-reductase was impaired in adult mice neonatally treated with DES [246]. Nuclear androgen receptor levels in these mice were decreased in dorsal and ventral prostate but not affected in the lateral lobe, and the numbers of androgen receptor-positive stromal cells was increased in all three lobes [263]. However, although the exact mechanisms of the carcinogenic effects of perinatal estrogen exposure for the prostate remain unclear, there appear to be lasting direct and indirect effects of this treatment on the mouse prostate. No abnormalities in circulating estrogen and androgen levels were found in boys that had been exposed to DES in utero [264]. Thus, the human relevance of the effects of perinatal estrogen exposure in mice remains uncertain at present.

5
Summary and Conclusion

There are no known exogenous hormonal or non-hormonal exposures that are associated with prostate cancer risk, with the exception of "exposure" to a western life style (including a high fat diet), to an African American "environment", and, perhaps, to venereal disease, unknown factors related to farming, and employment in armed services and the nuclear industry. None of these associations point to *specific* chemicals, hormones, or other factors. In view of the high frequency of this malignancy in Western countries, this lack of known specific risk factors is remarkable and may indicate that there are many exogenous risk factors for prostate cancer which are too ubiquitous and overlapping to be detectable by epidemiologists. It is also possible that there are strong endogenous determinants of prostate cancer risk which are "overwhelming" most exogenous risk factors in epidemiological analyses. Androgenic hormones and androgen receptor mechanisms are prime candidates to be such important strong endogenous factors, but the epidemiological evidence for this is weak. There are some indications that muscle mass is positively correlated with risk, perhaps reflecting exposure to endogenous androgens or anabolic steroids. Elevation of bioavailable and bioactive androgens in the circulation and in the target tissue as an important risk factor is biologically highly plausible, and the results of several animal model studies strongly support this notion. Some of these experiments indicate that substantial enhancement of prostate carcinogenesis can be achieved by only very small elevations of circulating testosterone. This finding, if also valid for humans, may explain why the epidemiological associations between circulating androgen levels and prostate cancer are at best weak. There is also epidemiological evidence for increased transactivation activity of the androgen receptor to be associated with increased prostate cancer risk, both at the population and individual levels. However, more research is needed to confirm and further define these associations in humans, and to dissect further the biological mechanisms that underlie the increased risk that is probably associated with elevated circulating androgen levels and increased androgen receptor sensitivity.

African American men have a twofold higher risk than European American men, which is probably related to unknown environmental and genetic factors that may act though modifying their hormonal status. Indeed, circulating levels of androgens and, in men under 50 years, estrogens appear to be higher in African American men than in European American men. Such hormonal factors perhaps act as early as in utero, because circulating levels of androgens and estrogens have been reported to be higher in young men and pregnant African American women than in European American women.

Familial aggregation of prostate cancer risk is consistently observed and confers a considerable increased risk, but explains 10% or less of all cases. Putative susceptibility loci have been identified on chromosome 1q and the X chromosome, but there are no indications that these are related to hormonal factors. The role of hormones in familial aggregation of prostate cancer risk is unclear, since circulating testosterone levels appear to be lower in men with a family history of prostate cancer than in other men.

Hormonal stimulation of prostatic epithelial cell proliferation enhances the susceptibility of the rat prostate to chemical carcinogens. Testosterone at near-physiological plasma concentrations is a weak complete carcinogen and a strong tumor promotor for the rat prostate. The weak complete carcinogenic activity of androgens can perhaps be explained by their very strong tumor promoting activity. The precise mechanism of the tumor-inducing and -promoting activities of androgens for the rat prostate remains unknown. It is unlikely that chronic stimulation of prostatic cell proliferation rates by androgens is involved. However, it is conceivable that prostatic epithelial cells carrying critical genetic alterations have a selective growth advantage over normal cells and do not respond to androgens by differentiation, as normal cells would, but by proliferation.

Chronic exposure to testosterone plus 17β-estradiol is strongly carcinogenic for the dorsolateral prostate of some rat strains, while testosterone alone is only weakly carcinogenic. The mechanism of the carcinogenic effect of estrogen plus androgen in the rat prostate is incompletely understood, but it appears to involve estrogen-generated oxidative stress and genotoxicity, in addition to androgen receptor- and estrogen receptor-mediated processes such as changes in sex steroid metabolism and receptor status. There is evidence for the presence of the enzyme aromatase in the human and rat prostate providing for a local source of estrogens, which in humans seems to increase in stromal cell activity with aging. Perinatal estrogen exposure is carcinogenic for the rodent male accessory sex glands. Hyperplastic and squamous metaplastic changes have been reported in human male genital tract tissues following prenatal DES exposure, suggesting that prenatal exposure to DES may also target the human prostate. The mechanisms of these prenatal estrogen effects are not clear but may involve permanently imprinted changes in hormone production and hormone sensitivity of prostatic tissues.

These observations lead to a multifactorial general hypothesis of prostate carcinogenesis in which androgens are strong tumor promotors acting via androgen receptor-mediated mechanisms to enhance the carcinogenic activity of endogenous genotoxic carcinogens, reactive estrogen metabolites, estrogen- and prostatitis-generated reactive oxygen species, and possibly unknown weak environmental carcinogens. In this concept, all of these processes are modulated by a variety of environmental factors such as diet and by genetic determinants such as hereditary susceptibility and polymorphic genes that encode for receptors and enzymes involved in the metabolism and action of steroid hormones. It is conceivable that environmental agents with hormonal activity are involved in prostate carcinogenesis, but their identity, role, and importance have not been defined.

Acknowledgement. This work was supported in part by NIH Grants No. CA58088 and CA 75293, and Center Grants No. CA13343 and ES 00260.

6
References

1. American Cancer Society (1999) Cancer facts and figures. American Cancer Society, Atlanta
2. Huff J, Cirvello J, Haseman J, Bucher J (1991) Environ Health Perspect 93: 247
3. Swirsky Gold L, Slone TH, Manley NB, Bernstein L (1991) Environ Health Perspect 93: 233
4. Kirkman H, Kempson RL (1982) Tumours of the testis and accessory male sex glands. In: Turusov VS (ed) Pathology of tumours in laboratory animals, vol III: tumours of the hamster. IARC, Lyon, p 175
5. Bosland MC (1987) Adenocarcinoma, prostate, rat. In: Jones TC, Mohr U, Hunt RD (eds) Genital system. Springer, Berlin Heidelberg New York, p 252
6. Bosland MC (1987) Adenoma, prostate, rat. In: Jones TC, Mohr U, Hunt RD (eds) Genital system. Springer, Berlin Heidelberg New York, p 261
7. Bosland MC (1987) Adenocarcinoma, seminal vesicle/coagulating gland, rat. In: Jones TC, Mohr U, Hunt RD (eds) Genital system. Springer, Berlin Heidelberg New York, p 272
8. Mitsumori K, Elwell MR (1988) Environ Health Perspect 77: 11
9. Ito N, Shirai T (1990) Tumours of the accessory male sex organs. In: Turusov V, Mohr U (eds) Pathology of tumours in laboratory animals, vol I: tumors of the rat. IARC, Lyon, p 21

10. Nomura AMY, Kolonel LN (1991) Am J Epidemiol 13: 200
11. Wingo PA, Ries LAG, Giovino GA, Miller DS, Rosenberg HM, Shopland DR, Thun MJ, Edwards BK (1999) J Natl Cancer Inst 91: 675
12. Landis SH, Murray T, Bolden S, Wingo PA (1999) CA Cancer J Clin 49: 8
13. Bosland MC (1988) Adv Cancer Res 50: 1
14. Meikle AW, Smith JA (1990) Urol Clin North Am 17: 709
15. Muir CS, Nectoux J, Staszewski J (1991) Acta Oncol 30: 133
16. Van der Gulden JWJ, Kolk JJ, Verbeek ALM (1992) J Occup Med 34: 402
17. Bosland MC (1994) The male reproductive system. In: Waalkes MP, Ward JM (eds) Target organ carcinogenicity. Raven Press, New York, p 339
18. Shibata A, Whittemore, AS (1996) Prostate 32: 65
19. Steinberg GD, Carter BS, Beaty TH, Childs B, Walsh PC (1990) Prostate 17: 337
20. Carter B, Beaty TH, Steinberg GD (1992) Proc Natl Acad Sci USA 89: 3367
21. Hayes RB, Liff J, Pottern LM, Greenberg R, Schoenberg J, Schwartz AG, Swanson GM, Silverman DT, Brown LM, Hoover RN (1995) Int J Cancer 60: 361
22. Whittemore AS, Kolonel LN, Wu AH, John EM, Gallagher RP, Howe GR, Burch JD, Hankin J, Drean DM, West DW, Paffenberger RS Jr (1995) J Natl Cancer Inst 87: 652
23. Dong JT, Isaacs WB, Isaacs JT (1997) Curr Opin Oncol 9: 101
24. Smith JR, Freije D, Carpten JD, Gronberg H, Xu J, Isaacs SD, Brownstein MJ, Bova GS, Guo H, Bujnovszky P, Nusskern DR, Dambers JE, Bergh A, Emanuelsson M, Kallioniemi OP, Walker-Daniels J, Baily-Wilson JE, Beaty TH, Meyers DA, Walsh PC, Trent JM, Isaacs WB (1996) Science 274: 1371
25. Gronberg H, Xu J, Smith JR, Carpten JD, Isaacs SD, Freije D, Bova GS, Danber JE, Bergh A, Walsh PC, Collins FS, Trent JM, Meyers DA, Isaacs WB (1997) Cancer Res 57: 4707
26. Gronberg H, Isaacs SD, Smith JR, Carpten JD, Bova GS, Freije D, Xu J, Meyers DA, Collins FS, Trent JM, Walsh PC, Isaacs WB (1997) J Am Med Assoc 278: 1251
27. Eeles RA, Durocher F, Edwards S, Teare D, Badzioch M, Hamoudi R, Gill S, Biggs P, Dearnaley D, Adern-Jones A, Dowe A, Shearer R, McLennan DL, Norman RL, Ghadirian P, Aprikanian A, Ford D, Amos C, King TM, Labrie F, Simard J, Narod SA, Easton D, Foulkes WD (1998) Am J Hum Genet 62: 653
28. Gibbs M, Chakrabarti L, Stanford JL, Goode EL, Kolb S, Schuster EF, Buckley VA, Shook M, Hood L, Jarvik GP, Ostrander EA (1999) Am J Hum Genet 64: 1087
29. Monroe KR, Yu MC, Kolonel LN, Coetzee GA, Wilkins LR, Ross RK, Henderson BE (1995) Nature Med 1: 827
30. Xu J, Meyers DA, Freije D, Isaacs SD, Wiley K, Nusskern D, Ewing C, Wilkins E, Bujnovszky P, Bova GS, Walsh P, Schleuter J, Matkainen M, Tammala T, Visakorpi T, Kallioniemi OP, Berry R, Schaid D, French A, McDonnell S, Schroeder J, Blute M, Thibodeau S, Trent JM, Isaacs WB (1998) Nature Genet 20: 175
31. Lehrer S, Fodor F, Stock RG, Stone NN, Eng C, Song HK, McGovern M (1998) Br J Cancer 78: 771
32. Mandel JS, Schuman LM (1987) J Gerontol 42: 259
33. Ross RK, Shimizu H, Paganini-Hill A, Honda G, Henderson BE (1987) J Natl Cancer Inst 78: 869
34. Honda GD, Bernstein L, Ross RK, Greenland S, Gerkins V, Henderson BE (1988) Br J Cancer 57: 326
35. Mills PK, Beeson WL, Phillips RL (1989) Cancer 64: 598
36. Hsing AW, McLaughlin JK, Schuman LM, Bjelke E, Gridley G, Wacholder S, Co Chien TH, Blot WJ (1990) Cancer Res 50: 6836
37. Hsing AW, McLaughlin JK, Hrubec Z, Blot WJ, Fraumeni JF (1991) Am J Epidemiol 133: 437
38. Hayes RB, Pottern LM, Swanson GM, Liff J, Schoenberg J, Greenberg R, Schwartz AG, Brown LM, Silverman DT, Hoover RN (1994) Cancer Causes Control 5: 221
39. Lumey LH (1996) Prostate 29: 249
40. Lumey LH, Pittman B, Zhang EA, Wynder EL (1997) Prostate 33: 195
41. Handelsman DJ, Conway AJ, Boylan LM, Turtle JR (1984) Int J Androl 7: 369

42. Meikle AW, Bishop DT, Stringham JT (1989) Genet Epidemiol 6: 399
43. Lumey LH, Pittman B, Wynder EL (1998) Prostate 36: 250
44. Hayes RB, Brown LM, Schoenberg J, Greenberg R, Silverman DT, Schwartz AG, Swanson GM, Benichou J, Liff J, Hoover RN, Pottern LM (1996) Am J Epidemiol 143: 692
45. Chopra IJ, Tulchinsky D, Greenway FL (1973) Ann Int Med 79: 198
46. Gordon GG, Olivo J, Rafii F, Southern L (1975) J Clin Endocrinol Metab 40: 1018
47. Krstev S, Baris D, Stewart PA, Hayes RB, Blair A, Dosemici M (1998) Am J Ind Med 34: 413
48. Krstev S, Baris D, Stewart PA, Dosemici M, Swanson GM, Greenberg RS, Schoenberg JB, Schwartz AG, Liff JM, Hayes RB (1998) Am J Ind Med 34: 421
49. IARC (1987) IARC Monographs on the evaluation of carcinogenic risks to humans, Supplement 7: Overall evaluations of carcinogenicity: an updating of IARC Monographs Vols 1–42. IARC, Lyon
50. Waalkes MP, Oberdörster G (1990) Biological effects of heavy metals. In: Foulkes ED (ed) Cadmium carcinogenesis, vol II. CRC Press, Boca Raton, p 129
51. Blair A, Zahm SH, Pearce NE, Heineman EF, Fraumeni JF (1992) Scand J Work Environ Health 18: 209
52. Checkoway H, Mathew RM, Shy CM, Watson JR, Tankersley WG, Wolfe SH, Smith JC, Fry SA (1985) Br J Ind Med 42: 525
53. Shimizu Y, Kato H, Schull WJ (1991) J Radiat Res 2: 54
54. Eatough JP, Henshaw DL (1990) Lancet 335: 1292
55. Bosland MC (1996) Hormonal factors in carcinogenesis of the prostate and testis in humans and animal models. In: Huff J, Boyd J, Barrett JC (eds) Cellular and molecular mechanisms of hormonal carcinogenesis: environmental influences. Wiley-Liss, New York, p 309
56. Kolonel LN (1996) Cancer Causes Control 7: 83
57. Kolonel LN, Nomura AMY, Cooney RV (1999) J Natl Cancer Inst 91: 414
58. Hayes RB, Ziegler RG, Gridley G, Swanson C, Greenberg R, Swanson GM, Schoenberg J, Silverman DT, Brown LM, Pottern LM, Liff J, Schwartz AG, Fraumeni JF, Hoover RN (1999) Cancer Epidemiol Biomarkers Prev 8: 25
59. Hankin JH, Zhao LP, Wilkens LR, Kolonel LN (1992) Cancer Causes Control 3: 17
60. National Research Council (1989) Diet and health: implications for reducing chronic disease risk. National Academy Press, National Academy of Sciences, Washington DC, p 314
61. Bosland MC, Oakley-Girvan I, Whittemore AS (1999) J Natl Cancer Inst 91: 489
62. Hill P, Wynder EL, Garbaczewski L, Garnes H, Walker ARP (1979) Cancer Res 39: 5101
63. Hill P, Wynder EL, Garbaczewski L, Garnes H, Walker ARP, Helman P (1980) Am J Clin Nutr 33: 1010
64. Hill P, Wynder EL, Garnes H, Walker ARP (1980) Prev Med 9: 657
65. Hill P, Wynder EL, Garbaczewski L, Walker ARP (1982) Cancer Res 42: 3864
66. Hamalianen E, Adlercreutz H, Puska P, Pietinen P (1984) J Steroid Biochem 20: 459
67. Dorgan JF, Judd JT, Brown C, Schatzkin A, Clevidence BA, Campbell WS, Nair PP, Franz C, Kahle L, Taylor PR (1996) Am J Clin Nutr 64: 850
68. Bosland MC (1992) J Cell Biochem Suppl 16H: 135
69. McCormick DL, Rao KNV, Steele VE, Lubet RA, Kelloff GJ, Bosland MC (1999) Cancer Res 59: 521
70. McCormick DL, Rao KNV, Dooley L, Steele VE, Lubet RA, Kelloff GJ, Bosland MC (1998) Cancer Res 58: 3282
71. Esquenet M, Swinnen JV, Heyns W, Verhoeven G (1996) Prostate 28: 182
72. Kuiper JW, Walden P, Bosland MC (1999) Proc Am Assoc Cancer Res 40: 304
73. Shen JC, Wang TT, Chang S, Hursting SD (1999) Mol Carcinogenesis 24: 160
74. Fuller PJ (1991) FASEB J 5: 3092
75. Schwartz DA, Norris JS (1992) Biochem Biophys Res Comm 184: 1108
76. Hall RE, Tilley WD, McPhaul MJ, Sutherland RL (1992) Int J Cancer 52: 778
77. Mettlin C, Natarajan N, Huben R (1990) Am J Epidemiol 132: 1056

78. Rosenberg L, Palmer JR, Zauber AG, Warshauer ME, Stolley PD, Shapiro S (1990) Am J Epidemiol 132: 1051
79. Perlman JA, Spirtas R, Kelaghan J (1991) Am J Epidemiol 134: 107
80. Spitz MR, Fueger JJ, Babaian RJ, Newell GR (1991) Am J Epidemiol 134: 108
81. Peterson DE, Remington PL, Anderson HA (1992) Am J Epidemiol 135: 324
82. Hayes RB, Pottern LM, Greenberg R, Schoenberg J, Swanson GM, Liff J, Schwartz AG, Brown LM, Hoover RN (1993) Am J Epidemiol 137: 263
83. Giovannuci E, Tosteson TD, Speizer FE, Vessey MP, Colditz GA (1992) N Engl J Med 326: 1392
84. Giovannuci E, Ascherio A, Rimm EB, Colditz GA, Stampfer JD, Willett WC (1993) J Am Med Assoc 269: 873
85. Giovannuci E, Ascherio A, Rimm EB, Colditz GA, Stampfer JD, Willett WC (1993) J Am Med Assoc 269: 878
86. Ross RK, Paganini-Hill A, Henderson BE (1983) Prostate 4: 333
87. Newell GR, Fueger JJ, Spitz MR, Sider JG, Pollack ES (1989) Am J Epidemiol 130: 395
88. Rosenberg L, Palmer JR, Zauber AG, Washauer ME, Strom BL, Harlap S, Shapiro S (1994) Am J Epidemiol 140: 431
89. John EM, Whittemore AS, Wu AH, Kolonel LN, Hislop TG, Howe GR, West DW, Hankin J, Dreon DM, The CZ (1995) J Natl Cancer Inst 87: 662
90. Zhu K, Stanford JL, Daling JR, McKnight B, Stergachis A, Brawer MK, Weiss NS (1996) Am J Epidemiol 144: 717
91. Platz EA, Yeole BB, Cho E, Jussawalla DJ, Giovannucci E, Ascherio A (1997) Int J Epidemiol 26: 933
92. Sidney S (1987) J Urol 138: 795
93. Sidney S, Quesenberry CP, Sadler MC, Guess HA, Lydick EG, Cattolica EV (1991) Cancer Causes Control 2: 113
94. Nienhuis H, Goldacre M, Seagroatt V, Gill L, Vessey M (1992) Brit Med J 304: 743
95. Bernal-Delgado E, Latour-Perez J, Pradas-Arnal F, Gomez-Lopes LI (1998) Fertil Steril 70: 191
96. Hayes RB (1995) J Natl Cancer Inst 87: 629
97. Howards SS, Peterson HB (1993) J Am Med Assoc 269: 913
98. Peterson HB, Howards SS (1998) Fertil Steril 70: 201
99. Mo ZN, Huang X, Zhang SC, Yang JR (1995) J Urol 154: 2065
100. Wieland RG, Hallberg MC, Zorn EM, Klein DE, Luria SS (1972) Fertil Steril 23: 779
101. Rosemberg E, Marks SC, Howard PJ, James LP (1974) J Urol 111: 626
102. Varma MM, Varma RR, Johanson AJ, Kowarski A, Migeon CJ (1979) J Clin Endocrinol Metab 40: 868
103. Johnsonbaugh RE, O'Connell K, Engel SB, Edison M, Sode (1975) Fertil Steril 26: 329
104. de la Torre B, Hedman M, Jensen F, Pederson PH, Dicksfalusy E (1983) Int J Androl 6: 125
105. Kobrinsky NL, Winter JSD, Reyes FI, Faiman C (1976) Fertil Steril 27: 152
106. Purvis K, Saksena SK, Cekan Z, Diczfalusy CK, Ginger J (1976) Clin Endocrinol 5: 263
107. Smith KD, Tcholakian RK, Chowdhury M (1979) Endocrine studies in vasectomized men. In: Lepow IH, Crozier R (eds) Vasectomy: immunologic and patho-physiologic effects in animals and men. Academic Press, New York, p 183
108. Whitby RM, Gordon RD, Blair BR (1979) Fertil Steril 31: 518
109. Alexander NJ, Free MJ, Paulsen CA, Bushbom R, Fulgham DL (1980) J Androl 1: 40
110. Skegg DCG, Mathews JD, Guillebaud J, Vessey MP, Biswas S, Ferguson KM, Kitchin Y, Mansfield MD, Sommersville IF (1976) Brit Med J 1: 621
111. Fincham SM, Hill GB, Hanson J, Wijayasinghe C (1990) Prostate 17: 189
112. Yoshida O, Oishi K, Ohno Y, Schroeder FH (1990) A comparative study on prostatic cancer in the Netherlands and Japan. In: Sasaki R, Aoki K (eds) Proceedings Monbusho 1989 International Symposium. Comparative study of etiology and prevention of cancer. Nagoya Press, Nagoya, p 73
113. Hayes RB, de Jong FH, Raatgever J, Bogdanovicz J, Schroeder FH, van der Maas P, Oishi K, Yoshida O (1992) Eur J Cancer Prev 1: 239

114. Tsitouras PD, Martin CE, Harman SM (1982) J Gerontol 37: 288
115. Giovannuci E, Leitzmann M, Speigelman D, Rimm EB, Colditz GA, Stampfer MJ, Willett WC (1998) Cancer Res 58: 5117
116. Adlercreutz H, Harkonen M, Kuoppasalmi K (1986) Int J Sports Med 7: 27
117. Keizer H, Janssen GME, Menheere P (1989) Int J Sports Med 10: S139
118. Demark-Wahnefried W, Lesko SM, Conaway MR, Robertson CN, Clark RV, Lobaugh B, Mathias BJ, Smith Strigo T, Paulson DF (1997) J Androl 18: 495
119. Severson RK, Grove JS, Nomura AMY, Stemmermann GN (1988) Brit Med J 297: 713
120. Landry GL, Primos WA (1990) Adv Pediat 37: 185
121. Kato I, Nomura A, Stemmermann GN, Chyou PH (1992) J Clin Epidemiol 45: 1417
122. Tymchuk M, Habib FK, Chisholm GD, Ross M, Tozawa K, Hayashi Y, Kohri K, Tanaka S (1998) Urol Res 24: 265
123. Sandeman TF (1975) Med J Aust 2: 571
124. Guinan PD, Sadoughi W, Alsheik H, Ablin RJ, Alrenga D, Bush IM (1976) Am J Surg 131: 599
125. Roberts JT, Essenhigh DM (1986) Lancet ii: 742
126. Reif J, Pearce N, Fraser J (1989) Lancet ii: 742
127. Jackson JA, Waxman J, Spiekerman AM (1989) Arch Intern Med 149: 2365
128. Ebeling DW, Ruffer J, Whittington R, Vanarsdalen K, Broderick GA, Malkowicz SB, Wein AJ (1997) Urology 49: 564
129. Andersson SO, Adami HO, Bergström R, Wide L (1993) Br J Cancer 68: 97
130. Carlstrom K, Stege R (1997) Br J Urol 79: 427
131. Signorelli LB, Tzonou A, Mantzoros CS, Lipworth L, Lagiou P, Hsieh CC, Stampfer M, Trichopoulos D (1997) Cancer Causes Control 8: 632
132. Wolk A, Andersson SO, Bergström R (1997) J Natl Cancer Inst 89: 820
133. Ross R, Bernstein L, Judd H, Hanisch R, Pike MC, Henderson BE (1986) J Natl Cancer Inst 76: 45
134. Ross RK, Bernstein L, Lobo RA, Shimizu H, Stanczyk FS, Pike MC, Henderson BE (1992) Lancet 339: 887
135. Ross RK, Pike MC, Coetzee GA, Reichardt JKV, Yu MC, Feigelson H, Stanczyk FS, Kolonel LN, Henderson BE (1998) Cancer Res 58: 4497
136. Mhatre AN, Trifiro MA, Kaufman M, Kazemi-Esfarjani P, Figlewicz D, Rouleau G, Pinsky L (1993) Nature Genet 5: 184
137. Chamberlain NL, Driver E, Miesfeld RL (1994) Nucleic Acids Res 22: 3181
138. Jenster G, van der Korput HAGM, Trapman J, Brinkmann AO (1995) J Biol Chem 270: 7341
139. Ahluwalia B, Jackson MA, Williams AO, Rao MS, Rajguru S (1981) Cancer 48: 2267
140. Hill P, Garbaczewski L (1984) Med Hypotheses 14: 29
141. Henderson BE, Bernstein L, Ross RK, Depue RH, Judd HL (1988) Br J Cancer 57: 216
142. Lookingbill PD, Demers LM, Wang C, Leung A, Rittmaster RS, Santen RJ (1991) J Clin Endocrinol Metab 72: 1242
143. De Jong FH, Oishi K, Hayes RB, Bogdanowicz JF, Raatgever JW, van der Maas PJ, Yoshida O, Schroeder FH (1991) Cancer Res 51: 3445
144. Ellis L, Nyberg H (1992) Steroids 57: 72
145. Wu AH, Whittemore AS, Kolonel LN, John EM, Gallagher RP, West DW, Hankin J, The CZ, Dreon DM, Paffenberger RS Jr (1995) Cancer Epidemiol Biomarkers Prev 4: 735
146. Santner SJ, Albertson B, Zhang GY, Zhang GH, Santulli M, Wang C, Demers LM, Shackleton C, Santen RJ (1998) J Clin Endocrinol Metab 83: 2104
147. Labrie F, Sugimoto Y, Luu-The V, Simard J, Lachance Y, Bachvarov D, Leblanc G, Durocher F, Paquet N (1992) Endocrinology 131: 1571
148. Thigpen AE, Davis DL, Milatovich A, Medonca BM, Imperato-Ginley J, Griffin JE, Francke U, Wilson JD, Russell DW (1992) J Clin Invest 90: 799
149. Davis DL, Russell DW (1993) Human Mol Genet 2: 820
150. Reichardt JKV, Makridakis N, Henderson BE, Yu MC, Pike MC, Ross RK (1995) Cancer Res 55: 3973

151. Makridakis N, Ross RK, Pike MC, Chang L, Stanczyk FZ, Kolonel LN, Shi CY, Yu MC, Henderson BE, Reichardt JKV (1997) Cancer Res 57: 1020
152. Labrie F, Simard J, Luu-The V, Belanger A, Pelletier G (1992) J Steroid Biochem Mol Biol 43: 805
153. Lunn RM, Bell DA, Mohler JL, Taylor JA (1999) Carcinogenesis 20: 1727
154. Devgan SA, Henderson BE, Yu MC, Shi CY, Pike MC, Ross RK, Reichardt JKV (1997) Prostate 33: 9
155. Irvine RA, Yu MC, Ross RK, Coetzee GA (1995) Cancer Res 55: 1937
156. Makridakis NM, Ross RK, Pike MC, Crocitto LE, Kolonel LN, Pearce CL, Henderson BE, Reichardt JK (1999) Lancet 354: 975
157. Carey AH, Waterworth D, Patel K, White D, Little J, Novelli P, Franks S, Williamson R (1994) Hum Mol Genet 3: 1873
158. Horton R, Hawks D, Lobo R (1982) J Clin Invest 69: 1203
159. Nomura A, Heilbrun LK, Stemmermann GN, Judd HL (1988) Cancer Res 48: 3515
160. Barrett-Connor E, Garland C, McPhillips JB, Tee Khaw K, Wingard DL (1990) Cancer Res 50: 169
161. Hsing AW, Comstock GW (1993) Cancer Epidemiol Biomarkers Prev 2: 27
162. Comstock GW, Gordon GB, Hsing AW (1993) Cancer Epidemiol Biomarkers Prev 2: 219
163. Nomura AMY, Stemmermann GN, Chyou PH, Henderson BE, Stanczyk GZ (1996) Cancer Epidemiol Biomarkers Prev 5: 621
164. Gann PG, Hennekens CH, Jing M, Loncope C, Stampfer MJ (1996) J Natl Cancer Inst 88: 1118
165. Guess HA, Friedman GD, Sadler MC, Stanczyk FZ, Vogelman JH, Imperato-McGinley, Lobo RA, Orentreich N (1996) Cancer Epidemiol Biomarkers Prev 6: 21
166. Vatten LJ, Usrin G, Ross RK, Stanczyk FZ, Lobo RA, Harvei S, Jellum E (1997) Cancer Epidemiol Biomarkers Prev 6: 967
167. Dorgan JF, Albanes D, Virtamo J, Heinonen OP, Chandler DW, Galmarini M, McShane LM, Barrett MJ, Tangrea J, Taylor PR (1998) Cancer Epidemiol Biomarkers Prev 7: 1069
168. Kantoff PW, Febbo PG, Giovannucci E, Krithivas K, Dahl DM, Chang G, Hennekens CH, Brwon M, Stampfer MJ (1997) Cancer Epidemiol Biomarkers Prev 6: 189
169. Giovannuci E, Stampfer MJ, Krithavas K, Brown M, Brufsky A, Talcott J, Hennekens CH, Kantoff PW (1997) Proc Natl Acad Sci (USA) 94: 3320
170. Ingles SA, Ross RK, Yu MC, Irvine RA, La Pera G, Haile RW, Coetzee GA (1997) J Natl Cancer Inst 89: 166
171. Stanford JL, Just JJ, Gibbs M, Wicklund KG, Neal CL, Blumenstein BA, Ostrander EA (1997) Cancer Res 57: 1194
172. Hardy DO, Scher HI, Bogenrieder T, Sabbatini P, Zhang ZF, Nanus DM, Catterall JF (1996) J Clin Endocrinol Metab 81: 4400
173. Hsing AW (1996) J Natl Cancer Inst 88: 1093
174. Heikkilä R, Aho K, Heliövaara M, Hakama M, Marniemi J, Reunanen A, Knekt P (1999) Cancer 86: 312
175. Febbo PG, Kantoff PW, Platz EA, Casey D, Batter S, Giovannucci E, Hennekens CH, Stampfer MJ (1999) Cancer Res 59: 5878
176. Wadelius M, Andersson SO, Johansson JE, Wadelius C, Rane A (1999) Pharmacogenetics 9: 635
177. Correa-Cerro L, Wöhr G, Häussler J, Berthon P, Drelon E, Mangin P, Fournier G, Cussenot O, Kraus P, Just W, Paiss T, Cantu JM, Vogel W (1999) Europ J Hum Genet 7: 357
178. Platz EA, Giovannucci E, Dahl DM, Krithivas K, Hennekens CH, Brown M, Stampfer MJ, Katntoff PW (1998) Cancer Epidemiol Biomarkers Prev 7: 379
179. Eaton NE, Reeves PN, Appelby PN, Key TJ (1999) Br J Cancer 80: 930
180. Hiramatsu M, Maehara I, Ozaki M, Harada N, Orikasa S, Sasano H (1997) Prostate 31: 118
181. Negri-Cesi P, Poletti A, Colciago A, Magni P, Martini P, Motta M (1998) Prostate 34: 283
182. Smith T, Chisholm GD, Habib FK (1982) J Steroid Biochem 17: 119
183. Brodie AM, Son C, King DA, Meyer KM, Inkster SE (1989) Cancer Res 49: 6551
184. Krieg M, Nass R, Tunn S (1993) J Clin Endocrinol Metab 77: 375

185. Meikle AW, Stanish WM, Taylor N, Edwards CW, Bishop CT (1982) Metabolism 31: 6
186. Meikle AW, Smith JA, West DW (1985) Prostate 6: 121
187. Zumoff B, Levin J, Strain GW, Rosenfeld RS, O'Connor J, Freed SZ, Kream J, Whitmore WS, Fukushima DK, Hellman L (1982) Prostate 3: 579
188. Hammond GL (1978) J Endocrinol 78: 7
189. Powell IJ (1997) Oncology 11: 599
190. Bosland MC (1992) J Cell Biochem Suppl 16H: 89
191. Bosland MC (1997) Urol Oncol 2: 103
192. McNeal JE, Redwine EA, Freiha FS, Stamey TA (1988) Am J Surg Pathol 12: 897
193. Price D (1963) Natl Cancer Inst Monogr 12: 1
194. Noble RL (1982) Int Rev Exp Pathol 23: 113
195. Pollard M, Luckert PH, Schmidt MA (1982) Prostate 3: 563
196. Pollard M, Luckert PH (1986) J Natl Cancer Inst 77: 583
197. Pour PM, Stepan K (1987) Cancer Res 47: 5699
198. Hoover DM, Best KL, McKenney BK, Tamura RN, Neubauer BL (1990) Cancer Res 50: 142
199. Bosland MC, Dreef-Van Der Meulen HC, Sukumar S, Ofner P, Leav I, Han X, Liehr JG (1992) Multistage prostate carcinogenesis: the role of hormones. In: Harris CC, Hirohashi S, Ito N, Pitot HC, Sugimura T, Terada M, Yokota J (eds) Multistage carcinogenesis. CRC Press, Boca Raton, p 109
200. Shirai T, Tamano S, Kato T, Iwasaki S, Takahashi S, Ito N (1991) Cancer Res 51: 1264
201. Pollard M, Luckert PH (1985) Prostate 6: 1
202. Pollard M, Luckert PH (1987) Prostate 11: 219
203. Pollard M, Luckert PH (1989) Prostate 15: 95
204. Pollard M, Luckert PH, Snyder DL (1989) Cancer Lett 45: 209
205. Drago JR (1984) Anticancer Res 4: 255
206. Leav I, Merk FB, Kwan PW, Ho SM (1989) Prostate 15: 23
207. Bosland MC, Ford H, Horton L (1995) Carcinogenesis 16: 1311
208. Ofner P, Bosland MC, Vena RL (1992) Toxicol Appl Pharmacol 112: 300
209. McLachlan JA (1981) Rodent models for perinatal exposure to diethylstilbestrol and their relation to human disease in the male. In: Herbst A, Bern HA (eds) Developmental effects of diethylstilbestrol (DES) in pregnancy. Thieme Verlag, Stuttgart, p 148
210. Arai Y, Mori T, Suzuki Y, Bern HA (1983) Int Rev Cytol 84: 235
211. Santti R, Newbold RR, Makela S, Pylkkänen L, McLachlan JA (1994) Prostate 24: 67
212. McLachlan JA, Newbold RR, Bullock BC (1975) Science 190: 991
213. Newbold RR, Bullock BC, McLachlan JA (1987) J Urol 138: 1446
214. Pylkkänen L, Santti R, Newbold RR, McLachlan JA (1991) Prostate 18: 117
215. Pylkkänen L, Santti R, Mäentausta O, Vihko R (1992) Prostate 20: 59
216. Vorherr H, Messer RH, Vorherr UF, Jordan SW, Kornfeld M (1979) Biochem Pharmacol 28: 1865
217. Pour PM (1983) Carcinogenesis 4: 49
218. Shirai T, Sakata T, Fukushima S, Ikawa E, Ito N (1985) Jap J Cancer Res 76: 803
219. Rivenson A, Silverman J (1979) Invest Urol 16: 468
220. Bosland MC, Prinsen MK, Kroes R (1983) Cancer Lett 18: 69
221. Shirai T, Fukushima S, Ikawa E, Tagawa T, Ito N (1986) Cancer Res 46: 6423
222. Ito N, Shirai T, Tagawa Y, Nakamura A, Fukushima S (1988) Cancer Res 48: 4629
223. Bosland MC, Prinsen MK (1990) Cancer Res 50: 691
224. Shirai T, Ikawa E, Imaida K, Tagawa Y, Ito N (1987) J Urol 138: 216
225. Takai K, Kakizoe T, Tobisu K, Ohtani M, Kishi K, Sato S, Aso Y (1988) J Urol 139: 1363
226. Shirai T, Imaida K, Iwasaki S, Mori T, Tada M, Ito N (1993) Jap J Cancer Res 84: 20
227. Waalkes MP, Rehm S, Riggs CW, Bare RM, Devor DE, Poirier LA, Wenk ML, Henneman JR, Balaschak MS (1988) Cancer Res 48: 4656
228. Waalkes MP, Rehm S, Riggs C, Bare RM, Devor DE, Poirier LA, Wenk ML, Henneman JR (1989) Cancer Res 49: 4282
229. Waalkes MP, Perantoni AO (1989) Toxicol Appl Pharmacol 101: 83

230. Waalkes MP, Rehm S, Perantoni AO, Coogan TP (1992) Cadmium exposure in rats and tumours of the prostate. In: Nordberg GF, Herber RFM, Alessio L (eds) Cadmium in the human environment: toxicity and carcinogenicity. IARC, Lyon, p 391
231. Hirose F, Takizawa S, Watanabe H, Takeichi N (1976) Jap J Cancer Res 67: 407
232. Takizawa S, Hirose F (1978) Jap J Cancer Res 69: 723
233. Driscoll SG, Taylor SH (1980) Obstet Gynecol 56: 537
234. Sugimura Y, Cunha GR, Yonemura CU, Kawamura J (1988) Human Pathol 19: 133
235. Henderson BE, Ross RK, Pike MC, Casagrande JT (1982) Cancer Res 42: 3232
236. Cunha GR, Foster B, Thomson A, Sugimura Y, Tanji N, Tsuji M, Terada N, Finch PW, Donjacour AA (1995) World J Urol 13: 264
237. Hayward SW, Rosen MA, Cunha GR (1997) Br J Urol 79(2): 18
238. Condon MS, Bosland MC (1999) In Vivo 13: 61
239. Ripple MO, Henry WF, Rago RP, Wilding G (1997) J Natl Cancer Inst 89: 40
240. Tuohimaa P (1980) Control of cell proliferation in male accessory sex glands. In: Spring-Mills E, Hafez ESE (eds) Male accessory sex glands. Elsevier/North-Holland Biomedical Press, Amsterdam, p 131
241. Orsilles MA, Depiante-Depaoli M (1998) Prostate 34: 270
242. De Marzo AM, Coffey DS, Neslon WG (1999) Urology 53: 29
243. Sukumar S, Armstrong B, Bruyntjes JP, Leav I, Bosland MC (1991) Mol Carcinogenesis 4: 362
244. Ho SM, Yu M, Leav I (1992) The conjoint actions of androgens and estrogens in the induction of proliferative lesions in the rat prostate. In: Li JJ, Nandi S, Li SA (eds) Hormonal carcinogenesis. Springer, Berlin Heidelberg New York, p 18
245. Marts SA, Padilla GM, Petrow V (1987) J Steroid Biochem 26: 25
246. Santti R, Newbold RR, McLachlan JA (1991) Reprod Toxicol 5: 149
247. Lau KM, Leav I, Ho SM (1998) Endocrinology 139: 424
248. Miyata E, Kawabe M, Sano M, Takesheda Y, Takahashi S, Shirai T (1997) Prostate 31: 9
249. Thompson CJ, Ho SM, Lane K, Leav I (1999) Proc Am Assoc Cancer Res 40: 382
250. Nakamura A, Shirai T, Ogawa K, Wada S, Fujimoto NA, Ito A (1990) Cancer Lett 53: 151
251. Lane KE, Leav I, Ziar J, Bridges RS, Rand WM, Ho SM (1997) Carcinogenesis 18: 1505
252. Yager JD, Liehr JG (1996) Annu Rev Pharmacol Toxicol 36: 203
253. Gladek A, Liehr JG (1989) J Biol Chem 264: 16,847
254. Liehr JG, Avitts TA, Randerath E, Randerath K (1986) Proc Natl Acad Sci USA 83: 5301
255. Han X, Liehr JG, Bosland MC (1995) Carcinogenesis 16: 951
256. Ho SM, Roy D (1994) Cancer Lett 84: 155
257. Lane KE, Ricci MJ, Ho SM (1997) Prostate 30: 256
258. Olinksi R, Zastawny TH, Foksinski M, Barecki A, Dizdaroglu M (1995) Free Radicals Biol Med 18: 807
259. Hietanen E, Bartsch H, Bereziat JC, Camus AM, McClinton S, Eremin O, Davidson L, Boyle P (1994) Eur J Clin Nutr 48: 575
260. Jean C, Andre JM, Berger MJ, De Turckheim M, Veyssiere G (1975) J Reprod Fertil 44: 235
261. Dalterio S, Bartke A, Steger R, Mayfield D (1985) Pharmacol Biochem Behavior 22: 1019
262. Edery M, Turner T, Dauder S, Young G, Bern H (1990) Proc Soc Exp Biol Med 194: 289
263. Prins GS (1992) Endocrinology 130: 2401
264. Ross RK, Garbeff P, Paganini-Hill A, Henderson BE (1983) Can Med Assoc J 128; 1197

Beneficial and Adverse Effects of Dietary Estrogens on the Human Endocrine System: Clinical and Epidemiological Data

Doris M. Tham

Stanford University School of Medicine, 300 Pasteur Drive, Room S-308, Stanford, California 94305-5208, USA
E-mail: *doris.tham@forsythe.stanford.edu*

Dietary estrogens, also known as phytoestrogens, represent a family of plant compounds which are of biological interest because they exhibit both *in vivo* and *in vitro* weak estrogenic and anti-estrogenic properties. Phytoestrogens appear to exert their physiological effects through a variety of possible mechanisms, such as their ability to bind to estrogen receptors and their actions on tyrosine kinases and growth factors. Phytoestrogens can be classified into three main categories consisting of isoflavones, lignans, and coumestans. A variety of commonly consumed foods contains appreciable amounts of these plant compounds which have been identified in various human body fluids, such as plasma, urine, bile, saliva, feces, breast milk, prostatic fluid and semen. Accumulating evidence from both clinical and epidemiological studies has suggested that dietary estrogens may potentially affect the human endocrine system. The existing evidence reviewed here will identify the current research in this area, which will include both the possible beneficial and adverse effects which dietary estrogens may have on the human endocrine system as it relates to breast, prostate and colon cancer, endogenous hormones, the menstrual cycle, menopausal symptoms, coronary heart disease, and osteoporosis. Moreover, the issue of infants fed soy-based formulas will be addressed.

Keywords: Isoflavone, Lignan, Cancer, Menopause, Coronary heart disease

1   Introduction   71

2   Metabolism of Dietary Estrogens   72

3   Sources and Typical Intake Levels of Dietary Estrogens   76

4   Biological Potencies of Dietary Estrogens   77

5   Potential Health Benefits of Dietary Estrogens   79
5.1   Breast Cancer   80
5.2   Prostate Cancer   83
5.3   Colon Cancer   85
5.4   Endogenous Hormones   86
5.5   Menstrual Cycle   88
5.6   Menopausal Symptoms   88
5.7   Coronary Heart Disease   90
5.8   Osteoporosis   93

The Handbook of Environmental Chemistry Vol. 3, Part M
Endocrine Disruptors, Part II
(ed. by M. Metzler)
© Springer-Verlag Berlin Heidelberg 2002

6   **Infant Soy Formula**   95

7   **Potential Adverse Effects of Dietary Estrogens**   98

8   **Conclusions**   100

9   **References**   101

Abbreviations

AOM	azoxymethane
CHD	coronary heart disease
DMBA	dimethylbenz[a]anthracene
E2	17β-estradiol
EGF	epidermal growth factor
ER	estrogen receptor
FSH	follicle-stimulating hormone
GC-MS	gas chromatography-mass spectrometry
HDL	high density lipoprotein
HPLC	high performance liquid chromatography
HRT	hormone replacement therapy
HO	hydroxyl
ISP	isolated soy protein
LP(a)	lipoprotein a
LDL	low density lipoprotein
LH	luteinizing hormone
NAF	nipple aspirate fluid
NCEP	National Cholesterol Education Program
NMU	N-methyl-N-nitrosourea
O-Dma	O-demethylangiolensin
PSA	prostate specific antigen
PTK	protein tyrosine kinase
RBA	relative binding affinity
SFBA	San Francisco Bay Area
SHBG	sex hormone-binding globulin
SMA	serum modified access
TGF	transforming growth factor
UK	United Kingdom
US	United States
VLDL	very low density lipoprotein

1
Introduction

Considerable effort has been directed towards identifying constituents in the diet that either prevent or contribute to human disease. The human diet contains several plant-derived, non-steroidal estrogenic compounds known as phytoestrogens [1]. By utilizing modern techniques such as high pressure liquid chromatography (HPLC) and gas chromatography-mass spectrometry (GC-MS), it has been possible to isolate and characterize biologically active compounds in plants. More than 300 plants contain phytoestrogens, but there is considerable variability in hormonal potency within each category of phytoestrogen, which include isoflavones, lignans, and coumestans. Certain compounds may be quite estrogenic while others may have no activity at all. Some of the biologically active compounds are not found in plant food itself, but are produced from plant precursor compounds by bacterial metabolism in the colon. These dietary estrogens mimic endogenous estrogens and may produce many of the same physiological effects [2].

Endogenous estrogens influence the growth, differentiation and functioning of many target tissues. These include tissues of the female and male reproductive systems such as the mammary gland, uterus, vagina, ovary, testes, epididymis, and prostate [1]. Estrogens also play an important role in bone maintenance, in the central nervous system, and in the cardiovascular system [1, 3–5]. Estrogens diffuse in and out of cells but are retained with high affinity and specificity in target cells by an intranuclear binding protein, termed the estrogen receptor (ER). Once bound by estrogen, the ER undergoes a conformational change allowing the receptor to interact with chromatin and modulate transcription of target genes [6–8]. Dietary estrogens, however, bind to ERs with a much lower affinity than does endogenous estrogen. The mechanisms of these phytoestrogens are complex because, like steroids, phytoestrogens may bind to one or more proteins, including hormone receptors, hormone binding proteins in serum and enzymes that metabolize steroid hormones. The extent of interaction of phytoestrogens with one or more of these proteins could modulate steroid hormone action to produce profound physiological effects [2]. For the most part, focus has been on understanding phytoestrogen binding to steroid receptors and determining if phytoestrogens act as either a steroid hormone agonist or antagonist in regulating transcription of steroid-dependent genes. In addition, phytoestrogens have also been shown to function through steroid-independent mechanisms.

Phytoestrogens are thought to function both as weak estrogens and as anti-estrogens. This hormonal duality is part of the reason these compounds seem to exert such a range of health effects. For example, the estrogenic effects of phytoestrogens could potentially reduce the risk of cardiovascular disease and osteoporosis, while their anti-estrogenic effects have been linked to protection from breast cancer. It may seem paradoxical that phytoestrogens can act as both an estrogen and an anti-estrogen, but this phenomenon is also demonstrated by drugs such as tamoxifen, which have estrogenic effects on the heart and bones, but anti-estrogenic effects on breast tissues [9]. The chemical structure of phytoestrogens is strikingly similar to that of endogenous estrogens (Fig. 1). This structural similarity allows phytoestrogens to bind to the ER. The most potent of the endogenous estrogens, 17β-estradiol (E2), is usually the measure against which estrogenic activity is compared. Even the most potent of the plant estrogens have only about 0.001 the potency of E2 [10]. Although relatively weak, phytoestrogens can potentially exert biological effects when consumed in large amounts [11]. The serum levels of these dietary estrogens in populations who consume phytoestrogen-rich foods can be as much as 10,000 times higher than serum levels of E2 [12]. Several lines of evidence (including human, animal and *in vitro* research) are beginning to accrue which suggests that phytoestrogens may begin to offer protection against a wide range of human conditions, including breast, prostate, and colon cancers, cardiovascular diseases, osteoporosis and menopausal symptoms. Epidemiological studies suggest that populations that consume a diet high in phytoestrogens have a lower risk of these diseases [13]. These dietary estrogens are of great interest because they could explain why diets containing large amounts of plant foods are associated with lower mortality and morbidity in adult life. Vegetarians, for example, have strikingly lower overall mortality rates than omnivores [14]. This chapter reviews the current literature regarding both the beneficial and adverse endocrinological effects of dietary estrogen consumption in humans in relation to breast, prostate and colon cancer, endogenous hormones, menopausal symptoms, the menstrual cycle, coronary heart disease, osteoporosis and infant soy formulas.

2
Metabolism of Dietary Estrogens

The major isoflavones, genistein and daidzein, commonly exist as inactive glycosides [11]. They are derived from precursors, biochanin A and formononetin, which are converted to genistein and daidzein, respectively, after breakdown by intestinal glucosidases [15]. Daidzein is further metabolized to the mammalian isoflavone metabolites, equol and O-demethylangiolensin (O-Dma) (Fig. 1) [15, 16]. Absorbed phytoestrogen metabolites undergo enterohepatic circulation and may be excreted in the bile as the 7-O-β-glucuronide conjugate [17–19], deconjugated by intestinal flora, reabsorbed, reconjugated by the liver and excreted in the urine [15, 20–22]. Not all humans produce equol from daidzein, presumably due to differences in colonic bacterial populations among individuals. Since the conversion of daidzein to equol occurs by the action of gut microflora, it has been proposed that the type of bacteria found in the colon of equol excreters is different from those of non-excreters. Moreover, it has been reported that an individual is consistently either an equol excreter or non-excreter, suggesting that the individual variability may be intrinsic or perhaps even genetically determined [23]. In a large study conducted by Lampe et al. [24], 30 men and 30 women consumed a soy protein beverage containing 22 mg genistein and 8 mg daidzein for 4 days as a supplement to their habitual diets. Approximately one-third of the participants excreted equol after 3 days of consuming the soy supplement. There appears to be no difference in equol excre-

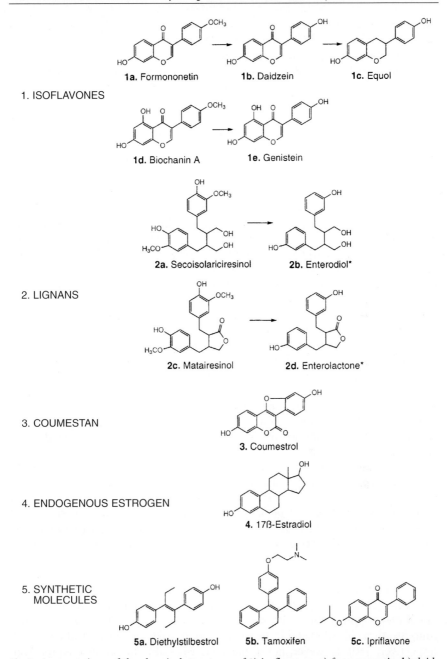

Fig. 1. A comparison of the chemical structures of **1**) isoflavones: **a**) formononetin, **b**) daidzein, **c**) equol*, **d**) biochanin A, **e**) genistein; **2**) lignans: **a**) secoisolariciresinol, **b**) enterodiol*, **c**) matairesinol, **d**) enterolactone*; **3**) coumestan: coumestrol; **4**) endogenous estrogen: 17β-estradiol; **5**) synthetic molecules: **a**) synthetic estrogen diethylstilbestrol, **b**) synthetic anti-estrogen tamoxifen, **c**) synthetic analogue ipriflavone. (*Mammalian phytoestrogens found in tissues and biological fluids which were derived from plant phytoestrogen precursors)

tion prevalence between men (43%) and women (27%). Daily excretion of daidzein, genistein and O-Dma was similar between equol excreters and non-excreters and between men and women [24].

The factors that contribute to the use of a particular isoflavone metaboliz-ing pathway (e.g., the capacity to produce equol) are unknown, but several ex-planations have been proposed. Setchell et al. [25] hypothesized that the com-position of the intestinal microflora, intestinal transit time and variability in the redox potential of the colon might contribute to variation in equol pro-duction in humans. The intestinal bacteria vary widely between individuals, and only selected strains are capable of hydrolyzing plant β-glucosides [26]. In addition, the relative stability of the composition of human intestinal mi-crofloral population may account for the consistent patterns of isoflavone ex-cretion. However, external and internal conditions influence microbial popula-tions and their activities and contribute to interindividual differences in bac-terial populations. Dietary intake is one of these factors which can mediate its influence directly through a change in substrate availability or indirectly through its effect on host metabolic functions [27]. Adlercreutz et al. [28] re-ported that equol excretion was positively associated with intake of fat and meat in a Japanese population and suggested that individuals consuming more fat and meat have an intestinal microfloral population capable of producing equol from daidzein. Kelly et al. [23] had observed that the patterns of iso-flavone excretion remain consistent over several years and postulated that some inherent factor probably plays a more significant role than diet in deter-mining isoflavone metabolism. Concentrations of the different phytoestrogen metabolites vary widely between individuals even when a controlled quantity of an isoflavone or lignan supplement is administered. Since dietary phyto-estrogen metabolism is predominantly determined by gastrointestinal flora, a variety of factors can modify metabolism such as antibiotic use, bowel disease, and gender [25, 29–31].

Phytoestrogens can be measured in urine, plasma, feces, prostatic fluid, se-men, bile, saliva, and breast milk [18, 32–38]. Both unconjugated (free) and con-jugated forms of isoflavones circulate in the human body. Many investigators have measured dietary estrogens in body fluids in correlation with dietary in-take of phytoestrogens. For example, urinary and blood levels of isoflavones have been correlated with an increase with supplements of soybean. Blood lev-els of isoflavones increase within 30 min of consumption of a soybean supple-ment, and begin to decline after 5 h ingestion, although elevated levels remain at 24 h [36]. However, only 7–30% of the ingested amount may be recovered in the urine [22, 23, 39]. Postmenopausal Australian women consuming a tradi-tional diet supplemented with soy flour or clover sprouts had the respective concentrations of equol, daidzein and genistein reach 43, 312, and 148 ng·ml^{-1}, respectively [40]. However, not all subjects were able to produce equol from daidzein. In a study conducted by Seow et al. [41], the distributions of dietary soy isoflavonoids (daidzein, genistein, and glycitein), urinary soy isoflavonoids and their metabolites (daidzein, genistein, glycitein, equol and O-Dma) among 147 Singapore Chinese of age 45–74 years were assessed. Among study subjects, there were statistically significant, dose-dependent associations between frequency of overall soy intake and levels of urinary daidzein and the sum of urinary daidzein, genistein and glycitein. In contrast, there were no associations between frequency of overall soy intake and levels of the two daidzein metabolites equol and O-Dma in urine. This study suggests that within the range of exposures experienced by Singapore Chinese, the urinary levels of daidzein or the sum of daidzein, genistein and glycitein obtained from a spot sample can serve as a biomarker of current soy consumption in epidemiological studies of diet-disease associations. Several studies have shown that Japanese men and women who consume a traditional diet have high levels of isoflavonoids in both urine and plasma [42]. Japanese women were found to excrete 10 times more daidzein and 20–30 times more equol and O-Dma than women in Boston and Helsinki [32]. In Western populations, urinary excretion of isoflavonoids appears highest among macrobiotic women, who consume mainly cereals, grains, legumes, and vegetables, followed by vegetarians, who in turn have higher levels than omnivores [32, 43]. Using a 3-day food diary, Adlercreutz et al. [28] demonstrated a significant correlation between urinary excretion of daidzein, equol and O-Dma with intakes of beans and pulses, soy products, and boiled soybeans in 19 Japanese men and women. Subjects were fed a soy beverage for a period of 2 weeks (~42 mg of genistein and ~27 mg of daidzein per day) and the measured plasma levels of genistein and daidzein ranged from $0.55–0.86$ μM, mostly as glucuronide and sulfate conjugates. Horn-Ross et al. [44] examined the racial/ethnic differences in urinary phytoestrogen levels in 50 Japanese, Caucasian, African American and Latina young women (ages 20–40 years) residing in the San Francisco Bay Area (SFBA). There was substantial variation in phytoestrogen levels along with racial/ethnic differences. The highest levels of coumestrol and lignans were observed in white women and the lowest levels in Latina and African American women. Genistein levels, however, were highest in Latina women. Other isoflavones levels did not differ significantly by race/ethnicity. Latina SFBA women, who were observed to have the highest urinary excretion of genistein and daidzein, reported consuming more beans and chili with beans. Although different varieties of beans contain different amounts and types of phytoestrogens [45], the urinary excretion of genistein and daidzein among these women is similar to that among persons consuming an experimental diet including garbanzo beans [46], which contains biochanin A, a precursor of genistein.

The lignans enterolactone and enterodiol form another class of phenolic compounds. Enterodiol and enterolactone are derived from the colonic microbial fermentation of seco isolariciresinol and matairesinol, respectively (Fig. 1) [15, 47]. These lignan precursors occur in the aleuronic layer of the grain close to the fiber layer. Seco isolariciresinol and matairesinol have glucose residues attached to the hydroxy (OH) groups. During fermentation, the colonic bacterial flora remove both glucose and methoxy groups to form the diphenols, enterodiol and enterolactone, which are structurally similar to E2 (Fig. 1). After absorption, these mammalian lignans also are excreted in urine. Urinary levels of lignans increase substantially when supplements of linseed are fed to human subjects [48, 49]. There is also a moderate increase in urinary excretion following vegetable supplementation [46]. Macrobiotic and other vegetarians, such as

Seventh Day Adventists, have the highest excretion values of lignans [29]. Postmenopausal Australian women consuming a traditional diet supplemented with linseed had combined levels of enterolactone and enterodiol which reached 500 ng·ml^{-1} [40].

3
Sources and Typical Intake Levels of Dietary Estrogens

Phytoestrogens are present naturally in many foods of plant origin (e.g., rice, rye, bran, pomegranates, apples, wheat, garlic, oats, coffee, fennel, licorice, barley, parsley, cherries, yeast, potatoes, soybeans, and soy-based products, including animal feeds). They have also been identified in alcoholic beverages. Phytoestrogens have been isolated from beer made from hops and bourbon made from corn [10, 50–55]. The type of isoflavone differs according to how the soybean is treated and its origin. Japanese soybean varieties contain different proportions of isoflavones to those grown in the United States (US), and fermented products, such as miso, contain more unconjugated isoflavones rather than daidzin and genistin [56]. Variation in phytoestrogen content can also occur because of genetic differences in plants such as soy varieties, location, season, infection with fungal diseases, and processing [57]. The main isoflavones of interest from a dietary perspective are genistein, daidzein, and glycitein (in minor amounts). Soy is essentially the only significant source of the isoflavones in foods with concentrations up to 1–3 mg·g^{-1} of daidzein, genistein, and glycitein [56–61]. Low levels of some isoflavones are also found in other legumes, such as lentils and beans (e.g., mung, haricot, broad, kidney, lima) and in products that contain most or all of the soybean (e.g., soy meal and flour, tofu and soy milk) [62]. Soy sauce, however, contains few isoflavones.

Lignans are obtained from high-fiber foods. Lignans are more widespread in plants where they form the building blocks of lignin in plant cell walls. Flaxseed (also known as linseed) is the richest source of lignan precursor compounds, but other good sources include other seeds (e.g., sesame and sunflower seeds), whole grain cereals, bran of rye, barley, wheat, fruits and vegetables that have a tendency to become woody or fibrous (e.g., carrots, broccoli, asparagus, and squash) [10, 15, 63].

Legumes are the only known dietary sources of coumestans. Soy sprout is a potent source of coumestrol, the major coumestan [10]. Coumestrol is also found in other sprouted legumes such as alfalfa and clover sprouts.

Americans are exposed to very little isoflavones in their diet, usually no more than 1–3 mg per day [64]. In some countries, women consume on the average 20–50 times more soy products per capita than Americans [64]. Soy contains significant amounts (1–3 mg·g^{-1}) of the isoflavone phytoestrogens genistein and daidzein [65]. Messina has estimated intakes of isoflavones to be approximately 30 mg per day in the Japanese population [64]. Asian populations, such as those in Japan, Taiwan, Indonesia and Korea, are estimated to consume 20–150 mg per day of isoflavones with a mean of about 40 mg from tofu and miso [29]. The median daily intake of soy protein among Chinese male and female subjects from Singapore were 2.0 and 2.1 g, respectively [41]. In Shanghai and

Tianjin, China, median intake levels among control subjects in two case-control studies of breast cancer were 3.5 and 2.8 g, respectively [66]. In an earlier case-control study of breast cancer in Singapore, a median daily level of 2.5 g was reported among control subjects [67].

4
Biological Potencies of Dietary Estrogens

Phytoestrogen are of biological interest because they exhibit both *in vivo* and *in vitro* weak estrogenic and anti-estrogenic actions through their binding to ERs [68]. The dual action of phytoestrogens as agonists and antagonists has resulted in a general confusion in the literature regarding the role these compounds may play in disease progression and/or protection. In common with many other weak estrogens, isoflavones have been shown to be anti-estrogens in model systems, competing for E2 at the receptor complex, yet failing to stimulate a full estrogenic response after binding to the nucleus [69]. This raises the possibility that they may be protective in hormone-related diseases, such as breast cancer [25]. *In vitro* studies have established that phytoestrogens are weakly estrogenic, since they have the ability to bind to mammalian ERs to a low degree. The relative potencies, determined by human cell culture bioassays (compared with E2, to which an arbitrary value of 100 was given) are: coumestrol 0.202, genistein 0.084, equol 0.061, daidzein 0.013 and formononetin 0.0006 [70]. The biological potency of these compounds, in animal and *in vitro* models, varies considerably and may depend on many factors including target tissue, functional state of the target tissue, species, age of the subject, route of delivery, dose, length of exposure, and metabolism. Furthermore, it is important to note that these wide ranges are derived from studies of cell-culture systems or animal models, rather than from direct effects in humans.

The human, rat and mouse ER exists as two subtypes, $ER\alpha$ and $ER\beta$, which differ in the *C*-terminal ligand-binding domain and in the *N*-terminal transactivation domain. Kuiper et al. [71] investigated the estrogenic activity of phytoestrogens in competition binding assays with $ER\alpha$ or $ER\beta$ protein and in a transient gene expression assay. In most instances, the relative binding affinities (RBA) of phytoestrogens are at least 1000-fold lower than that of E2. Some phytoestrogens, such as coumestrol and genistein, compete more strongly with E2 for binding to $ER\beta$ than to $ER\alpha$. Phytoestrogens can stimulate the transcriptional activity of both ER subtypes at concentrations of $1-10$ nM. The ranking of the estrogenic potency of phytoestrogens for both ER subtypes in the transactivation assay is different; that is $E_2 >$ coumestrol $>$ genistein $>$ daidzein $>$ biochanin A $>$ formononetin $=$ ipriflavone for $ER\alpha$ and $E_2 >$ genistein $=$ coumestrol $>$ daidzein $>$ biochanin A $>$ formononetin $=$ ipriflavone for $ER\beta$. The binding affinity of coumestrol to $ER\beta$ is 7-fold higher in comparison to $ER\alpha$, whereas genistein has a 20- to 30-fold higher binding affinity for $ER\beta$. The exact position and number of the hydroxy substituents on the isoflavone molecule seems to determine the ER binding affinity. For example, the isoflavone genistein has a particular high binding affinity for $ER\beta$, but elimination of one hydroxy group (e.g., daidzein, biochanin A) or two hydroxy groups (e.g., formononetin) causes a great loss in binding affinity (Fig. 1). Anti-estrogenic activity of the phyto-estrogens could not be detected in this particular study. However, the estrogenic potency of phytoestrogens in this study is significant, especially for ERβ, and they may trigger many of the biological responses that are evoked by the phys-iological estrogens. Interestingly, ERβ shows a different anatomical distribution from ERα, being expressed more prominently in tissues such as the brain, ovary, uterus, prostate, lung and urinary tract [72]. ERβ is also expressed in breast cells, although apparently weakly [73].

In vitro research has played an important role in determining the biological potencies of dietary estrogens. In cultured cells, both proliferative and anti-pro-liferative effects have been ascribed to genistein [74–77]. Wang et al. [68] ob-served that genistein stimulated estrogen-responsive pS2 mRNA expression, and this effect could be inhibited by tamoxifen. Thus, the estrogenic effect of genistein would appear to be a result of an interaction with the ER. At lower concentrations (10^{-8} to 10^{-6} M), genistein stimulated growth of ER-positive cells, but at higher concentrations ($>10^{-5}$ M), genistein inhibited growth. Genistein failed to stimulate the proliferation of ER-negative cells at low con-centrations. The biphasic effects of genistein on growth at lower concentrations appeared to be via the ER pathway, while the effects at higher concentrations were independent of the ER. These effects might appear to be involved in the mechanism by which genistein contributes to the decreased risk in hormone-dependent cancers associated with ingestion of soy products. However, this may be unlikely since the human circulating level of genistein does not exceed 2.5×10^{-5} M [78]. Even with increased intake of soy products, circulating levels of genistein normally do not exceed 10^{-6} M. It is more likely that the effects of dietary genistein on tumorigenesis are mediated through the ER, for which it has a relatively high affinity. Taken together, the results support the idea that genistein, at physiologically achievable concentrations, may interfere with the action of estrogen through direct competition for binding to ER and by reduc-ing ER expression. These effects could be considered anti-estrogenic, consistent with the proposed cancer-preventive effects of soy diet. In addition, prolonged experimental exposure to phytoestrogens may result in the reduction of ER ex-pression and hence lead to decreased responsiveness to endogenous estrogens since prolonged exposure to genistein resulted in a decrease in ER mRNA level, as well as a decrease in response to stimulation by E2 [68].

Although there have been many interesting studies on the effects of isoflavones on biochemical targets in tissue culture experiments, the concen-trations used by investigators have exceeded 10 µM in most cases. Based on simple pharmacokinetic calculations involving daily intakes of isoflavones, ab-sorption from the gut, distribution to peripheral tissues, and excretion, it is un-likely that blood isoflavone concentrations, even in high soy consumers, could be greater than 1–5 µM [17]. Although the plasma genistein levels achievable with soy food feedings are unlikely to be sufficient to inhibit the growth of ma-ture, established breast cancer cells by chemotherapeutic-like mechanisms, these levels are sufficient to regulate the proliferation of epithelial cells in the breast and thereby may cause a chemopreventive effect. For breast cancer and prostate cancer, delivery of genistein via the blood supply is the only route available physiologically. The effective concentration of genistein in blood will therefore depend on plasma protein binding. Physiological steroids are 98% bound to plasma proteins, thereby leaving less than 2% available for uptake into tissues.

There are several other anti-cancer effects of isoflavones which do not involve ER mechanisms. Genistein is known to inhibit protein tyrosine kinases (PTKs), which are responsible for phosphorylating proteins required for the regulation of cell function, including cell division [75]. These enzymes appear to be necessary for epidermal growth factor function and the action of other growth factors, which indicate the anti-proliferative potential of this isoflavonoid. Since many of the protein products of human oncogenes were found to take part in growth factor signaling through PTKs, a rationale was developed for the role of genistein in cancer prevention [79]. Genistein can inhibit cell cycle progression in tumor cells independent of action at the ER [80–82]. Genistein has also been shown to inhibit the DNA repair enzyme topoisomerase [83], and to act as an antioxidant, thus potentially preventing apoptosis, or programmed cell death, a protective mechanism induced in cells that have been damaged in order to prevent the proliferation of harmful mutations and possible cancer [84]. Alternative mechanisms of action of genistein have also been postulated. They include the effects of reactive oxygen species [85], and the expression of DNA transcription factors c-fos and c-jun [86]. Recently, genistein's regulation of the level of transforming growth factor β (TGFβ), an inhibitor of the G_1/S phase of cell cycle control, has been identified [87]. It has also been shown to inhibit ras gene expression in a rat pheochromocytoma cell line [88]. In addition, genistein has been shown to inhibit angiogenesis, the formation of new blood vessels, which is an abnormal event that occurs as a part of the growth and expansion of malignant tumors [76]. It has been pointed out that many of these effects have also been shown with very high concentrations, and not in cells treated with the levels likely to be achieved in plasma of human subjects eating foods containing phytoestrogens [17]. For example, 100 μM concentrations were needed for a significant suppression of angiogenesis, although proliferation was inhibited at 5 μM levels [76]. This compares with plasma levels of about 0.4 μM in Japanese men and women fed on dietary supplements of isoflavones [40, 78, 89] and peak levels of about 4 μM in women receiving 126 mg isoflavones as soy milk, rising to 12 μM when receiving 480 mg isoflavones per day [22, 90].

5
Potential Health Benefits of Dietary Estrogens

Both clinical and epidemiological data suggest that dietary estrogens may have a beneficial effect on the human endocrine system. Breast cancer, prostate cancer, colon cancer, menopausal symptoms, heart disease, and osteoporosis share a common epidemiology in that they are rare in Far Eastern populations eating traditional diets containing soybean products compared with Western populations. However, with Westernization and loss of traditional eating patterns, the pattern of disease incidence is also changing in these countries. Cross-sectional

studies have shown higher phytoestrogen levels in the urine and plasma of populations at lower risk of these diseases [28, 91]. This section will focus on the
beneficial role which phytoestrogens may play in breast cancer, prostate cancer,
colon cancer, endogenous hormones, the menstrual cycle, menopausal symptoms, cardiovascular disease, and osteoporosis.

5.1
Breast Cancer

Breast cancer is the most common cancer of women living in Western popula-
tions [92]. The incidence of disease and death from breast cancers in the US
and Western Europe substantially exceeds rates observed in most of Southeast
Asia [93–95]. Incidence and mortality rates of breast cancer in Chinese and
Japanese women in Asia are reported to be significantly lower than those of
Chinese and Japanese women living in the US. The age-adjusted death rates
from breast cancer are 2- to 8-fold lower in Asian countries than in the US and
Western Europe [96]. This difference in breast cancer incidence has been correlated with differences in dietary patterns [93]. A large body of literature, including epidemiological, human, animal, and *in vitro* data, supports a beneficial role for dietary estrogens, particularly as a protective agent against breast
cancer. Migration studies have indicated that the differences in these rates are
dissipated after one generation following emigration from Southeastern Asia
to the US [97]. Breast cancer rates of North American-born Japanese and
"early" Japanese immigrants are almost identical to those of Caucasian North
Americans, whereas "late" immigrants have an incidence rate intermediate between the former groups and that of Japanese residing in Japan [97]. Such findings support the role for environmental conditions in the etiology of breast
cancer. Hirayama [98] reported a significant graded inverse association in
Japanese women between risk of breast cancer and consumption of miso (soybean paste soup). A diet high in soy products conferred a low risk of breast
cancer in premenopausal women in Singapore [67]. Wu et al. [99] have report
that the risk of breast cancer was lowered in association with the number of
servings of tofu consumed per week. A review of existing epidemiological data
on the role of soy in the risk of cancer revealed some evidence of a preventive
effect [64].

Genistein has been the phytoestrogen of great interest in the reduction of
breast cancer risk because it has been shown to exert both proliferative (estrogenic) and anti-proliferative (anti-estrogenic) effects in human cell lines [68,
100]. In the human ER-positive MCF-7 breast cancer cell line, these effects are
biphasic and concentration-dependent with stimulation of cell growth occurring at low concentration of genistein (10^{-5} to 10^{-8} M) and dose-dependent
inhibition at higher concentrations (10^{-4} to 10^{-5} M) [68, 75, 84, 100–108].
Sathyamoorthy et al. [100] have demonstrated a similar stimulatory effect with
daidzein, equol and enterolactone at 10^{-6} M concentrations. The anti-proliferative effects of genistein described above occurred in both ER-positive and ER-
negative cell lines and thus appear not to be mediated by the ER activity alone
[68]. Extrapolation of cell culture studies to humans is questionable, especially

in terms of tumor growth inhibition. A high soy diet results in an approximate plasma level of $1-5$ µM of genistein [40, 109], whereas the studies describing an anti-proliferative effect indicate a minimum cell culture concentration of $10-100$ µM of genistein [64, 70]. Dees et al. [110] demonstrated that dietary estrogens at low concentrations do not act as anti-estrogens, but stimulate human breast cancer cells to enter the cell cycle. This paradoxical role of the growth-promoting effects of genistein on ER-positive breast cancer cells *in vitro* and the apparent preventive and tumor growth-inhibitory actions of genistein *in vivo* maybe most plausibly explained by genistein's ability to mediates its action via ER as an estrogen agonist while also cross-talking with other ER-independent cellular mechanisms at higher concentration to inhibit cell proliferation induced by genistein through ER pathways. Collectively, these data support the hypothesis that the actions of genistein on ER and on inhibition of cell proliferation are a result of distinctly different cellular mechanisms.

Isoflavones have been shown to exhibit anti-carcinogenic activity *in vivo*. Laboratory animals fed soy-fortified diets have demonstrated protective effects as measured by a decrease in breast tumor proliferation, tumor number, incidence, metastases and an increase in latency, after stimulation with direct-acting, e.g., *N*-methyl-*N*-nitrosourea (NMU) and indirect-acting, e.g., dimethyl-benz[*a*]anthracene (DMBA) tumor-inducing agents [111–113]. In addition, pre-pubertal genistein-treated rats developed fewer mammary gland terminal end buds, with significantly fewer cells in the S-phase of the cell cycle, and more lobules than controls at 50 days of age [114]. Interestingly, emerging evidence from animal studies suggests that short-term exposure to dietary isoflavones neonatally or pre-pubertally decreases carcinogen-induced differentiated cells in the mammary gland [114]. These studies support a concept derived from other epidemiological investigations that the protective effect of the Asian diet occurs early in life [115]. Perhaps the early administration of genistein resulted in a more mature gland with less-susceptible structures to later initiation by the chemical carcinogen [111, 112]. This may explain why the epidemiological studies, which in essence focus on the adult consumption of soy, are relatively unimpressive. Recently, Yan et al. [116] investigated the effect of dietary supplementation of flaxseed, the richest source of lignans, on experimental metastasis of murine melanoma cells in C57BL/6 mice. The median number of tumors in mice fed the flaxseed-supplemented diets were not only lower than that of the controls, but the addition of flaxseed to the diet also caused a dose-dependent decrease in the tumor cross-sectional area and the tumor volume. These results provide experimental evidence that flaxseed reduces metastasis and inhibits the growth of the metastatic secondary tumors in animals. Dietary supplementation of soybeans, a rich source of isoflavone phytoestrogens, was also able to reduce experimental metastasis of melanoma cells in mice [117]. From these results, it has been suggested that flaxseed and soybeans may be a useful nutritional adjuvant to prevent metastasis in cancer patients.

Human data regarding phytoestrogens and breast cancer are limited. Two early prospective investigations on breast cancer were the studies in Hawaii by Nomura et al. [118] and by Hirayama in Japan [98]. Both studies found that the fermented soybean paste, miso, appeared to have a protective effect against breast cancer in premenopausal women. Using a case-control study design, Lee et al. [67] reported a significantly lower risk of breast cancer in premenopausal Chinese women in Singapore who consumed soy. Many studies thus far have attempted to correlate an increase in consumption of dietary estrogens with a decrease in breast cancer risk by measuring urinary excretion of phytoestrogens. Maskarinec et al. [119] conducted a cross-sectional study to investigate the associations between urinary isoflavone excretion and self-reported soy intake of 102 women of Caucasian, Native Hawaiian, Chinese, Japanese and Filipino ancestry. Japanese women excreted more daidzein, genistein, and glycitein than did Caucasian women, whereas Caucasian women excreted slightly more coumestrol than any other group. Equol was excreted at the highest rate by Chinese women, whereas all other women excreted equol at a very low rate. These results demonstrate differential intestinal absorption by ethnic groups. Dietary soy protein and isoflavone intakes during the previous 24 h were positively related to urinary isoflavone excretion. The strong correlation between urinary isoflavone excretion and self-reported soy intake validates the dietary history questionnaire which can now be used in studies exploring dietary risk factors for breast cancer. A study from California [44] has also reported urinary isoflavone excretion by ethnic group. In this study, no statistically significant difference among various ethnic groups for isoflavone excretion was found, but the small group of Japanese women (n = 5) excreted slightly more isoflavones than did other ethnic groups. Their results also showed that Japanese women excreted very low levels of coumestrol compared with Caucasian women. Ingram et al. [120] conducted a case-control study to assess the association between phytoestrogen intake (as measured by urinary excretion) and the risk of breast cancer. Women with newly diagnosed early breast cancer and controls had their urine analyzed for the isoflavone phytoestrogens daidzein, genistein and equol, and for the lignans enterodiol, enterolactone and matairesinol. After adjustment for age at menarche, parity, alcohol intake, and total fat intake, high excretion of both equol and enterolactone was associated with a substantial reduction in breast cancer risk.

Currently, there are no reported toxic effects in humans from eating soy protein. Asian women, who for centuries have consumed soy as a staple of the diet, have not experienced any evident adverse effects on their hormonal and reproductive systems. It remains to be determined if dietary estrogens are beneficial or, as suggested by some *in vitro* studies, constitute an additional carcinogenic risk factor for tissues where proliferation is controlled by estrogens. If the amount of estrogens that can be derived from the dietary sources does not contribute a level high enough to suppress ER-positive cell growth, dietary estrogens may increase the risk of breast cancer. The molecular effects of dietary-derived estrogen on the ER-positive breast cancer cells appear to be complex. The effects of dietary estrogens may be concentration-dependent and may interact with synthetic and natural estrogens. It may be premature at this time to suggest dietary changes that significantly alter the amount of dietary-derived estrogens until additional research can fully elucidate the effects they have on the reproductive tissues in terms of dose, tissue-specific effects, and potential interaction with other estrogenic compounds.

5.2
Prostate Cancer

Prostate cancer is the most common hormone-related cancer in men and the incidence in the United Kingdom (UK) has been rising rapidly by about 3–4% per year [92]. High-fat and high-meat diets are currently linked to increased risk of the disease, and like breast cancer, it is comparatively rare in Far Eastern populations consuming soybean. The incidence of and mortality from prostate cancer is very much lower in Asian men in comparison to men from the West [121]. In addition, men who adopt a vegetarian diet are also at lower risk from prostate cancer [122]. The diets of Asian and vegetarian men are not only much lower in fat than the traditional diet of omnivorous Western man, but they are also a rich source of weak dietary estrogens [13]. It has been estimated that the traditionally-eating Chinese man consumes, on average, 35 times more soy than the North American man [56]. Lifetime exposure to the isoflavonoids in soy may play a significant role in the low incidence of prostate cancer in Asian men. Mills et al. [122] reported that increased consumption of beans, lentils and peas, tomatoes, and dried fruits was associated with significantly decreased prostate cancer risk in 14,000 Seventh Day Adventist men. Hirayama [98] reported that the protective effects from prostate cancer were correlated with the consumption of green leafy vegetables. A prospective study of men of Japanese ancestry living in Hawaii has demonstrated a decreased prostate cancer risk in those who consume rice and tofu [123], and a study of Japanese immigrants to North America reported an increase in the incidence rate of prostate cancer with younger age at immigration, supporting the hypothesis that environmental changes, including diet, can impact on cancer risk even in later life. In men, phytoestrogen levels have been evaluated in plasma, prostatic fluid and semen. Morton et al. [36] measured isoflavonoid and lignan levels in the prostatic fluid of men from the UK, Portugal, Hong Kong and China. The mean levels of the isoflavone equol observed in the prostatic fluid from men in Hong Kong (29.2 ng·ml^{-1}) and China (8.5 ng·ml^{-1}) was 17 times higher than those from Britain and 5 times higher than those from Portugal. Similarly, the prostatic fluid of men from Hong Kong and China also had the highest mean concentrations of the soy-derived isoflavone, daidzein at 70 and 24.2 ng·ml^{-1}, respectively, which was 15 times greater than those from Europe. The highest mean concentration of the lignans, enterodiol (13.5 ng·ml^{-1}) and enterolactone (162 ng·ml^{-1}) were found in the prostate fluid of men from Portugal. The plasma levels of free non-conjugated enterolactone have been reported to range from 3.1 to 6.3 ng·ml^{-1} where total levels were much higher, between 11 and 38 ng·ml^{-1} [33]. In human semen, the values were approximately 5-fold greater [33]. Adlercreutz et al. [78] has reported plasma concentrations of isoflavones in Japanese males to be 7–110 times higher than in Finnish males. Urinary excretion of genistein is 40 times higher in the Japanese than in Caucasian populations [28]. The higher levels of isoflavones in Japanese men relative to American and European men have been correlated to Japan's low mortality rate for prostate cancer [78]. These epidemiological data provide evidence of an inverse relationship between soy ingestion and the development of clinically significant prostate cancer [123], suggesting that phytoestrogens may play a protective role against the development of prostate cancer [42].

The physiological significance of the inhibitory effects of genistein on prostate cancer has been evaluated in animal and *in vitro* experiments. In human prostate cancer cell lines, high concentrations of genistein and biochanin A inhibit the growth of androgen-dependent and -independent cells. However, fairly high concentrations of genistein were required to achieve this effect [124]. These are much higher doses than can be achieved with the human diet. Using the two soybean isoflavones genistein and daidzein and their glucosides genistin and daidzin, Onozawa et al. [125] tested for their effects on cell growth and apoptosis of LNCaP human prostatic cancer cells. Among these isoflavones, genistein was found to inhibit the growth of LNCaP most effectively with an IC_{50} value of 40 µM. Inhibition of cell growth by genistein was accompanied by the suppression of DNA synthesis and the induction of apoptosis. Expression of prostate specific antigen (PSA) in LNCaP cells was also significantly reduced by treatment with genistein. These results suggest that genistein might primarily influence prostate cancer development by reducing tumor growth. In a study conducted by Dalu et al. [126], genistein fed to rats at 1.0 mg·g^{-1} was able to down-regulate EGF and ErbB2/Neu receptors in the rat prostate, with no toxicity to the host. The levels of circulating total genistein (252–2,712 pmol·ml^{-1}) in these rats are comparable to those in Asians consuming a traditional diet containing soy products with an average of 276 pmol·ml^{-1}; one Japanese subject had circulating levels of genistein exceeding 2,400 pmol·ml^{-1} [78]. Genistein's inhibition of the EGF signaling pathway suggests that this phytoestrogen may be useful in both protecting against and treating prostate cancer.

There are very few studies which have evaluated humans and the effects dietary estrogens can have on the progression of prostate cancer. Genistein and biochanin A were found to inhibit 5α-reductase in the lysate of human genital skin fibroblast and prostatic tissue [127]. Like isoflavones, lignans also inhibit 5α-reductase [127]. Genistein and other isoflavonoids were found to inhibit the activity of 3- and 17β-hydroxysteroid dehydrogenase [128]. Thus far, there has been one unique case of prostate cancer regression in a 66-year-old patient diagnosed with high-grade adenocarcinoma of the prostate gland. The subject took phytoestrogens in a pill form (160 mg per day) for 1 week before undergoing radical prostatectomy. Stephans [129] reported an increase in apoptosis or cell death, in the resected prostate gland. The potential benefits of soy constituents, phytoestrogens, has been speculated in the treatment of prostate cancer. Some investigators have moved beyond simple speculation and have begun clinical trials addressing the role of phytoestrogens in patients with prostate cancer. Rosenthal et al. [130] have commenced a large clinical study examining the effects of soy ingestion in men with rising serum PSA levels after radical prostatectomy and in men at high risk of recurrent disease after radical prostatectomy. Eventually, Rosenthal's group would like to undertake a true chemoprevention study of soy protein in patients at high risk for developing clinically significant prostate cancer. Clinical trials of soybean in men with abnormal PSA are also reported to be in progress [131]. Further human studies are needed to evaluate the role of phytoestrogens in prostate cancer.

5.3
Colon Cancer

Cancer is one of the main causes of mortality in the industrialized world, yet the incidence of colon cancer varies worldwide. Bowel cancer is the second most common cancer in the UK, after breast cancer in women and lung cancer in men. Far Eastern populations, such as China and Japan, used to have a much lower incidence [132]. Up to the 1960s, prostate cancer was rare in Japan and Far East countries. The apparent protection of an Asian heritage is speculated to probably be dietary which can be lost upon adoption of a Western style diet [118, 133]. Soy is a particular component of Far Eastern diets, not present in the traditional Western diet. There have been several case-control studies examining a role for soybean in protection against colo-rectal cancer, in China, Japan and in Japanese migrants to the US [64]. However, there is not a clear relationship because studies have yielded non-significant results.

Many hypotheses have suggested lignans to be the active chemoprotective agent, but there has been little investigation of lignan intake or excretion in relation to bowel cancer [134, 135]. Many speculations which have suggested lignans' role in the protection against colon cancer have been extrapolated from the beneficial effects of a well-balanced diet, which is high in dietary fiber. Slattery et al. [136] evaluated the diet diversity, diet composition and risk of colon cancer in an incident population-based study of 1,993 cases and 2,410 controls. Total diet diversity was not associated with colon cancer. However, eating a diet with greater diversity of meats, poultry, fish and eggs, was associated with a 50% increase in risk among all men with a slightly stronger association for younger men and men with distal tumors. A diet with a greater number of refined grain products was also associated with increased risk among men. Women who ate a diet with a more diverse pattern of vegetables were approximately at a 20% lower risk than women who had the least diverse diet in vegetables. In addition, a high intake of dietary fiber has been thought to reduce the risk of colorectal cancer and adenoma [137]. However, in a recent large prospective study with a follow-up extending to 16 years, Fuchs et al. [138] found no association between the intake of dietary fiber and the risk of colorectal cancer.

Colon cancer does not have a strong association with hormone status, but there are a number of other possible mechanisms whereby lignans or isoflavones could be involved in the etiology of bowel cancer. For example, there may be a role for isoflavones to inhibit endogenous N-nitrosation that occurs when meat is consumed via suppression of inducible NO synthase [139]. In animals, aberrant crypts are accepted early markers of colon cancer which can be induced by standard chemical carcinogens. One study has shown that linseed is able to reduce the numbers of these aberrant crypts formed in such animal models, and another study has shown that genistein has the same effect [140, 141]. Genistein was found to inhibit azoxymethane (AOM)-induced formation of aberrant crypts in the colon of male F344 rats [142]. These studies provided a rationale for the evaluation of genistein in the long-term pre-clinical colonic tumor assay for a closer assessment of the potential utility of this agent. In addition, Sung et al. [143] determined that, at 100 μM concentrations, lignans significantly reduced the proliferation of colon tumor cells. The growth was not affected by the presence of E2, implying that these cells are not estrogen-sensitive. Although evidence for the presence of ER in the colon in inconsistent [144], tamoxifen, a synthetic anti-estrogen, has demonstrated growth inhibitory effects on colon tumor cells [145]. In contrast, Rao et al. [146] demonstrated that the administration of genistein significantly increased non-invasive and total adenocarcinoma multiplicity in the colon, compared to the control diet, but had no effect on the colon adenocarcinoma incidence or on the multiplicity of invasive adenocarcinoma. The results of this investigation emphasize that the biological effects of genistein may be organ specific, inhibiting cancer development in some sites yet showing no effect or enhancing effects on the tumorigenesis at other sites, such as the colon. The exact mechanism(s) of colon tumor enhancement by genistein remains to be elucidated. The data regarding dietary estrogens and colon cancer are still limited and need to be investigated further.

5.4
Endogenous Hormones

Steroidal estrogens which circulate in blood are associated with serum proteins. E2 is primarily bound with high affinity to glycoproteins, such as α-fetoprotein in mice and rats or sex hormone-binding globulin (SHBG) in humans, and with low affinity to serum albumin [147, 148]. Of the total serum E2 in adults, typically only 1–3% is able to pass into cells and bind to intracellular receptors [148 –151]. When the protein-bound and free fractions of E2 are near steady state, the free fraction is the concentration that determines receptor occupancy and ultimately the level of response. Such carrier proteins have at least two roles in hormone action, i.e., sequestering of the hormone which reduces metabolism by enzymes, and modulation of the concentration of hormone that is available to the target cell [152, 153]. The activity of carrier proteins would be affected by dietary estrogens which may displace the steroid from its carrier. Of great importance, although often overlooked, is that dietary estrogens which do not bind to serum proteins will escape this mechanism to limit cell uptake. A number of phytoestrogens, including genistein and coumestrol, bind poorly to SHBG [101]. These phytoestrogens, that show less binding to serum proteins than E2, may have a greater proportion of their total concentration in serum available to interact with intracellular ERs and this would increase their effective estrogenic activity in serum. Nagel et al. [154] developed the relative binding affinity-serum modified access (RBA-SMA) assay to compare the competition of unlabeled phytoestrogens with [^3H]-E2 in the presence and absence of serum to determine the effect of serum on the access of phytoestrogens to ERs in intact cells. They found that several phytoestrogens, including coumestrol, equol, genistein, and daidzein, showed greater access to ERs than E2 in the presence of adult serum, indicating that the activity of these dietary estrogens could be underestimated in serum-free or low serum assays. Conversely, the phytoestrogen, biochanin A, showed decreased access relative to E2, and the activity of these compounds could be overestimated when the effects of serum are not taken into account.

It has been suggested that phytoestrogens would exert their effects via the stimulation of SHBG, reducing the proportion of free estrogens circulating in plasma. This observation is based on cross-sectional comparisons and some studies with cell lines [42]. However, cross-sectional comparisons are difficult because SHBG can be affected by many factors, such as changes in body weight. In controlled intervention studies, there appears to be no effect of soybean on SHBG levels in premenopausal women [39, 155, 156]. There is also no effect in premenopausal women receiving lignans [157]. Hence, existing evidence to date suggests that a protective effect of phytoestrogens is unlikely to be brought about by lowering the levels of free estrogens in plasma, since they do not seem to have a direct effect on serum E2 or SHBG levels. Nagata et al. [12] examined a cross-sectional relationship of soy product intake to serum concentrations of E2 and SHBG in 50 healthy premenopausal Japanese women. The intakes of soy products was inversely correlated with E2 on days 11 and 22 of the cycle after controlling for age, body mass index, cycle length and intakes of total energy, fat and crude fiber. No significant correlation was observed between soy product intake and SHBG. These results suggest that the consumption of soy products lowers the risk of developing breast cancer by modifying estrogen metabolism. The results of this cross-sectional study also reflect the association of the hormonal status with the usual diet of the subjects over periods longer than that described in the intervention studies. Shoff et al. [158] conducted a cross-sectional study examining the relations between consumption of phytoestrogen-containing foods and serum sex hormones and SHBG in a population-based sample of postmenopausal women. Partial correlations between hormones and intake of phytoestrogen-containing foods were computed with adjustment for age, body mass index, years since menopause, and total energy intake. Number of standard servings per week of whole grain products from the dark bread group was inversely associated with total testosterone. Although not statistically significant, other hormones displayed similar inverse associations with dark bread consistent with a common metabolic pathway. Although the magnitude of association was small, the data are consistent with the possibility that consumption of some phytoestrogen-containing food may affect levels of testosterone in postmenopausal women. The lack of associations between hormones and other phytoestrogen-containing foods is likely due to infrequent consumption and/or lower levels of phytoestrogens in particular food items. Additionally, interindividual variability in serum phytoestrogen concentration due to differences in gut microflora (those responsible for phytoestrogen conversion to estrogen-like compounds) may attenuate associations of dietary intake and steroid hormone levels. Isoflavones and lignans are thought to interfere with intestinal microflora in their metabolic pathway [23, 159] which would have an effect on the reabsorption of E2 and secondarily, their levels in the blood. Adlercreutz and associates [31] found a significant inverse correlation between the urine excretion of equol and the plasma percent free E2 in Finnish women. In addition, the urinary excretion of the lignan metabolite, enterolactone was significantly and inversely correlated with percent free E2. Their findings for equol and enterolactone might reflect the effect of isoflavones and lignans on E2 metabolism. It is more likely that isoflavones and their metabolites may alter the intestinal steroid hormone metabolism, which affects E2 concentration since such a mechanism is postulated for the role of fiber intake in the prevention of breast cancer [160].

5.5
Menstrual Cycle

Soy intake exerts pronounced physiological effects on the menstrual cycle length (which has been associated with breast cancer risk) and serum concentrations of follicle-stimulating hormone (FSH) and luteinizing hormone (LH) in premenopausal. Cassidy et al. [155] monitored the effect of a diet rich in isoflavones on the menstrual cycle of premenopausal women. The results from their study demonstrated that a high intake of soy protein (60 g per day containing 45 mg isoflavones) increased the length of the follicular phase and/or delayed menstruation by 2–3 days. No effects were shown with supplements from which isoflavones had been removed. The midcycle surges of LH and FSH were also significantly suppressed during the soy-diet period. Another study with a soy drink caused an 'erratic' elevation throughout the cycle [156]. Twelve ounces of soy milk 3 times per day decreases serum E2 and luteal phase serum progesterone in 22- to 29-year-old females [161]. A cyclic pattern of lignan excretion has been observed during the menstrual cycle in humans [162]. In another study, flax seed supplementation (10 g per day) increases luteal phase duration, but with no difference in follicular phase length in normally cycling women [157]. Soya milk supplements induced a 3-day increase in menstrual cycle length [161]. These observations suggest that the agonist-antagonist effect of the phytoestrogens present in soy may be responsible for these hormonal modifications of the cycle. An increase in the menstrual cycle length reduces the lifetime exposure to estrogens [163]. This, as well as the concomitant lengthening of the follicular phase, would be beneficial in lowering breast cancer risk since the mitotic activity of the breast tissue is reported to be 4-fold higher in the luteal phase than during the follicular phase. What is of greater interest in the context of these findings is that two of the women who had the highest urinary equol excretion showed the largest increase in follicular phase length [39]. These effects of soybean in women are coupled with the fact that breast cancer rates are low and menstrual cycle lengths longer in Far Eastern populations where the traditional diet is rich in soybean [164].

5.6
Menopausal Symptoms

One of the most disruptive and classic aspects of the menopause is the hot flash. In Western societies, it is the most common symptom of the menopause, although the prevalence is much lower in Japan [165, 166]. This and the rarity of the problem in soybean-consuming populations have prompted investigations to determine whether phytoestrogens have a similar effect. Hormone replacement therapy (HRT) generally alleviates the hot flashes, as well as the vaginitis occurring at the menopause due to atrophy. More recently, postmenopausal

HRT has been seen as a specific treatment for symptoms in the short-term and preventative therapy in the long-term [167]. Postmenopausal hormone therapy has both benefits and hazards, including a decreased risk of osteoporosis and cardiovascular disease and an increased risk of breast and endometrial cancer [168, 169]. In a prospective study conducted by Grodstein et al. [170], the relation between postmenopausal hormones and mortality was examined to provide a balanced assessment of the risks and benefits of hormone use. On average, mortality among women who use postmenopausal hormones is lower than among non-users. However, the survival benefit diminishes with longer duration of use and is lower for women at low risk for coronary disease. In a systematic review of scientific and lay literature, the strongest controlled study data of alternative treatment for menopausal symptoms supports phytoestrogens for their role in diminishing menopausal symptoms related to estrogen deficiency and for possible protective effects on bones and the cardiovascular system [171]. Seidl and Stewart [172] conducted a qualitative study of women's experiences with alternative treatments for symptoms attributed to menopause. Women perceived alternative treatments to be safe and somewhat effective because of their "natural" origin. Factors influencing the use of alternative therapies included personal control over health, confidence in advice from non-physicians, perceived pressure from physicians to use HRT, and physicians disinterest and frequently negative attitudes towards alternatives. Despite medical evidence for HRT, the women interviewed were mostly against HRT, predominantly for fear of cancer. The scientifically proven benefits of HRT (e.g., short-term symptom relief, cardiovascular and bone protection) were not seen to outweigh the potential risks and side effects of HRT.

Hot flashes are related to the fall in circulating estrogen, rather than absolute levels, and are associated with surges of gonadotropins from the pituitary, e.g., LH. As ovarian activity declines during menopause, serum estrone and E2 fall and remain low. It has been hypothesized that phytoestrogen consumption, due to ER binding affinity, may exert an estrogenic effect in the postmenopausal woman. Japanese women are reported to have a lower frequency of hot flashes compared with postmenopausal Western women, which has been attributed to their high phytoestrogen consumption [173]. Two studies have demonstrated that phytoestrogens, either as soybean or lignans, lower plasma gonadotropins after the menopause, although FSH rather than LH is lowered [39, 174]. However, two other studies showed no effects [175, 176]. Several studies have examined the effects of phytoestrogens on menopausal symptoms. One study has reported an improvement with 45 g raw soybean flour per day, but an improvement occurred also with white wheat flour which contains little phytoestrogens [176]. Significant decreases in hot flashes have also been demonstrated with the consumption of 45 g linseed [177], 80 g soy protein drink [178], and 80 g of tofu and miso plus 10 g linseed [179]. Several studies have investigated the effects of phytoestrogen supplements on vaginal cytology. One study reported an increase in vaginal cell proliferation, which is a sensitive and specific indicator of estrogen response, using a mixture of soybean clover and linseed [174]. Significant effects on vaginal cytology have also been reported in studies using 45 g of soy grit enriched bread [177] and 80 g of tofu and miso plus

10 g linseed [179]. In another study, 91 postmenopausal women were studied after ingesting a diet containing soy-based foods (165 mg isoflavones) or a usual diet for 4 weeks. Although anticipated, no estrogenic effects were seen in the liver or pituitary as measured by no change in SHBG or gonadotropin levels. The overall vaginal maturation index did not differ between the groups, but the percentage of vaginal superficial cells increased slightly [175]. No effects were found, however, in another intervention in which postmenopausal women were asked to add 45 g raw soybean flour to their diet [176]. There is inconsistency in the data pertaining to menopausal symptoms and phytoestrogens, and this variability may be due to differing study designs (particularly with respect to duration of exposure), phytoestrogen sources, individual variability in response or non-response, the natural resolution of symptoms over time, and populations studied [180]. Overall, however, more studies show a beneficial effect of these compounds than no effect. At present it would appear that an increase in phytoestrogens in the diet may assist women with mild menopause symptoms to a small degree, which may be an indication of estrogenic activity with reversal of menopausal atrophy.

5.7
Coronary Heart Disease

Coronary heart disease (CHD) is a multifactorial disease, for which the main established risk factors are raised serum cholesterol, raised blood pressure and smoking. The proportion of both men and women who are hypertensive steadily increases with age. Compared with men, serum cholesterol levels are lower in women up to the age of 50 years. After the menopause, levels of serum cholesterol in women exceed those of men. In women, therefore, the relative importance of CHD as a cause of death steadily increases with age, whereas in men, its importance declines after 55–64 years of age [181]. CHD accounts for 23% of deaths in women in the UK, and 30% of deaths in men, although rates have been falling since the late 1970s. Rates are low in Far East countries, such as Japan, and also declining [181, 182]. Postmenopausal estrogen replacement has been shown to decrease lipoprotein (a) [Lp(a)] [183, 184]. The synthetic anti-estrogen, tamoxifen, has also been shown to beneficially alter serum lipid and Lp(a) concentrations [184, 185]. In addition, there may be similar beneficial effects of estrogen on low density lipoprotein (LDL) and high density lipoprotein (HDL) cholesterol concentrations in men [186–188].

The importance of lowering serum cholesterol in reducing the risk of CHD and total mortality is now well established. It has been estimated that a 1% lowering of plasma cholesterol translates on a population basis to a 2–3% reduction in risk of CHD [181]. Decreases in serum cholesterol concentrations of ~5.5 mM, even within subjects with average total cholesterol concentrations, can reduce cardiovascular disease risk [189]. Cholesterol reduction can be achieved by a reduction in the saturated fatty acid content of the diet and by drugs. Both estrogens in HRT and the anti-estrogen tamoxifen can lower LDL cholesterol although the decrease in LDL is complemented by an increase in HDL levels with HRT [190, 191]. Much of the effect of soybean in lowering serum cholesterol has been attributed to the phytoestrogens [192, 193]. There is evidence to support the hypothesis that phytoestrogen consumption contributes to the lower incidence of cardiovascular disease in Asian countries and in vegetarians and that phytoestrogens may be cardioprotective [13]. However, the mechanism is uncertain, since genistein is reported to both up- and down-regulate LDL receptors, and some of the products used in studies reporting cholesterol-lowering effects contained unexpectedly low levels of isoflavones [194]. In addition, because of its effects on tyrosine kinases, genistein may have a role in the suppression of the cellular processes which lead to thrombus formation and eventually, atherosclerosis. In cell lines, genistein has been found to inhibit the proliferation brought about by platelet-derived growth factor in the artery wall, and to interfere with release of inflammatory cytokines from macrophages. It also inhibits platelet aggregation and acts as a thromboxane receptor agonist [195, 196]. Some of these effects are brought about at low levels. Genistein is also able to suppress the release of the endothelial relaxing factor NO via its effect on inducible NO synthase [174, 197].

Animal studies have demonstrated that the concentration of cholesterol in blood is lowered by the consumption of soy protein rather than animal protein [198]. A study of male and female Rhesus monkeys (*Macaca mulatta*) has demonstrated that an isoflavone-containing soy protein supplementation results in a hypocholesterolemic effect when compared with a soy diet depleted of phytoestrogen [199]. Both male and female animals receiving the soy diet were observed to have LDL cholesterol and very low density lipoprotein (VLDL) cholesterol values 30–40% lower than controls. HDL cholesterol ratio increased by 50% in female and 20% in male animals. To distinguish the relative contributions of the protein moiety versus the alcohol-extractable phytoestrogens for cardiovascular protection, Anthony et al. [200] studied young male cynomolgus macaques fed a moderately atherogenic diet and randomly assigned to three groups for 14 months. The groups differed only in the source of dietary protein, which was either casein/lactalbumin, soy protein with the phytoestrogen intact (soy+), or soy protein with the phytoestrogens mostly extracted (soy–). Animals fed soy+ had significantly lower total and LDL plus VLDL cholesterol concentrations compared with the other two groups. The soy+ animals had the highest HDL cholesterol concentrations, the casein group had the lowest and the soy– group was intermediate. Coronary artery atherosclerotic lesions were smallest in the soy+ group, largest in the casein group and intermediate in the soy– group. It could not be determined whether the beneficial effects seen in the soy– group relate to the protein itself or to the remaining traces of phytoestrogens. The beneficial effects of soy protein on atherosclerosis appear to be mediated primarily by the phytoestrogen component. In a study conducted by Wagner et al. [201], ovarectomized female monkeys fed soy protein compared with casein consumption resulted in a significant improvement in plasma lipid and lipoprotein concentration, and a decrease in arterial lipid peroxidation. This study also demonstrated that E2 reduced the number of CHD risk factors, and the combination of both soy and E2 resulted in a significant interactive effect with regard to aortic cholesteryl ester content. Another study examined the effects of soy phytoestrogens on coronary vascular reactivity in 22 rhesus monkeys with pre-existing diet-induced atherosclerosis. The monkeys were randomized to a soy-enriched or non-enriched diet for 6 months. The soy-enriched diet enhanced the dilator responses of atherosclerotic coronary arteries to acetylcholine in female rhesus monkeys [202]. These animal studies further encourage studies in humans in this area.

Dietary soy phytoestrogens may provide cardioprotective benefits for humans via a direct effect on lipids. Many human clinical trials have examined the effects of dietary estrogens on serum lipids. Meta-analysis of controlled clinical trials examining soy protein consumption and serum lipid concentrations found that consumption of 47 g soy protein daily, significantly decreased serum concentrations of total cholesterol (~9%), LDL cholesterol (~13%) and triglycerides in 34 of 38 studies [192]. Cassidy et al. [39] observed a 9% reduction in total cholesterol in a small study of normolipemic premenopausal women given a 60 g soy protein supplement. In another study, men consuming 1 L of soy drink daily, reduced their elevated cholesterol and LDL cholesterol by 9.3% and 11.3%, respectively [203]. A soybean protein diet in subjects with type II hyperlipoproteinemia may lower cholesterol on average by 20% [204]. Consumption of 25 g soy protein-enriched bread resulted in a decreased total serum cholesterol and increased HDL cholesterol in hypercholesterolemic men [203]. A study of normolipidemic postmenopausal women supplemented with a 40 mg phytoestrogen pill demonstrated a 22% increase in HDL cholesterol and no significant change in other parameters [205]. Another study also observed a significant increase in HDL cholesterol on subjects consuming soy milk [206]. In a cross-sectional study conducted by Nagata et al. [207], the relationship between soy products and serum cholesterol concentration in a community in Japan was examined and a significant trend was observed for decreasing total cholesterol concentration with an increasing intake of soy products in men after controlling for age, smoking status and intake of total energy, total protein and total fat. This negative trend was also noted in women after controlling for age, menopausal status, body mass index and intake of total energy and vitamin C. In another study, 66 hypercholesterolemic, free-living, post-menopausal women were investigated during a 6-months parallel-group, double-blind trial with 3 interventions [208]. After a control period of 14 days, all subjects were randomly assigned to 1 of 3 dietary groups (all with 40 g protein): a National Cholesterol Education Program (NCEP) Step 1 diet with protein from casein and nonfat dry milk (control), an NCEP Step 1 diet with protein from isolated soy protein (ISP) containing moderate amounts of isoflavones (ISP56), or an NCEP Step 1 diet with protein from ISP containing high amounts of isoflavones (ISP90). Non-HDL cholesterol (LDL and VLDL) in both the ISP56 and ISP90 groups was significantly reduced compared to the control group, whereas total cholesterol was not changed. HDL cholesterol significantly increased in both the ISP56 and ISP90 groups, whereas the ratio of total to HDL cholesterol decreased significantly in both groups compared with the control. Mononuclear cell LDL receptor messenger RNA concentrations significantly increased in subjects consuming ISP56 or ISP90 compared with the control. These results indicate that soy protein with different amounts of isoflavones may decrease the risk of cardiovascular disease via improved blood lipid profiles, and that the mechanism by which apolipoprotein B-containing lipoproteins were depressed may be via alterations in LDL receptor quantity or activity. Low-fat, low-cholesterol diets similar to the NCEP Step 1 diet have been shown to reduce serum cholesterol concentrations in humans by as much as 14% [209]. The enhanced effects on plasma lipoproteins observed with the addition of soy protein to an NCEP Step 1 diet provide additional evidence that dietary protein influences the risk of CHD in humans. It is reasonable to postulate that soy protein, by some not yet understood mechanism, may also increase hepatic LDL receptor mRNA concentrations to enhance LDL receptor activity. Reductions in LDL-cholesterol concentrations with soy protein by way of increased LDL receptor activity in human blood monocytes was demonstrated by Lovati et al. [210], whereas Angelin et al. [211] documented the effects of estrogen on the up-regulation of human hepatic LDL receptors. It has also been hypothesized that isoflavonoid antioxidants derived from soy could be incorporated into lipoproteins and could possibly protect them against oxidation, which is regarded as atherogenic. Six healthy volunteers received 3 soy bars containing 12 mg genistein and 7 mg daidzein daily for 2 weeks [212]. Compared with baseline values, lag phase of LDL oxidation curves were significantly prolonged by a mean of 20 min during soy intake, indicating a reduced susceptibility to oxidation. These results suggest that intake of soy-derived antioxidants such as genistein and daidzein may provide protection against oxidative modification of LDL. Hodgson et al. [213] conducted a study to determine if isoflavonoids could improve serum lipids in 46 men and 13 postmenopausal women. One tablet containing 55 mg of isoflavonoids (predominantly in the form of genistein) or one placebo tablet was taken daily with the evening meal for 8 weeks. After adjustment for baseline values, no significant differences in post-intervention serum lipid and Lp(a) concentrations between groups were identified. Further adjustment for age, gender, and weight did not alter the results. In addition, changes in urinary isoflavonoids were not significantly correlated with changes in serum lipids and Lp(a). This study does not support the hypothesis that isoflavonoid phytoestrogens can improve the serum lipids, at least in subjects with average serum cholesterol concentrations. The effect of dietary soy protein on serum cholesterol concentration has been examined in humans in many clinical trials, but the results have not been consistent. Some factors which may affect the outcome of these clinical trials include the study design, type of phytoestrogen used, dosage and pretreatment cholesterol level. Although some studies reported a significant decrease in plasma or serum cholesterol concentration in some hypercholesterolemic subjects as a result of soy-protein diets [203, 204, 214], most studies of normocholesterolemic subjects have shown little difference in effects on plasma or serum cholesterol concentration between soy protein and control diets [215–220].

5.8
Osteoporosis

Osteoporosis is defined as a condition in which the amount of bone per unit volume is decreased, but the composition remains unchanged. The bone becomes porous due to an imbalance in forming and resorbing bone cells, causing structural failure and predisposition to fracture. Osteoporosis in women is particularly associated with the menopause, since the loss of estrogen accelerates bone loss. Osteoporosis is a major health care problem leading to a high incidence of vertebral, radial and mainly hip fractures that are causes of morbidity and mortality in an aging population. HRT prevents this loss, at least up to age 75, if taken for several years early on in the postmenopausal period [221]. HRT prevents the lowering in bone density related to the postmenopausal hypoestrogenism [222, 223]. In addition, tamoxifen is known to prevent bone loss [224]. However, the doses of estrogen required to prevent the postmenopausal bone loss are generally higher than those required to cure the clinical subjective symptoms [225]. Thus, when moderate to high estrogen dose administration is contraindicated, lower estrogen dosages may be sufficient to control subjective symptoms of menopause, but an additional anti-osteoporotic agent must be considered to prevent or cure the postmenopausal osteopenia. Dietary factors have been investigated for their effect in achieving peak bone mass and preventing bone loss in later life. The rates of osteoporosis differ within populations with a lower incidence in Asian women than Western women [226]. The hormonal effects of phytoestrogens, coupled with the comparative rarity of the disease in populations consuming soybean, has prompted investigations of their effects in osteoporosis. In animal models, dietary soybean prevents significant bone loss in ovarectomized rats by increasing formation which then exceeds resorption [227]. Thus far, data are limited on the effect of phytoestrogens in humans. Postmenopausal women randomized to receive casein, soy protein with either 1.39 mg total isoflavones g^{-1} protein ('ISP') or 2.25 mg total isoflavone g^{-1} protein ('ISP+') for six months demonstrated increased bone mineral content and density with 'ISP+' compared with controls [228]. One study has demonstrated that 45 g soy grits added to bread increased bone mineral content by 5.4% after 3 months [177].

A synthetic analogue, ipriflavone (7-isopropoxy-isoflavone), is known to be effective in inhibiting bone resorption in postmenopausal women, although its action is not thought to involve direct action with ERs [229]. Ipriflavone is a synthetic isoflavone-derived compound able to modulate the oxidative phosphorylation and to exert a direct inhibitory effect on osteoclastic activity [230–232]. In different experimental models of osteoporosis, ipriflavone can inhibit bone resorption in animals [233]. In humans, oral ipriflavone has been shown to maintain or even increase bone density, to inhibit osteoclast recruitment and function, and to prevent bone loss at the distal radius in osteoporotic postmenopausal women, mainly through an inhibition of bone resorption, when administered at the dose of 600 mg per day [234–241]. The same effect can be obtained with the combined administration of low dose (400 mg per day) ipriflavone with low dose (0.3 mg per day) conjugated estrogens [241]. In another study, the observed ipriflavone effects on biochemical markers of bone turnover are somewhat controversial [242]. Although changes were not statistically significant, the study showed increased levels of bone metabolic markers after 1 year of treatment. The gain (~3%) was greater in those subjects who received both ipriflavone and HRT. Furthermore, their *in vitro* study results suggest that ipriflavone and E2 have different mechanisms of action on cell proliferation and differentiation in culture human vertebrae-derived cells. Ipriflavone has mild stimulatory effects on osteoblasts, in contrast to E2 which inhibits cell proliferation and differentiation. Even though ipriflavone potentiates the effect of estrogen *in vivo* [243], their study suggests that ipriflavone does not interact with the ER in human vertebrae-derived cells. Cross-talk between ipriflavone and E2 should be considered and it is essential to understand the signal pathway to elucidate the mechanism of action of ipriflavone on osteoblast activity and bone formation in future studies. Since the positive effect was more pronounced after 1 year, the possibility of a long-term ipriflavone treatment must be taken into consideration in the future. Despite its structural similarity to natural phytoestrogens, ipriflavone has been shown to be devoid of any estrogenic effect in postmenopausal women [244].

6
Infant Soy Formula

For more than 60 years, soy-based infant formulas have been fed to millions of infants worldwide and studied in controlled clinical research. The first description of the use of a soy formula in the Modern World as a cow's milk substitute appeared in 1909, whereas the use of soybean in infant feeding goes back to Eastern countries where it was employed for alimentation as early as in 82 BC [245]. The safety of isoflavones in soy-based products, including infant formulas, has been questioned recently owing to reports of possible endocrine effects in animals and in cultured cells. Based on model systems and pharmacological principles, isoflavones consumed by infants fed soy formula could theoretically contribute estrogenic activity (perhaps in addition to endogenous estrogen), suppress endogenous estrogen levels or have no effect. Among humans, the highest of all phytoestrogen doses appears to be provided to infants exposed to soy formula, and this exposure occurs during development, often the most sensitive life-stage for induction of toxicity [246]. Consumption of phytoestrogen-containing soy products by women produces demonstrable estrogenic responses at phytoestrogen doses about 5-fold lower than those in soy infant formulas given to infants [39, 246, 247].

Given the relatively broad choice of infant foods becoming available, exposure to dietary isoflavones during the first year of life is virtually ubiquitous. The isoflavones in soy products are in the form of glycosides whereas those in breast milk are in the form of glucuronides. The bioavailability of these compounds may be a function of the conjugating group as well as the gut microflora of the individual, but the absorption and subsequent metabolism of these compounds by infants has not been fully investigated. The possibility that infants may be exposed to phytoestrogens at concentrations greater than those found in breast milk is cause for concern, given the evidence that hormonal imbalance early in life can affect the sexual development of some animal species [248–252]. We know little about the bioavailability of isoflavones in breast milk and soy-containing infant formulas. Most of the isoflavones in breast milk and soy formulas are in the conjugated (bound) form [253], which may be less biologically active than unconjugated (free) forms. Thus, biological activity may not be assumed from the presence of isoflavones alone. Franke and Custer [34] demonstrated that lactating women consuming a diet with a high phytoestrogen content will produce breast milk containing the two isoflavones genistein and daidzein at ~20 $\mu g \cdot L^{-1}$, although there is a large variation between individuals. This is in agreement with the published GC-MS data of Morton et al. [254] who reported isoflavone concentrations of ~4 $\mu g \cdot L^{-1}$ in the breast milk of Hong Kong women. A typical infant weighing 7 kg and consuming 0.8 L of breast milk per day should thus have an upper range of isoflavone consumption of ~0.4 μg $\cdot kg^{-1}$ body weight per day. Nguyenle et al. [255], using HPLC, reported concentrations of the glycoside conjugates equivalent to ~9,000 $\mu g \cdot L^{-1}$ for daidzein and 24,000 $\mu g \cdot L^{-1}$ for genistein, averaged over 4 different formulas. Dwyer et al. [62] used GC-MS and found lower values, but still in the range of 200–1,400 μg $\cdot L^{-1}$ for daidzein and 600–3,100 $\mu g \cdot L^{-1}$ for genistein. Even considering these low values, infant formulas clearly contain at least 10-fold the amount of phytoestrogen found in breast milk and some data indicate that the difference may be as much as 1000-fold. This means that the phytoestrogen intake of the infant described above would increase from a minimum of 90 $\mu g \cdot kg^{-1}$ body weight per d, up to 4,000 $\mu g \cdot kg^{-1}$ body weight per day when fed infant formulas. This is compared with the intake of an adult consuming a daily ration of 50 g of soy protein, (e.g., 700 $\mu g \cdot kg^{-1}$ body weight per day of isoflavones), an amount known to cause hormonal effects in premenopausal women [39]. Knight et al. [256] investigated the phytoestrogen content of different foods, formulas and drinks that may be consumed by infants during their first year of life in an attempt to define levels of exposure on different feeding regimens. HPLC was used to determine the levels of genistein, daidzein, biochanin A, formononetin and equol in samples purchased from Australian supermarkets. All foods tested contained isoflavones, at varying levels. Casein-based infant formulas contained between 0.001 and 0.03 $mg \cdot L^{-1}$. Soy-based infant formulas ranged from 17.2 to 21.9 mg L^{-1} with the values detected in yogurt at similar levels to that of cow's milk. For comparison, the soy-based beverages (which are not recommended for use under 12 months of age) contained levels of isoflavones from 22.9 to 71.5 $mg \cdot L^{-1}$. In addition, clover contains isoflavonoids and may, therefore, represent, via milk, a source of isoflavonoids in the human diet. King et al. [257] measured cows' milk samples obtained from 76 farms in 3 Australian states. Concentrations in all samples were found to be extremely low. The mean daidzein concentration was <5 $ng \cdot ml^{-1}$. Mean genistein concentrations ranged from just detectable (~2 $ng \cdot ml^{-1}$) in Victorian samples collected during summer to 20–30 $ng \cdot ml^{-1}$ in samples from all states collected during spring when isoflavonoid-containing clover is most dominant. Mean equol concentrations ranged from 45±10 $ng \cdot ml^{-1}$ in Victorian farms samples collected during summer to 293±52 $ng \cdot ml^{-1}$ in Western Australian samples collected in spring. The mean concentrations of genistein and equol in post-pasteurization samples collected in spring were approximately double those for samples collected in autumn. Pasteurization had no effect on isoflavonoid concentrations. The concentrations of isoflavonoids in Australian cows' milk are low and therefore are unlikely to have any pronounced biological effect in human consumers.

Soy-based infant formulas continue to be a safe, nutritionally complete feeding option for most infants [258]. Clinical cross-reactivity to legumes is very rare in children, therefore explaining why soy allergy is so uncommon [259]. These products provide essential nutrients required for normal growth and development. Iacono et al. [260] performed a double-blind crossover study comparing cow's milk with soy milk in 65 children with chronic constipation. Forty-five of the 65 children (68%) had a response while receiving soy milk. Anal fissures and pain with defecation resolved. None of the children who received cow's milk had a response. In young children, chronic constipation can be a manifestation of intolerance to cow's milk. Soy protein formulas are used for different conditions including cow's milk allergy, lactose and galactose intolerance and in the management of severe gastroenteritis. Some studies show that feeding soy protein formulas for the first 6 months of life significantly reduces the prevalence of atopic diseases in high risk babies [261]. It is also worth noting that infants in soy-consuming communities, such as Japan, China and Seventh Day Adventists, typically are weaned onto soy products between 6 and 12 months of age and then receive isoflavone-containing foodstuffs (e.g., tofu, miso, tempeh) on a long-term basis from that time, a period which embraces the bulk of their reproductive development. A question that needs to be answered is whether it is more important to have eaten phytoestrogen-rich foods as a child than as an adult. It appears that regulatory pathways for synthesis and breakdown of hormones or sensitivity to hormonal control can be set very early in life, even during fetal development, by exposures to certain chemicals, and have lifelong effects.

Additionally, no increased incidence of endocrine effects has been documented in infants in Asian populations whose traditional diet includes large amounts of soy products. There have been no reports of abnormal pubertal development in adolescents who received soy-based formulas as infants. Growth is normal and no changes in timing of puberty or infertility rates have been reported in humans who consumed soy formulas as infants. Similarly, there have been no reports of infertility in adults who consumed soy-based formula as infants. There have been no reports of infants fed soy-based formulas developing breast buds. If isoflavones are absorbed in significant amounts and yet do not cause breast development or increased growth rate, both of which are accepted as early indicators of estrogenic effect in human children, it is very unlikely that they cause any other acute or delayed adverse endocrine effects. The lack of reported effects on millions of infants consuming isoflavones in soy formulas suggest there are no biological or clinical effects. The lack of any apparent reproductive and endocrinological dysfunction in these communities compared to Western communities argues against a detrimental impact of these compounds.

Although large doses of isoflavones have estrogenic effects in some animals, no parallel effects in human infants have been reported. Furthermore, it may not be appropriate to extrapolate animal or cultured-cell observations to human infants who consume modern soy-based infant formulas because it is well documented that effects of isoflavones are dependent on numerous factors, including species, age, dose, duration of exposure, metabolism and individual variability. In addition, the neonatal rodent and postnatal human are not at equivalent morphological stages of development [262] and the neonatal rodent does not model the infant human. The considerably different timetable of developmental milestones in laboratory animals in contrast to humans must also be considered. The maturation of rodent reproductive tracts is ~70 times more rapid compared with humans suggesting a far greater degree of susceptibility of the immature reproductive tissues of rodent to possible adverse effects of hyper-estrogenicity. Cross species differences in the effects of estrogenic compounds are also well known. For example, tamoxifen is estrogenic in mice, anti-estrogenic in frogs and chickens and may be estrogenic and anti-estrogenic in humans, depending on tissue specificity [263]. Toxicological studies of estrogenic isoflavones in animals and in particular those claiming to show a detrimental effect of reproductive tract development in infant animals must be viewed cautiously. At present, the beneficial effects of soy-based formulas or of milk from mothers consuming phytoestrogens are speculative, but the same is true for potential risks.

7
Potential Adverse Effects of Dietary Estrogens

Concern has been expressed that some phytoestrogens may disrupt the developing endocrine system similarly to the effects of other endogenous estrogens [264, 265]. Much of this concern has stemmed from animal research. There are well-described examples of phytoestrogen-containing plants inhibiting fertility via estrogenic activity in animals. For instance, sheep grazing on Australian pastures containing a particular type of clover rich in formononetin, which is converted to daidzein in the rumen during fermentation, developed a widespread infertility in the 1940s [248, 266]. Other examples are the "moldy corn syndrome" in pigs and cattle fed corn contaminated by *Fusarium* sp., which produces the estrogenic β-resorcyclic acid lactone, zearalenone [267], and the inhibition of reproduction of California quail by phytoestrogens produced by plants growing in dry conditions [266]. The use of soybean in captive cheetah in Cincinnati zoo was also shown to be responsible for an infertility syndrome, reversed by its removal from the feed [268]. The phytoestrogen, coumestrol, which is ~30 times more effective than genistein in mice, is known to cause estrogen-related disorders in animals which seem to have a cumulative effect. Whitten et al. [269] demonstrated the effect of phytoestrogens on the sexual differentiation of gonadotropin function by examining neonatal exposure of pups through milk of rat dams fed coumestrol ($100\ \mu g \cdot g^{-1}$) during the critical period of the first 10 postnatal days or throughout the 21 days of lactation. In females, exposure to coumestrol throughout the period of lactation produced growth suppression and an acyclic condition in early adulthood resembling the premature anovulatory syndrome. When the period of treatment was restricted to the first 10 postnatal days, however, no effects on vaginal cyclicity were seen. The 10-day exposure period produced more marked effects in males, resulting in transitory reductions in body weight in weaning males and reductions in mount and ejaculation frequency and prolongation of the latencies to mount and ejaculate. Testicular weights and plasma testosterone levels did not differ among treatment groups suggesting that the deficits in male sexual behavior were not due to deficits in adult gonadal function. These data provide evidence that lactational exposure to phytoestrogen diets can alter neuroendocrine development in both female and male rats. In addition, feeding flaxseed, the richest source of the mammalian lignan precursor secoisolariciresinol diglycoside, to rats during a hormone-sensitive period has demonstrated reproductive effects [270]. The female offspring had shortened anogenital distance, greater uterine and ovarian relative weights, earlier age and lighter body weight at puberty, lengthened estrous cycle, and persistent estrus, whereas the males had reduced postnatal weight gain and greater sex gland and prostate relative weights, suggesting estrogenic effects. These examples in animals suggest that the phytoestrogen content of soy products and other dietary products may induce unintended adverse effects on reproduction and development in humans. A general argument can be made that the long history of apparent safe use of soy argues that it is not toxic. To date, there are no long-term studies in humans in which a possible association between soy exposure and toxicity has been systematically and rigorously explored. Given the prevalence of soy exposure and the possible health benefits, it is appropriate to include adverse effects in any future large-scale, long-term epidemiological studies. Because reproductive and developmental toxicity have been demonstrated in animals and humans with a wide variety of estrogens, and phytoestrogen exposure has been shown to induce reproductive and developmental toxicity in experimental animals and livestock, these endpoints should receive particular attention.

Despite the hypothesized beneficial effects of phytoestrogens in human cancer, two reports suggest that caution may be necessary at this stage. In one study, 29 women took 60 g soybean (containing 45 mg isoflavones) for 14 days and demonstrated a significant increase in the proliferation rate of breast lobular epithelium [271]. In an earlier study, Petrakis et al. [156] evaluated the influence of the long-term ingestion of commercial soy protein isolate on breast secretory activity. It was hypothesized that the features of nipple aspirate fluid (NAF) of non-Asian women would be altered so as to resemble those previously found in Asian women. Both pre- and post-menopausal white women ingested 38 g of soy protein isolate containing 38 mg of genistein daily. Unfortunately, the findings did not support their hypothesis. Compared to baseline values, a 2- to 6-fold increase in NAF volume ensued during ingestion of soy protein isolate in all premenopausal women. This 6-months pilot study indicates that prolonged consumption of soy protein isolate has a stimulatory effect on premenopausal female breast, characterized by increased secretion of breast fluid, the appearance of hyperplastic epithelial cells and elevated levels of plasma E2. Compared with white and African-American women, hyperplastic, atypical epithelial cells and apocrine metaplasia were less frequently found in NAF from Chinese and Japanese women [63, 272–275].

Safety concerns regarding soy-based formulas have been raised despite no apparent deleterious effects. Unfortunately, there is very little known in regard to the toxicity of estrogens in human infants. Longer-term studies to assess the potential benefits or adverse effects of phytoestrogen exposure early in life are needed. Soy formula feedings in early life have been associated with the development of auto-immune thyroid disorders [276]. The soybean and its products have been considered goitrogenic in humans and animals. Goiter and hypothyroidism were reported in infants receiving soy-containing formula [277–279] although iodine supplementation of the formula has reversed this problem [280]. Several investigators have reported induction of goiter in iodine-deficient rats maintained on a soybean diet [281–284]. Furthermore, Kimura et al. [281] reported the induction of thyroid carcinoma in rats fed an iodine-deficient diet containing 40% defatted soybean diet. Genistein and daidzein were found to inhibit the thyroid peroxidase-catalyzed iodination of tyrosine at concentrations that approach the total isoflavone levels previously measured in plasma from humans consuming soy products [285]. Because inhibition of thyroid hormone synthesis can induce goiter and thyroid neoplasia in rodents, delineation of anti-thyroid mechanisms for soy isoflavones may be important for extrapolating goitrogenic hazards identified in chronic rodent bioassays to humans consuming soy products. It is difficult for humans to consume the amounts of isoflavones from natural soy foods to reach the toxicological levels that induce pathological effects recorded in animals. However, a trend towards isoflavone supplements (e.g., in pill form) will facilitate patient intake and the potential dangerous effects of mega-dosing are a concern. Careful studies of the soy infant formula-exposed population should be undertaken, as it is a well-identified group and phytoestrogen doses can be estimated with some accuracy. Such studies should include not only infants currently consuming soy infant formulas, but older children, adolescents and adults previously exposed. They should incorporate estrogenic and thyroid hormone related endpoints, as well as a wide variety of other endpoints of toxicity.

8
Conclusions

Our understanding of dietary estrogens and their physiological impact on the human endocrine system is broadening as research in this area is expanding with both clinical and epidemiological studies. These dietary estrogens are not only structurally similar to endogenous hormones, but have demonstrated their ability to bind to ERs and have both estrogenic and anti-estrogenic effects. The growing interest in this area suggests that these dietary estrogens may confer significant health benefits related to hormonally-related diseases and conditions such as breast, prostate and colon cancer, menopausal symptoms, the menstrual cycle, osteoporosis, and coronary heart disease. These compounds may play a significant role in the molecular processes concerned with the pathogenesis of these diseases with a real possibility that they can exercise a restraining influence on their development. The possibility still exists that the association between risk of disease and phytoestrogen intake is not casual. Despite some of the inconclusive findings, several putative mechanisms that could account for the hypothesized preventive effects of phytoestrogens have been proposed. Extrapolating from human cell lines and animal studies has its limitations. More human studies are needed to establish the role of phytoestrogens as estrogen agonists and antagonists and of their actions on tyrosine kinases and other growth factors. In addition, larger and longer-term studies are needed to more thoroughly document clinical effects and to examine the target effects on responsive tissue. Although epidemiological, human and animal data suggest that phytoestrogens may play a protective role in disease progression, the adverse effects observed in animals and in cancer cell lines must be interpreted with caution because of the difficulty in correlating their phytoestrogen exposure to the *in vivo* tissue levels in humans. The determination of the risks and benefits of human phytoestrogen exposure must be addressed for both dietary exposure and potential pharmaceutical treatments. Potentially, dietary estrogens may be an important factor in not only offering protection of the human endocrine system, but also in providing an alternative therapy against disease.

9
References

1. Korach KS, Migliaccio S, Davis VL (1994) Estrogens. In: Munson PL (ed), Principles of Pharmacology-Basic Concepts and Clinical Applications. Chapman and Hall, New York, p 809
2. Baker ME (1995) Proc Soc Exp Biol Med 208: 131
3. Turner RT, Riggs BL, Spelsberg TC (1994) Endocrine Rev 15: 275
4. Farhat MY, Lavigne MC, Ramwell PW (1996) FASEB J 10: 615
5. Iafrati MD, Karas RH, Aronovitz M, Kim S, Sullivan TR Jr, Lubahn DB, O'Donnell TF Jr, Korach KS, Mendelsohn ME (1997) Nature Med 3: 545
6. Jensen EV (1995) Ann NY Acad Sci 761: 1
7. Beato M, Herrlich P, Schutz G (1995) Cell 83: 851
8. Tsai MJ, O'Malley BW (1995) Ann Rev Biochem 63: 451
9. Osborne CK (1998) N Engl J Med 339: 1609
10. Price KR, Fenwick GR (1985) Food Addit Contam 2: 73
11. Axelson M, Sjovall J, Gustafsson BE, Setchell KD (1984) J Endocrinol 102: 49
12. Nagata C, Kabuto M, Kurisu Y, Shimizu H (1997) Nutr Cancer 29: 228
13. Adlercreutz H (1990) Scand J Clin Lab Invest 201: 3
14. Thorogood M, Mann J, Appleby P, McPherson K (1994) Brit Med J 308: 1667
15. Setchell KDR, Adlercreutz H (1988) Mammalian lignans and phytoestrogen. Recent studies on their formation, metabolism and biological role in health and disease. In: Rowland I (ed), Role of the Gut Flora in Toxicity and Cancer. Academic Press, London, p 315
16. Borriello SP, Setchell KD, Axelson M, Lawson AM (1985) J Appl Bacteriol 58: 37
17. Barnes S, Sfakianos J, Coward L, Kirk M (1996) Adv Expt Med Biol 401: 87
18. Sfakianos J, Coward L, Kirk M, Barnes S (1997) J Nutr 127: 1260
19. Axelson M, Setchell KD (1981) FEBS Lett 123: 337
20. Adlercreutz H, Fotsis T, Bannwart C, Hamalainen E, Bloigu S, Ollus A (1986) J Steroid Biochem 24: 289
21. Joannou GE, Kelly GE, Reeder AY, Waring M, Nelson C (1995) J Steroid Biochem Mol Biol 54: 167
22. Xu X, Harris KS, Wang HJ, Murphy PA, Hendrich S (1995) J Nutr 125: 2307
23. Kelly GE, Nelson C, Waring MA, Joannou GE, Reeder AY (1993) Clin Chim Acta 223: 9
24. Lampe JW, Karr SC, Hutchins AM, Slavin JL (1998) Proc Soc Exp Biol Med 217: 335
25. Setchell KD, Borriello SP, Hulme P, Kirk DN, Axelson M (1984) Am J Clin Nutr 40: 569
26. Winter J, Moore LH, Dowell VR, Bokkenheuser VD (1989) Appl Environ Microbiol 55: 1203
27. Rao A (1995) Effects of dietary fiber on intestinal microflora and health. In: Kritchevsky D, Bonfield C (eds), Dietary Fiber in Health and Disease. Eagan Press, St. Paul, p 257

28. Adlercreutz H, Honjo H, Higashi A, Fotsis T, Hamalainen E, Hasegawa T, Okada H (1991) Am J Clin Nutr 54: 1093
29. Kelly GE, Joannou GE, Reeder AY, Nelson C, Waring MA (1995) Proc Soc Exp Biol Med 208: 40
30. Kirkman LM, Lampe JW, Campbell DR, Martini MC, Slavin JL (1995) Nutr Cancer 24: 1
31. Adlercreutz H, Hockerstedt K, Bannwart C, Bloigu S, Hamalainen E, Fotsis T, Ollus A (1987) J Steroid Biochem 27: 1135
32. Adlercreutz H, Fotsis T, Bannwart C, Wahala K, Makela T, Brunow G, Hase T (1986) J Steroid Biochem 25: 791
33. Dehennin L, Reiffsteck A, Jondet M, Thibier M (1982) J Reprod Fertil 66: 305
34. Franke AA, Custer LJ (1996) Clin Chem 42: 955
35. Morton MS, Chan PS, Cheng C, Blacklock N, Matos-Ferreira A, Abranches-Monteiro L, Correia R, Lloyd S, Griffiths K (1997) Prostate 32: 122
36. Morton MS, Matos-Ferreira A, Abranches-Monteiro L, Correia R, Blacklock N, Chan PS, Cheng C, Lloyd S, Chieh-ping W, Griffiths K (1997) Cancer Lett 114: 145
37. Kurzer MS, Lampe JW, Martini MC, Adlercreutz H (1995) Cancer Epidemiol Biomarkers Prev 4: 353
38. Zava DT, Dollbaum CM, Blen M (1998) Proc Soc Exp Biol Med 217: 369
39. Cassidy A, Bingham S, Setchell KD (1994) Am J Clin Nutr 60: 333
40. Morton MS, Wilcox G, Wahlqvist ML, Griffiths K (1994) J Endocrinol 142: 251
41. Seow A, Shi CY, Franke AA, Hankin JH, Lee HP, Yu MC (1998) Cancer Epidemiol Biomarkers Prev 7: 135
42. Adlercreutz H (1995) Environ Health Perspect 103: 103
43. Herman C, Adlercreutz H, Goldin BR, Gorbach SL, Hockerstedt KAV, Watanabe S, Hamalainen EK, Markkanen MH, Makela TH, Wahala KT, Hase TA, Fotsis T (1995) J Nutr 125: 757S
44. Horn-Ross PL, Barnes S, Kirk M, Coward L, Parsonnet J, Hiatt RA (1997) Cancer Epidemiol Biomarkers Prev 6: 339
45. Franke AA, Custer LJ (1994) J Chromatography B: Biomed Appl 662: 47
46. Hutchins AM, Lampe JW, Martini MC, Campbell DR, Slavin JL (1995) J Am Diet Asso 95: 769
47. Axelson M, Sjovall J, Gustafsson BE, Setchell KD (1982) Nature 298: 659
48. Setchell KD, Lawson AM, Conway E, Taylor NF, Kirk DN, Cooley G, Farrant RD, Wynn S, Axelson M (1981) Biochem J 197: 447
49. Lampe JW, Martini MC, Kurzer MS, Adlercreutz H, Slavin JL (1994) Am J Clin Nutr 60: 122
50. Rosenblum ER, Campbell IM, Van Thiel DH, Gavaler JS (1992) Alcohol Clin Exp Res 16: 843
51. Rosenblum ER, Stauber RE, Van Thiel DH, Campbell IM, Gavaler JS (1993) Alcohol Clin Exp Res 17: 1207
52. Gavaler JS, Rosenblum ER, Van Thiel DH, Eagon PK, Pohl CR, Campbell IM, Gavaler J (1987) Alcohol Clin Exp Res 11: 399
53. Gavaler JS, Imhoff AF, Pohl CR, Rosenblum ER, Van Thiel DH (1987) Alcohol Alcoholism Suppl 545
54. Gavaler JS (1993) J Am Coll Nutr 12: 349
55. Gavaler JS, Rosenblum ER, Deal SR, Bowie BT (1995) Proc Soc Exp Biol Med 208: 98
56. Coward L, Barnes NC, Setchell KDR, Barnes S (1993) J Agric Food Chem 41: 1961
57. Eldridge A (1982) J Nutr 30: 353
58. Franke AA, Custer LJ, Cerna CM, Narala K (1995) Proc Soc Exp Biol Med 208: 18
59. Franke AA, Cooney RV, Custer LJ, Mordan LJ, Tanaka Y (1998) Adv Expt Med Biol 439: 237
60. Tsukamoto C, Shimada S, Igita K, Kudou S, Kokubun M, Okubo K, Kitamura K (1995) J Agric Food Chem 43: 1184
61. Wang HJ, Murphy PA (1994) J Agric Food Chem 42: 1666
62. Dwyer JT, Goldin BR, Saul N, Gualtieri L, Barakat S, Adlercreutz H (1994) J Am Diet Assoc 94: 739

63. Thompson LU, Robb P, Serraino M, Cheung F (1991) Nutr Cancer 16: 43
64. Messina MJ, Persky V, Setchell KD, Barnes S (1994) Nutr Cancer 21: 113
65. Barnes S, Kirk M, Coward L (1994) J Agric Food Chem 42: 2466
66. Yuan JM, Wang QS, Ross RK, Henderson BE, Yu MC (1995) Br J Cancer 71: 1353
67. Lee HP, Gourley L, Duffy SW, Esteve J, Lee J, Day NE (1991) Lancet 337: 1197
68. Wang TT, Sathyamoorthy N, Phang JM (1996) Carcinogenesis 17: 271
69. Tang BY, Adams NR (1980) J Endocrinol 85: 291
70. Markiewicz L, Garey J, Adlercreutz H, Gurpide E (1993) J Steroid Biochem Mol Biol 45: 399
71. Kuiper GG, Lemmen JG, Carlsson B, Corton JC, Safe SH, van der Saag PT, van der Burg B, Gustafsson JA (1998) Endocrinology 139: 4252
72. Kuiper GG, Gustafsson JA (1997) FEBS Lett 410: 87
73. Dotzlaw H, Leygue E, Watson PH, Murphy LC (1997) J Clin Endocrinol Metab 82: 2371
74. Miksicek RJ (1993) Mol Pharmacol 44: 37
75. Akiyama T, Ishida J, Nakagawa S, Ogawara H, Watanabe S, Itoh N, Shibuya M, Fukami Y (1987) J Biol Chem 262: 5592
76. Fotsis T, Pepper M, Adlercreutz H, Fleischmann G, Hase T, Montesano R, Schweigerer L (1993) Proc Natl Acad Sci USA 90: 2690
77. Okura A, Arakawa H, Oka H, Yoshinari T, Monden Y (1988) Biochem Biophys Res Commun 157: 183
78. Adlercreutz H, Markkanen H, Watanabe S (1993) Lancet 342: 1209
79. Barnes S, Peterson TG, Coward L (1995) J Cell Biochem Suppl 22: 181
80. Constantinou A, Huberman E (1995) Proc Soc Exp Biol Med 208: 109
81. Barnes S, Peterson TG (1995) Proc Soc Exp Biol Med 208: 103
82. Rauth S, Kichina J, Green A (1997) Br J Cancer 75: 1559
83. Kondo K, Tsuneizumi K, Watanabe T, Oishi M (1991) Cancer Res 51: 5398
84. Pagliacci MC, Smacchia M, Migliorati G, Grignani F, Riccardi C, Nicoletti I (1994) Eur J Cancer 30A: 1675
85. Wei H, Bowen R, Cai Q, Barnes S, Wang Y (1995) Proc Soc Exp Biol Med 208: 124
86. Wei H, Barnes S, Wang Y (1996) Oncol Report 3: 125
87. Kim H, Peterson TG, Barnes S Am J Clin Nutr (in press)
88. Nakafuku M, Satoh T, Kaziro Y (1992) J Biol Chem 267: 19448
89. Adlercreutz H, Fotsis T, Lampe J, Wahala K, Makela T, Brunow G, Hase T (1993) Scand J Clin Lab Invest 215: 5
90. Xu X, Wang HJ, Murphy PA, Cook L, Hendrich S (1994) J Nutr 124: 825
91. Adlercreutz H, Fotsis T, Heikkinen R, Dwyer JT, Woods M, Goldin BR, Gorbach SL (1982) Lancet 2: 1295
92. Coleman MP, Esteve J, Damieki P, Arslan A, Renard H (1993) Trends in cancer incidence and mortality. IARC Scientific Publications, Lyon, France
93. Dunn JE (1975) Cancer Res 35: 3240
94. Gray GE, Pike MC, Henderson BE (1979) Br J Cancer 39: 1
95. Greenwald P (1989) Principles of cancer prevention: diet and nutrition. In: DeVita VTJ, Hellman S, Rosenberg SA (eds), Cancer: Principles and Practice of Oncology. JB Lippincott, Philadelphia, p 167
96. Parker SL, Tong T, Bolden S, Wingo PA (1996) CA Cancer J Clin 46: 5
97. Shimizu H, Ross RK, Bernstein L, Yatani R, Henderson BE, Mack TM (1991) Br J Cancer 63: 963
98. Hirayama T (1986) A large-scale cohort study on cancer risks by diet with special reference to the risk reducing effects of green-yellow vegetable consumption. In: Hayashi Y (ed), Diet, Nutrition and Cancer. Japanese Scientific Society Press, Tokyo, p 41
99. Wu AH, Ziegler RG, Horn-Ross PL, Nomura AM, West DW, Kolonel LN, Rosenthal JF, Hoover RN, Pike MC (1996) Cancer Epidemiol Biomarkers Prev 5: 901
100. Sathyamoorthy N, Wang TT, Phang JM (1994) Cancer Res 54: 957
101. Martin PM, Horwitz KB, Ryan DS, McGuire WL (1978) Endocrinology 103: 1860
102. Drane HM, Patterson DS, Roberts BA, Saba N (1980) Food Cosmet Toxicol 18: 425

103. Peterson G (1995) J Nutr 125: 784S
104. Willard ST, Frawley LS (1998) Endocrine 8: 117
105. Hoffmann R (1995) Biochem Biophys Res Commun 211: 600
106. Clark JW, Santos-Moore A, Stevenson LE, Frackelton AR Jr. (1996) Int J Cancer 65: 186
107. Verma SP, Salamone E, Goldin B (1997) Biochem Biophys Res Commun 233: 692
108. Makela S, Davis VL, Tally WC, Korkman J, Sala L, Vinko R, Korach KS (1994) Environ Health Perspect 192: 572
109. Barnes S (1995) J Nutr 125: 777S
110. Dees C, Foster JS, Ahamed S, Wimalasena J (1997) Environ Health Perspect 105: 633
111. Lamartiniere CA, Moore JB, Brown NM, Thompson R, Hardin MJ, Barnes S (1995) Carcinogenesis 16: 2833
112. Lamartiniere CA, Moore J, Holland M, Barnes S (1995) Proc Soc Exp Biol Med 208: 120
113. Barnes S, Grubbs C, Setchell KD, Carlson J (1990) Prog Clin Biol Res 347: 239
114. Murrill WB, Brown NM, Zhang JX, Manzolillo PA, Barnes S, Lamartiniere CA (1996) Carcinogenesis 17: 1451
115. Colditz GA, Frazier AL (1995) Cancer Epidemiol Biomarkers Prev 4: 567
116. Yan L, Yee JA, Li D, McGuire MH, Thompson LU (1998) Cancer Lett 124: 181
117. Yan L, Yee JA, McGuire MH, Graef GL (1997) Nutr Cancer 28: 165
118. Nomura A, Henderson BE, Lee J (1978) Am J Clin Nutr 31: 2020
119. Maskarinec G, Singh S, Meng L, Franke AA (1998) Cancer Epidemiol Biomarkers Prev 7: 613
120. Ingram D, Sanders K, Kolybaba M, Lopez D (1997) Lancet 350: 990
121. Dhom G (1991) Epidemiology of Hormone-Dependent Tumors. Raven Press, New York
122. Mills PK, Beeson WL, Phillips RL (1989) Cancer 49: 1857
123. Severson RK, Nomura AM, Grove JS, Stemmermann GN (1989) Cancer Res 49: 1857
124. Peterson G, Barnes S (1993) Prostate 22: 335
125. Onozawa M, Fukuda K, Ohtani M, Akaza H, Sugimura T, Wakabayashi K (1998) Jap J Clin Oncol 28: 360
126. Dalu A, Haskell JF, Coward L, Lamartiniere CA (1998) Prostate 37: 36
127. Evans BA, Griffiths K, Morton MS (1995) J Endocrinol 147: 295
128. Keung WM (1995) Biochem Biophys Res Commun 215: 1137
129. Stephens FO (1997) Med J Aust 167: 138
130. Rosenthal MA, Taneja S, Bosland MC (1998) Med J Aust 168: 467
131. Barnes D, Urban W, Grizzel L, Coward L, Kirk M, Weiss H, Irwin W (1996) Proceedings from the Second International Symposium on the Role of Soy in Preventing and Treating Chronic Disease. Brussels, Belgium
132. Doll R, Fraumeni JFJ, Muir CS (eds) (1994) Trends in Cancer Incidence and Mortality. Cold Spring Harbor Laboratory Press, Plainview
133. Trichopoulos D, Yen S, Brown J, Cole P, MacMahon B (1984) Cancer 53: 187
134. Setchell KD, Lawson AM, Borriello SP, Harkness R, Gordon H, Morgan DM, Kirk DN, Adlercreatz H, Anderson LC, Axelson M (1981) Lancet 2: 4
135. Adlercreutz H (1984) Gastroenterology 86: 761
136. Slattery ML, Berry TD, Potter J, Caan B (1997) Cancer Causes Control 8: 872
137. Burkitt DP (1971) Cancer 28: 3
138. Fuchs CS, Giovannucci EL, Colditz GA, Hunter DJ, Stampfer MJ, Rosner B, Speizer FE, Willett WC (1999) N Engl J Med 340: 169
139. Bingham S, Pignatelli B, Pollock J, Ellul A, Mallaveille C, Gross G, Runswick S, Cummings JH, O'Neill IK (1996) Carcinogenesis 17: 515
140. Jenab M, Thompson LU (1996) Carcinogenesis 17: 1343
141. Steele VE, Pereira MA, Sigman CC, Kelloff GJ (1995) J Nutr 125: 713S
142. Pereira MA, Barnes LH, Rassman VL, Kellof GV, Steele VE (1994) Carcinogenesis 15: 1049
143. Sung MK, Lautens M, Thompson LU (1998) Anticancer Res 18: 1405
144. Wobbes T, Beex LVAM, Koenders AMT (1984) Dis Colon Rectum 27: 591
145. Lointier P, Wildrick DM, Boman BM (1992) Anticancer Res 12: 1523

146. Rao CV, Wang CX, Simi B, Lubet R, Kelloff G, Steele V, Reddy BS (1997) Cancer Res 57: 3717
147. Sheehan D, Young M (1979) Endocrinology 104: 1442
148. Dunn JF, Nisula BC, Rodbard D (1981) J Clin Endocrinol Metab 53: 58
149. Ekins R, Edwards P, Newman B (1982) The role of binding-proteins in hormone delivery. In: Albertini A, Ekin RP (eds), Free Hormones in Blood. Elsevier Biomedical Press, New York, p 3
150. Mendel C (1989) Endocrine Rev 10: 232
151. Hammond GL, Nisker JA, Jones LA, Siiteri PK (1980) J Biol Chem 255: 5023
152. Arnold SF, McLachlan JA (1996) Environ Health Perspect 104: 1020
153. Sheehan DM, Branham WS (1987) Teratogen Carcinogen Mutagen 7: 411
154. Nagel SC, vom Saal FS, Welshons WV (1998) Proc Soc Exp Biol Med 217: 300
155. Cassidy A, Bingham S, Setchell KD (1995) Br J Nutr 74: 587
156. Petrakis NL, Barnes S, King EB, Lowenstein J, Wiencke J, Lee MM, Miike R, Kirk M, Coward L (1996) Cancer Epidemiol Biomarkers Prev 5: 785
157. Phipps WR, Martini MC, Lampe JW, Slavin JL, Kurzer MS (1993) J Clin Endocrinol Metab 77: 1215
158. Shoff SM, Newcomb PA, Mares-Perlman JA, Klein BE, Haffner SM, Storer BE, Klein R (1998) Nutr Cancer 30: 207
159. Chang YC, Nair MG, Nitiss JL (1995) J Nat Prod 58: 1901
160. Rose DP (1990) Nutr Cancer 13: 1
161. Lu LJ, Anderson KE, Grady JJ, Nagamani M (1996) Cancer Epidemiol Biomarkers Prev 5: 63
162. Setchell KDR, Adlercreutz H (1979) J Steroid Biochem 11: 15
163. Henderson B, Ross RK, Judd HL, Krailo MD, Pike MC (1985) Cancer 56: 1206
164. Treloar AE, Boynton RE, Behn BG, Brown BW (1970) Int J Fertil 12: 77
165. Kronenberg F (1994) Hot Flashes. In: Lobo RA (ed), Treatment of the Postmenopausal Woman. Raven, New York, p 97
166. Knight DC, Eden JA (1995) Maturitas 22: 167
167. McKinney KA, Thompson W (1998) Drugs 56: 49
168. Grodstein F, Stampfer M (1995) Prog Cardiovascular Dis 38: 199
169. Grady D, Rubin SM, Petitti DB, Fox CS, Black D, Ettinger B, Ernster VL, Cummings SR (1992) Ann Int Med 117: 1016
170. Grodstein F, Stampfer MJ, Colditz GA, Willett WC, Manson JE, Joffe M, Rosner B, Fuchs C, Hankinson SE, Hunter DJ, Hennekens CH, Speizer FE (1997) N Engl J Med 336: 1769
171. Seidl MM, Stewart DE (1998) Can Family Phys 44: 1299
172. Seidl MM, Stewart DE (1998) Can Family Phys 44: 1271
173. Lock M (1986) Culture Med Psych 10: 23
174. Wilcox G, Wahlqvist ML, Burger HG, Medley G (1990) Br Med J 301: 905
175. Baird DD, Umbach DM, Lansdell L, Hughes CL, Setchell KD, Weinberg CR, Haney AF, Wilcox AJ, McLachlan JA (1995) J Clin Endocrinol Metab 80: 1685
176. Murkies AL, Lombard C, Strauss BJ, Wilcox G, Burger HG, Morton MS (1995) Maturitas 21: 189
177. Dalais FS, Rice GE, Wahlqvist ML, Murkies AL, Medley G, Strauss BJG (1998) Climacteric 1: 124
178. Harding C, Morton M, Gould V (1996) Proceedings from the Second International Symposium on the Role of Soy in Preventing and Treating Chronic Disease. Brussels, Belgium
179. Brezezinski A, Adlercreutz H, Shaoul R (2001) Menopause (in press)
180. Lu LJ, Lin SN, Grady JJ, Nagamani M, Anderson KE (1996) Nutr Cancer 26: 289
181. Department of Health (1994) Nutritional Aspects of Cardiovascular Disease. Report on Health and Social Subjects. H.M. Stationary Office, London
182. Kesteloot H, Sasaki S, Zhang X, Joossens JV (1995) Acta Cardiologica 50: 343
183. Kim CJ, Jang HC, Cho DH, Min YK (1994) Arterioscler Thromb 14: 275
184. Shewmon DA, Stock JL, Rosen CJ, Heinluoma KM, Hogue MM, Morrison A, Doyle EM, Ukena T, Weale V, Baker S (1994) Arterioscler Thromb 14: 1586

185. Love RR, Wiebe DA, Newcomb PA, Cameron L, Leventhal H, Jordan VC, Feyzi J, DeMets DL (1991) Ann Int Med 115: 860
186. Berglund L, Carlstrom K, Stege R, Gottlieb C, Eriksson M, Angelin B, Henriksson P (1996) J Clin Endocrinol Metab 81: 2633
187. Eriksson M, Berglund L, Rudling M, Henriksson P, Angelin B (1989) J Clin Invest 84: 802
188. Ohshige A, Ito M, Koyama H, Maeda T, Yoshimura T, Okamura H (1996) Artery 22: 115
189. Sacks FM, Pfeffer MA, Moye LA, Rouleau JL, Rutherford JD, Cole TG, Brown L, Warnica JW, Arnold JM, Wun CC, Davis BR, Braunwald E (1996) N Engl J Med 335: 1001
190. Prentice R, Thompson D, Clifford C, Gorbach S, Goldin B, Byar D (1990) J Natl Cancer Inst 82: 129
191. Wolfe BM, Huff M (1995) Metabolism 44: 410
192. Anderson JW, Johnstone BM, Cook-Newell ME (1995) N Engl J Med 333: 276
193. Setchell KD (1985) Naturally occurring non steroidal estrogen of dietary origin. In: McLachlan JA (ed), Estrogens in the Environment. Elsevier, New York, p 69
194. Sirtori CR, Gianazza E, Manzoni C, Lovati MR, Murphy PA (1997) Am J Clin Nutr 65: 166
195. Sargeant P, Farndale RW, Sage SO (1993) J Biol Chem 268: 18151
196. Nakashima S, Koike T, Nozawa J (1991) Mol Pharmacol 39: 475
197. Raines EW, Ross R (1995) J Nutr 125: 624S
198. Carroll KK, Kurowska EM (1995) J Nutr 125: 594S
199. Anthony MS, Clarkson TB, Hughes CL Jr, Morgan TM, Burke GL (1996) J Nutr 126: 43
200. Anthony MS, Clarkson TB, Bullock BC, Wagner JD (1997) Arterioscler Thromb Vasc Biol 17: 2524
201. Wagner JD, Cefalu WT, Anthony MS, Litwak KN, Zhang L, Clarkson TB (1997) Metab Clin Exp 46: 698
202. Honore EK, Williams JK, Anthony MS, Clarkson TB (1997) Fertil Steril 67: 148
203. Bakhit RM, Klein BP, Essex-Sorlie D, Ham JO, Erdman JW Jr, Potter SM (1994) J Nutr 124: 213
204. Sirtori CR, Agradi E, Conti F, Mantero O, Gatti E (1977) Lancet 1: 275
205. Eden JA, Knight DC, Howes JB (1996) Eighth International Congress on the Menopause. Sydney, Australia
206. Kurowska EM, Jordan J, Spence JD, Wetmore S, Piche LA, Radzikowski M, Dandona P, Carroll KK (1997) Clin Invest Med 20: 162
207. Nagata C, Takatsuka N, Kurisu Y, Shimizu H (1998) Nutr Cancer 128: 209
208. Baum JA, Teng H, Erdman JW, Jr., Weigel RM, Klein BP, Persky VW, Freels S, Surya P, Bakhit RM, Ramos E, Shay NF, Potter SM (1998) Am J Clin Nutr 68: 545
209. Ramsay LE, Yeo WW, Jackson PR (1991) Brit Med J 303: 953
210. Lovati MR, Manzoni C, Canavesi A, Sirtori M, Vaccarino V, Marchi M, Gaddi G, Sirtori CR (1987) J Clin Invest 80: 1498
211. Angelin B, Olivecrona H, Reihner E, Rudling M, Stahlberg D, Eriksson M, Ewerth S, Henriksson P, Einarsson K (1992) Gastroenterology 103: 1657
212. Tikkanen MJ, Wahala K, Ojala S, Vihma V, Adlercreutz H (1998) Proc Natl Acad Sci USA 95: 3106
213. Hodgson JM, Puddey IB, Beilin LJ, Mori TA, Croft KD (1998) J Nutr 128: 728
214. Verillo A, De Teresa A, Giarrusso PC, La Rocca S (1985) Atherosclerosis 54: 321
215. Giovannetti PM CK, Wolfe BM (1986) Nutr Res 6: 609
216. Mendis S, Kumarasunderam R (1990) Br J Nutr 63: 547
217. Meinertz H, Faergeman O, Nilausen K, Chapman MJ, Goldstein S, Laplaud PM (1988) Atherosclerosis 72: 63
218. Raaij JMA, Katan MB, Hautvast JGAJ, Hermus RJJ (1981) Am J Clin Nutr 34: 1261
219. Raaij JMA, Katan MB, West CE, Hautvast JGAJ (1982) Am J Clin Nutr 35: 925
220. Sacks FM, Breslow JL, Wood PG, Kass EH (1983) J Lipid Res 24: 1012
221. Felson DT, Zhang Y, Hannan MT, Kiel DP, Wilson PW, Anderson JJ (1993) N Engl J Med 329: 1141
222. Wallach S, Henneman P (1959) JAMA 171: 1637
223. Notelovitz M (1993) Fertil Steril 59: 707

224. Fentiman IS, Fogelman I (1993) Eur J Cancer 29A: 485
225. Lindsay R, Hart DM, Clark DM (1984) Obstet Gynecol 63: 759
226. Abelow BJ, Holford TR, Igsogna KL (1992) Calcif Tiss Int 50: 14
227. Arjmandi BH, Alekel L, Hollis BW, Amin D, Stacewicz-Sapuntzakis M, Guo P, Kukreja SC (1996) J Nutr 126: 161
228. Erdman J (1996) Proceedings from the Second International Symposium on the Role of Soy in Preventing and Treating Chronic Disease. Brussels, Belgium
229. Petilli M, Fiorelli G, Benvenuti S, Frediani U, Gori F, Brandi ML (1995) Calcif Tiss Int 56: 160
230. Reginster JY (1993) Bone Miner 23: 223
231. Cheng SL, Zhang SF, Nelson TL, Warlow PM, Civitelli R (1994) Calcif Tiss Int 55: 356
232. Benvenuti S, Tanini A, Frediani U, Bianchi S, Masi L, Casano R, Bufalino L, Serio M, Brandi ML (1991) J Bone Min Res 6: 987
233. Bonucci E, Ballanti P, Martelli A, Mereto E, Brambilla G, Bianco P, Bufalino L (1992) Calcif Tiss Int 50: 314
234. Agnusdei D, Adami S, Cervetti R, Crepaldi G, Di Munno O, Fantasia L, Isaia GC, Letizia G, Ortolani S, Passeri M et al. (1992) Bone Miner 19: S43
235. Adami S, Bufalino L, Cervetti R, Di Marco C, Di Munno O, Fantasia L, Isaia GC, Serni U, Vecchiet L, Passeri M (1997) Osteoporosis Int 7: 119
236. Gambacciani M, Spinetti A, Cappagli B, Taponeco F, Felipetto R, Parrini D, Cappelli N, Fioretti P (1993) J Endocrinol Invest 16: 333
237. Gambacciani M, Spinetti A, Piaggesi L, Cappagli B, Taponeco F, Manetti P, Weiss C, Teti GC, La Commare P, Facchini V (1994) Bone Miner 26: 19
238. Valente M, Bufalino L, Castiglione GN, D'Angelo R, Mancuso A, Galoppi P, Zichella L (1994) Calcif Tiss Int 54: 377
239. Agnusdei D, Crepaldi G, Isaia GC, Mazzuoli GF, Ortolani S, Passeri M, Bufalino L, Gennari C (2000) Calcif Tiss Int (in press)
240. Melis GB, Paoletti AM, Bartolini R, Tosti Balducci M, Massi GB, Bruni V, Becorpi A, Ottanelli S, Fioretti P, Gambacciani M (1992) Bone Miner 19 Suppl 1: S49
241. Gambacciani M, Ciaponi M, Cappagli B, Piaggesi L, Genazzani AR (1997) Maturitas 28: 75
242. Choi YK, Han IK, Yoon HK (1997) Osteoporosis Int 7: S174
243. Agnusdei D, Gennari C, Bufalino L (1995) Osteoporosis Int 5: 462
244. Melis GB, Paoletti AM, Cagnacci A, Bufalino L, Spinetti A, Gambacciani M, Fioretti P (1992) J Endocrinol Invest 15: 755
245. Ruhrah J (1909) Arch Pediatr 26: 496
246. Irvine CHG, Fitzpatrick M, Robertson I, Woodhams D (1995) N Zealand Med J 108: 208
247. Irvine CH, Fitzpatrick MG, Alexander SL (1998) Proc Soc Exp Biol Med 217: 247
248. Kaldas RS, Hughes CL Jr (1989) Reprod Toxicol 3: 81
249. Whitten PL, Russell E, Naftolin F (1992) Steroids 57: 98
250. Whitten PL, Naftolin F (1992) Steroid 57: 56
251. Levy JR, Faber KA, Ayyash L, Hughes CL Jr (1995) Proc Soc Exp Biol Med 208: 60
252. Whitten PL, Lewis C, Naftolin F (1993) Biol Reprod 49: 1117
253. Setchell KD, Zimmer-Nechemias L, Cai J, Heubi JE (1997) Lancet 350: 23
254. Morton MS, Leung SSF, Davies DP, Griffiths K, Evans BAJ (1996) Proceedings from the Second International Symposium on the Role of Soy in Preventing and Treating Chronic Disease. Brussels, Belgium
255. Nguyenle T, Wang E, Cheung AP (1995) J Pharmacol Biomed Anal 14: 221
256. Knight DC, Eden JA, Huang JL, Waring MA (1998) J Ped Child Health 34: 135
257. King RA, Mano MM, Head RJ (1998) J Dairy Res 65: 479
258. Klein KO (1998) Nutr Rev 56: 193
259. Bernhisel-Broadbent J, Sampson HA (1989) J Allergy Clin Immunol 83: 435
260. Iacono G, Cavataio F, Montalto G, Florena A, Tumminello M, Soresi M, Notarbartolo A, Carroccio A (1998) N Engl J Med 339: 1100
261. Cantani A, Lucenti P (1997) Pediatr Allergy Immunol 8: 59

262. Ojeda SR, Andrews WW, Advis JP, White SS (1980) Endocrine Rev 1: 228
263. Jordan VC, Koch R, Bain RR (1985) Prolactin synthesis by cultured rat pituitary cells: an assay to study estrogens, antiestrogens and their metabolites in vitro. In: Lachlan JM (ed), Estrogens in the Environment II. Influences on Development. Elsevier, New York, p 221
264. Sheehan DM (1995) Proc Soc Exp Biol Med 208: 3
265. Medlock KL, Branham WS, Sheehan DM (1995) Proc Soc Exp Biol Med 208: 67
266. Shutt DA (1976) Endeavour 35: 110
267. Sheehan DM, Branham WS, Medlock KL, Shanmugasundaram ER (1984) Teratology 29: 383
268. Setchell KD, Gosselin SJ, Welsh MB, Johnston JO, Balistreri WF, Kramer LW, Dresser BL, Tarr MJ (1987) Gastroenterology 93: 225
269. Whitten PL, Lewis C, Russell E, Naftolin F (1995) Proc Soc Exp Biol Med 208: 82
270. Orcheson LJ, Rickard SE, Seidl MM, Thompson LU (1998) Cancer Lett 125: 69
271. McMichael-Phillips D, Harding C, Morton M (1996) Proceeding from the Second International Symposium on the Role of Soy in Preventing and Treating Chronic Disease. Brussels, Belgium
272. Sobell J, Block G, Koslowe P, Tobin J, Andres R (1989) Am J Epidemiol 130: 173
273. Horn-Ross PL (1995) Cancer Causes Control 6: 567
274. Block G, Woods M, Potosky A, Clifford C (1990) J Clin Epidemiol 43: 1327
275. Block G, Subar AF (1992) J Am Diet Assoc 92: 969
276. Fort P, Moses N, Fasano M, Goldberg T, Lifshitz F (1990) J Am Coll Nutr 9: 164
277. Hydovitz J (1960) N Engl J Med 262: 351
278. Shepard TH, Pyne GE, Kirschvink JF, McLean M (1960) N Engl J Med 262: 1099
279. Pinchera A, MacGillivray MH, Crawford JD, Freeman AG (1965) N Engl J Med 265: 83
280. Kay T (1998) J Trop Pediatr 44: 251
281. Kimura S, Suwa J, Ito M, Sato H (1976) Gann 67: 763
282. Nordisiek F (1962) Proc Soc Exp Biol Med 110: 417
283. Block JR, Mandl RH, Howard HW, Bauer CD, Anderson DW (1961) Arch Biochem Biophys 93: 15
284. McCarrison R (1933) Indian J Med Res 21: 179
285. Divi RL, Chang HC, Doerge DR (1997) Biochem Pharmacol 54: 1087

Mechanism-Based Carcinogenic Risk Assessment of Estrogens and Estrogen-Like Compounds

Ben A. T. Willems[1], Christopher J. Portier[2], George W. Lucier[2]

[1] 1418 Maple St. Apt B, Santa Monica, CA 90405, USA
E-mail: ben.willems@accupuncture.com
[2] Laboratory of Computational Biology and Risk Analysis, National Institute of Environmental Health Sciences, Mail Drop A3-06, P.O. Box 12233, Research Triangle Park, NC 27709, USA
E-mail: portier@niehs.nih.gov

Estrogens play an important role in mammalian reproductive cycles and a disturbance of this cycle can cause adverse effects like carcinogenicity. A mechanistic modeling approach is helpful in attempting to unravel the entanglement of genomic, non-genomic, and feedback signals involved in endocrine signaling. To make more balanced decisions on possible human health risks put forth by endocrine active compounds (e.g., estrogens and estrogen-like agents), a multidisciplinary approach is necessary. Besides traditional epidemiological studies, accumulation of mechanistic data is essential for all classes of agents. Combining all available information is a positive move towards a more complete risk assessment process.

Keywords. Endocrine disruptors, Estrogen, Risk assessment, Mechanistic modeling

1	**Introduction**	111
2	**Biological Signaling Pathways of Estrogen**	112
2.1	Estrogen Metabolism and Fluctuation	112
2.2	Genomic Processes	112
2.3	Non-Genomic Processes	113
2.4	Estrogen Receptor Regulation	114
3	**Endocrine Active Compounds Interactions and Adverse Endpoints**	115
3.1	Adverse Endpoints	116
3.2	Agonists/Antagonists	116
3.3	Xenoestrogens	117
3.4	Phytoestrogens	118
4	**Carcinogen Risk Assessment**	119
4.1	Change in Perspective	119
4.2	Modes of Action	119
4.3	Change in Risk Assessment Approach	120
5	**Mechanistic Model Development for Estrogens**	122
5.1	Model Types	122
5.2	Estrus Cycle Model	123
5.2.1	Hypothalamus System	123

The Handbook of Environmental Chemistry Vol. 3, Part M
Endocrine Disruptors, Part II
(ed. by M. Metzler)
© Springer-Verlag Berlin Heidelberg 2002

5.2.2   Hypothalamus/Pituitary Portal System 125
5.2.3   Pituitary 125
5.2.4   Ovaries 125

6   Discussion 126

7   References 128

Abbreviations

AC	adenylate cyclase
cAMP	cyclic adenosine monophosphate
AP	activator protein
ATP	adenosine triphosphate
BM	benchmark dose
CL	corpora lutea
CRE	cAMP response element
DDT	dichlorodiphenyltrichloroethane
DES	diethylstilbestrol
DNA	deoxyribonucleic acid
E2	17β-estradiol
EAC	endocrine active compound
ECM	estrus cycle model
EGF	epidermal growth factor
EPA	Environmental Protection Agency
ER	estrogen receptor
ERE	estrogen response element
FSH	follicle-stimulating hormone
GnRH	gonadotropin-releasing hormone
IGF-I	insulin-like growth factor I
Inh	inhibin
LH	luteinizing hormone
LMS	linearized multistage model
MOE	margin of exposure
mRNA	messenger ribonucleic acid
NADPH	nicotinamide adenine dinucleotide phosphate, reduced form
NOAEL	no observable adverse effect level
P	progesterone
PAH	polycyclic aromatic hydrocarbon
PBPD	physiologically based pharmacodynamic
PBPK	physiologically based pharmacokinetic
PCB	polychlorinated biphenyl
PDE	phosphodiesterase
PKA	protein kinase A
QSAR	quantitative structure-activity relationship
T	testosterone
TCDD	2,3,7,8-tetrachlorodibenzo-p-dioxin

1
Introduction

Carcinogenesis has historically been viewed as a disease resulting from genetic mutations occurring in a particular order with possible selective stimulation of these mutated cells to grow. This multistage theory of carcinogenesis has formed the basis for much of our understanding of the cancer process and has stimulated many of the experimental methods used to assess cancer risks. Only recently have researchers begun to focus on the effects of secondary pathways on the initiation, progression, and promotion of carcinogenesis. One area receiving considerable attention is the impact of modifications of endocrine hormones on cancer risks.

Genomic and non-genomic endocrine signaling pathways are extensively present in the body and function in a complicated manner. Receptors, ligands, enzymes, proteins, and catalysts work together closely within the same cell and between different cells and organ systems. By means of this direct and indirect signaling, a more or less homeostatic state is preserved. At this moment, it is still not clear on how this closely regulated system is affected by an exogenous agent.

Increasing interest in this area of research arose after epidemiological studies identified a significant increase in the incidence of hormone-dependent diseases including cancers of the breast, prostate, and testis, and suggested that environmental factors may contribute to this increased incidence.

Development of cancers can be influenced by exposure to estrogens or estrogenic drugs. This has been demonstrated through experimental initiation/promotion studies [1], epidemiology studies [2], and efficacy of hormone agonists in treating cancers. It is possible that estrogen-like compounds such as diethylstilbestrol, DDT, dioxins, and bisphenol A could yield similar results. Widespread exposure and the commercial significance of many of these agents have made endocrine-disrupting chemicals a contentious health concern and environmental issue.

Focused studies on the potential for toxicity from endocrine active compounds is a fairly new field of scientific research. Much of the research in this area has derived from mainly two factors that developed exponentially in the last few decades. The first is molecular biology. There have been substantial gains in our understanding of receptor binding, interaction, signaling pathways, and compounds involved in these processes. This research has explored down to the level of receptor-types, genetic background, genomic and non-genomic interactions. The second factor is an enormous increase in "desk" computing power, making it possible for a scientist to run complicated biomathematical models, to analyze and test effects resulting from a particular intervention. The challenge for the coming years is to combine these two aspects. This combination is only possible when both sides work together. Molecular biologist and toxicologists need to understand why mathematical modeling is important and helps in exploring new directions of their own research field. And mathematicians must make great strives to fully understand the biology, focusing on mechanistic/physiologically based models. Only through the integration of information can a true, mechanism-based approach to risk assessment be achieved. What follows is a discussion of this approach, weighing heavily on our experience of modeling the estrogen cycle in mammalian systems.

2
Biological Signaling Pathways of Estrogen

The steroid hormone estrogen is distributed by the blood in a mostly bound fashion to high-affinity plasma proteins, like sex hormone-binding globulin and corticosteroid-binding globulin, or low affinity, low specificity proteins like albumin and orosomucoid [3, 4]. However, the concentration of unbound estrogen is important, as only "free" estrogen is able to trigger an effect at the target tissue, i.e., tissue characterized by the presence of cytoplasmic and/or nuclear estrogen receptors (ERs). The way estrogens and estrogen-like compounds are believed to exert their influence on these target cells is described below.

2.1
Estrogen Metabolism and Fluctuation

Estrogens are formed from androgens by a widely distributed mono-oxygenase enzyme system called aromatase, containing NADPH-cytochrome c reductase and cytochrome P-450. Estradiol (i.e., 17β-estradiol, E2), by far the most potent endogenous estrogen, is synthesized from testosterone by aromatase present in the ovaries, placenta, and various other estrogen target tissues such as the brain, prostate, uterus, and mammary gland. In turn E2 can be metabolized to multiple hydroxylated products by enzymes of the P450 family, or broken down to sulfates, glucuronides, or fatty acid esters. The latter three pathways are also reversible, creating active estrogens [5, 6].

In the intact female mammalian system the fluctuation in estrogen levels is mostly influenced by the estrous cycle. The mechanisms involved in this cycle are complex and require the signaling interaction of many hormones. To address the involvement of estrogen a simplified description is given below.

Following synthesis in the ovaries under the influence of gonadotropic hormones, estrogen is carried to the brain and stimulates the hypothalamus in synthesizing and excreting pulses of gonadotropin releasing hormone (GnRH). Together with GnRH, estrogen also has an effect on the pituitary gland; while GnRH pulses induce production and release of luteinizing hormone (LH) and follicle-stimulating hormone (FSH), estrogen inhibits the production of these gonadotropins. Thus, by generating changes in LH and FSH concentrations, estrogen has an indirect influence on the growth and time of ovulation of the follicles residing in the ovaries, and subsequently on its own synthesis [7].

2.2
Genomic Processes

The most common and established pathway for estrogen as it arrives at the target cell is to enter the cell by diffusion or in some cases by active uptake [8, 9].

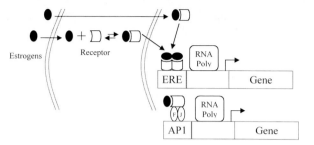

Fig. 1. Genomic estradiol action; RNA Poly = RNA polymerase; F = Fos; J = Jun

Subsequently, estrogen binds cytoplasmatic or nuclear ERs (see Fig. 1), which are members of a large superfamily of receptor proteins that share a similar configuration and functionality, and play an important role in cell differentiation, growth and metabolism [10–13].

After the ligand binds the receptor this complex forms a dimer and then binds to specific DNA sequences, called estrogen response elements (EREs), of a responsive gene. These estrogen-ER complexes are also thought to bind to other nuclear sites, called activator proteins (AP) and require the transcription factors Fos and Jun [14] (see Fig. 1). Binding of the estrogen-ER complex to DNA can alter the transcription of a gene by an RNA polymerase to produce messenger RNA (mRNA), which in turn is translated to the corresponding protein by the cytoplasmic ribosomes. The ligand-receptor complex activates, represses, or modifies the level of gene expression, causing a change in the levels of specific proteins, and, as a result, altering cell function, growth and/or differentiation [8–9, 12]. After dissociation of estrogen from the receptor, or detachment of the complex from the DNA acceptor site, gene transcription will terminate. Estrogen may diffuse out of the cell and be metabolized in the liver to less or non-active forms, or bind once again to another receptor. Altogether, the effects of this genomic response are distinguished by a comparatively long latency and duration of action, on the scale of minutes to hours to days [15].

2.3
Non-Genomic Processes

The second, less recognized, route of estrogen action is via a non-transcriptional signaling pathway which has been observed in the brain, pituitary, vascular smooth muscle cells, and breast cells [16–18]. Both short-term latency and duration, varying from milliseconds to minutes, are typical for non-transcriptional estrogen action. This possible alternative mechanism to the genomic route is described by Moss et al. [15] and Zakon [19], containing both extra- and intracellular non-genomic effects (see Fig. 2).

Extracellular estrogen can bind specific receptors, and subsequently trigger a G-protein-coupled mechanism. The activated G-protein is linked to the enzyme adenylate cyclase (AC), thereby causing the cyclase to produce the second messenger cyclic-AMP (cAMP). In turn this cAMP will activate protein

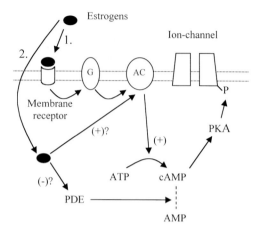

Fig. 2. Possible mechanism for non-genomic pathways of estrogen signaling, where G is a membrane G-protein, AC is adenylate cyclase, ATP is adenosine triphosphate, AMP is adenosine monophosphate, cAMP is cyclic-AMP, PKA is protein kinase A, and PDE is phosphodiesterase

kinase A (PKA), an enzyme which catalyzes the transfer of a phosphate group from adenosine triphosphate (ATP) to a specific protein. Moss et al. [15] suggest that phosphorylation by PKA prolongs the opening of specific ion-channels enhancing the membrane conductance. The second mechanism, involving intracellular estrogen, is not exactly known. But estrogen could reduce the activity of phosphodiesterase (PDE), thus decreasing the breakdown of cAMP to AMP, or assist in releasing more cAMP through interaction with AC.

Nevertheless these two pathways are not completely separate from the transcriptional route as PKA or cAMP might also influence cAMP response elements (CRE) and hence the transcription of genes. Another possible example of "crosstalk" is the stimulation of ER transcriptional activity by growth factors such as epidermal growth factor (EGF) and insulin-like growth factor I (IGF-I) [20].

2.4
Estrogen Receptor Regulation

Besides the estrogen concentration, the response level of a tissue to estrogen is also dependent on both the state of the ERs (active vs. inactive) and ER concentration. After dissociation of estrogen the ER can either be recycled directly to its active form, reside in a refractory state and become activated later, or degrade to an inactive form [9]. It is known that estrogen pretreatment increases the concentration of ERs in several target tissues and that it is associated with an increased sensitivity to subsequent estrogen treatments [21]. Our understanding of the complexity of estrogen and estrogen receptor interaction increased considerably after Simerly and Young [22] and Shupnik et al. [23] pointed out that ER expression, induced by estrogen, is down-regulated instead of up-regulated in tissues like the hypothalamus and uterus, creating a paradox. However, the recent discovery of a second ER isoform, called ER-β [24] (the traditional ER is currently called ER-α), might explain this difference. Because ERs work as dimers, the complexity is increased since α-α homodimers, β-β homodimers, and α-β heterodimers can form [25]. Barkhem et al. [13] showed that the two ER isoforms respond similar to some ligands but that there are also receptor specific responses. This latter action was confirmed by Paech et al. [14]. Different levels of expression of these ER isoforms in different tissues could be an explanation for the differences in response between target tissues.

3
Endocrine Active Compounds Interactions and Adverse Endpoints

Endocrine active compounds (EACs), or sometimes called endocrine disrupting chemicals, are a wide class of compounds defined by the EPA as: "Exogenous agents that interfere with the production, release, transport, metabolism, binding action or elimination of natural hormones in the body responsible for the maintenance of homeostasis and the regulation of developmental processes". Despite the "negative" title, endocrine disrupting agents with an anti-estrogenic action can be used in clinical applications, especially for fertility control or treatment of certain hormone-dependent cancers. The main anti-estrogen being used to date is tamoxifen. This non-steroidal EAC has turned out to be effective in the treatment of breast cancer because of its partial agonist effect in decreasing the action of estrogen in breast tissue [26]. It also has several beneficial side effects like maintaining bone-density and reducing blood cholesterol levels. However, while tamoxifen works as an antagonist in breast cancer cells, it has an agonistic effect in endometrial cancer cells. Even though there exists a long list of agents that could be considered EACs, only agents with an estrogenic or anti-estrogenic capacity will be discussed in this chapter. Synthetic steroids and phytoestrogens fall into this category and may act as a substitute for the endogenous estrogen at

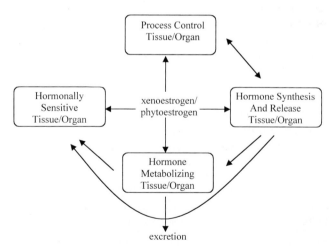

Fig. 3. Minimal endocrine system. The arrows illustrate the actions of estrogens on the described organs, plus the interaction between endocrine organs [86]

the plasma protein level or at the target tissue, thus changing the availability of this natural hormone, and influence the response in the cells [27].

3.1
Adverse Endpoints

During the last 5–10 years, research has shown that signaling is the predominant mechanism for toxic effects rather than a direct role of reactive chemicals. As a result of this shift, subtle biological effects like altered biological signals versus direct effects like mutations have attracted greater attention [28]. In that light the U.S. EPA considers endocrine disruption as a step that could possibly cause toxic effects, and encourages studies to define dose-response relationships across multiple endpoints, describe mechanisms of action, and determine the complete organism's response to an environmental exposure [29]. The minimal system needed to study alterations in endocrine pathways is displayed in Fig. 3, where a xenoestrogen or phytoestrogen is able to interfere at any level.

The bioavailability of EACs, their effect on endogenous hormone metabolism, their interaction with hormone receptors and the interaction with cell-signaling pathways, each have a distinct outcome. Whether this outcome, e.g., increased transcription of a particular gene, is considered as an (adverse) effect or not requires a more defined description of variability in unexposed populations [28] partially resulting from different sensitivities to agents at different ages. Secondly, besides intensity, frequency and duration of the exposure, timing of the exposure should also be taken into account. Third, both primary and secondary interference can disturb endocrine function [29]. Consideration of all of the above will allow discrimination between a genuine (adverse) effect, as a result of the compound's impact, and the boundaries of normal homeostatic endocrine functioning. Furthermore, the outcomes can function as precursors, individually or combined, for intermediate and long-term endpoints. This poses a second, more philosophical question of which marker best describes an adverse (health) effect. In the long term there are only two actual adverse effects; increased incidence or earlier onset of illness and/or death. See Fig. 4 for an overview of (adverse) effects of endocrine disruption.

3.2
Agonists/Antagonists

Estrogenic compounds that are rapidly cleared from target tissues are generally referred to as short-acting estrogens and can display both agonist and antagonis-

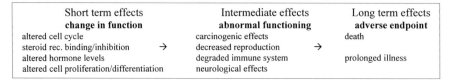

Short term effects change in function	Intermediate effects abnormal functioning	Long term effects adverse endpoint
altered cell cycle	carcinogenic effects	death
steroid rec. binding/inhibition →	decreased reproduction →	
altered hormone levels	degraded immune system	prolonged illness
altered cell proliferation/differentiation	neurological effects	

Fig. 4. Overview of (adverse) effects and endpoints due to EACs.

tic properties [9]. When they are administered by pellet implant (assumed constant blood and tissue levels), they act consistently. When administered as single exposures, opposite effects of short-acting estrogens can occur. For example, estriol administered in a continuous fashion, will act as a full agonist because the estrogen receptors are continuously occupied [30, 31]. However, in short-term exposures, estriol acts as an antagonist through competitive interaction with E2 for the ER complexes at the nuclear acceptor sites [32] and is capable of stimulating metabolic activity of the human breast cell (MCF-7) in culture [33]. Compared to E2, estriol shows a more rapid loss as it dissociates quicker from the receptor. The degree of antagonism observed with any estrogen antagonist is related to the dissociation constant (Kd) of the hormone from the receptor [34]. Furthermore, steroid derivatives like estriol cyclopentyl ether are able to extend the biological half-life of estrogen and thus are analogous to the hormone implant system [35].

For non-steroidal anti-estrogens like tamoxifen and clomiphene, the relation between the dissociation constant, Kd, and the degree of antagonism does not seem to hold [36, 37]. Here the level of antagonism, and even the change to agonistic effects, is dependent on the species, organ, and tissue [37, 38]. Also, experimental conditions can determine the effect of a compound. For example, an estrogen antagonist, ICI 164,384, lacked agonist activity *in vivo* [26] although it has shown agonistic effects in yeast cell experiments [39].

3.3
Xenoestrogens

Besides tamoxifen, clomiphene, and ICI 164,384, there are many synthetic chemicals expressing estrogenicity and/or anti-estrogenicity. Even chemicals that do not posses a hormone-like design, such as DDT and Kepone, may still display hormonal activity. Although DDT was banned in many countries in the early 1970s, this chemical recently gained attention once again following linkage of "feminized" sex-characteristics in male alligators in Lake Apopka to DDT exposure [40]. Also Kepone was shown to have weak estrogenic activity resulting in possible environmental effects, e.g., decreased sperm count in men exposed to the chemical after a 1975 spill of the compound [8]. A third synthetic estrogen, which had clinical applications, is diethylstilbestrol (DES). DES was used as early as 1948 to prevent miscarriages in women. But in the 1970s daughters of women who had taken DES sometimes expressed vaginal carcinomas [8]. It also caused malformations of the male reproductive tract. DES has, compared to E2, little affinity for extracellular proteins. Thus, at an equivalent concentration in the blood, a greater fraction of DES is unbound to protein than is E2 allowing for greater diffusion into cells making DES a more efficient estrogen than the natural hormone [8]. Similar effects have been shown for o,p'-DDT [41].

Certain non-estrogenic compounds (e.g., polycyclic aromatic hydrocarbons or PAHs), can be metabolized into agents with estrogenic activity by adding a hydroxy group. The estrogenicity of polychlorinated biphenyls (PCBs) is also enhanced by hydroxylation. But some PCB congeners also have estrogenic effects by acting as a substrate for the cytochrome P450 enzymes, thus inhibiting estrogen metabolism. The situation becomes even more complex as PCB congeners can express anti-estrogen behavior by binding to the Ah-receptor, inducing cytochrome P450 1A1 (CYP1A1) and cytochrome P450 1B1 (CYP1B1) synthesis. These cytochromes catalyze the metabolism of E2 to non-active forms [42] and their induction by xenobiotics will result in lower natural hormone levels. This latter pathway may also account for the anti-estrogenicity of 2,3,7,8-tetrachlorodibenzo-p-dioxin (TCDD) observed in rodent mammary and uterus and in human breast cancer cell lines [43].

The situations described above only focus on the impact of a single compound on the endocrine system. Mixtures of chemicals with estrogenic activity have received considerable study following a report that four weakly estrogenic pesticides, dieldrin, endosulfan, toxaphene, and chlordane, when combined in different combinations, results in dramatic synergy [44]; this report could not be validated [45] and was later retracted [46]. Synergism is clearly possible and can result from several mechanisms. For example, multiple chemicals may combine to form an estrogen-like compound; various xenoestrogens and natural estrogens may bind to different receptor subunits and thereupon form a functional receptor dimer; the receptor may have two or more interactive binding sites [8, 47] or combinations of metabolism and gene expression from multiple xenobiotics.

Combining these effects with the persistency of some chemicals in the body and in the environment may result in a major impact on the human endocrine system.

3.4
Phytoestrogens

More than 300 species of plants, across more than 16 families are known to contain estrogenic substances [48], also called phytoestrogens, which can be divided into three main classes: isoflavones, lignans, and coumestans. Most of these agents are non-steroidal in structure and a lot less potent than E2 (10^{-3} to 10^{-5} times less potent) [49]. In general lignans and isoflavonoids (i.e., isoflavones and coumestans) seem to stimulate the synthesis of sex hormone-binding globulin in the liver and thus reduce the effect of E2 by lowering the percentage of free hormone [50, 51].

The phytoestrogen that is of foremost interest at this moment is genistein. Its action is biphasic causing estrogenic effects by inducing cell-growth at lower concentrations (10^{-5} to 10^{-8} M), and anti-estrogenic effects, inhibiting cell-growth, at higher concentrations (10^{-4} to 10^{-5} M) [52]. At low concentrations, genistein, competes with E2 for the ER binding sites and stimulates the expression of specific mRNA markers recognized for estrogen activity [52]. Daidzein and equol (isoflavones), and enterolactone (lignan) show a similar effect [53]. The anti-proliferative effects of genistein do not seem caused by the estrogen receptor [52], but rather by interfering with the kinase pathways [49].

Phytoestrogens not only exert their activity by influencing the sex hormone metabolism and biological activity but also have an effect on intracellular enzymes, protein synthesis, growth factor action, tumor cell proliferation, and angiogenesis [54]. And like xenoestrogens, phytoestrogens can express both estrogenic and anti-estrogenic effects [55].

Overall more human studies are needed to verify the role of xenoestrogens or phytoestrogens as extrapolating from *in vitro* and *in vivo* studies for such a complicated system poses serious difficulties.

4
Carcinogen Risk Assessment

Regulatory agencies need input from the scientific community to develop adequate policies with regard to potentially hazardous agents, such as EACs. The biological sciences aspect of the risk assessment process includes hazard identification (qualitative), providing information on effects caused in exposed individuals, and dose-response assessment (quantitative), linking the size of an exposure to the response. Estrogenic compounds in the environment have the ability to cause a variety of (adverse) effects in humans and wildlife. The focus here will be on the carcinogenic risk assessment for these agents.

4.1
Change in Perspective

Cancer risk assessment of chemical exposure always depended, and still relies, on animal carcinogenicity bioassays carried out at doses generally higher than those routinely encountered by humans. In the absence of detailed understanding of the mechanism driving an effect, default assumptions are made in the risk assessment process. Examples of defaults include assuming the same biological processes between humans and animals, assuming that time of exposure and interindividual variance make no difference to a species' susceptibility for a compound and/or assuming a specific shape for dose-response below the experimental/observational range [56]. Generally, the information used for a risk assessment comes from epidemiological studies, acute and chronic animal toxicity studies, teratology and mutagenicity animal studies, and *in vitro* studies. Increasingly, the scientific community is endorsing mechanism-based toxicology for cancer risk assessment.

The scientific evidence is generally used to classify an agent into one or more categories of toxicity such as that used by the US National Toxicology Program for their Report on Carcinogens, and the International Agency for Research on Cancer Monograph Series. In these classification schemes, mechanistic research plays a key role in strengthening or weakening a causal relationship between cancer and exposure. When the mechanisms involved in carcinogenicity resulting from EACs are better understood, the prediction of whether a particular EAC has carcinogenic potential will improve, and be less dependent on arbitrary assumptions.

4.2
Modes of Action

Carcinogens are able to work via many different pathways to increase the mutation frequency in DNA and/or increase the number of target cells, both resulting in a higher probability for a tumor to develop. Direct and indirect geno-toxicity (altering metabolism, inhibition of repair mechanisms) have an effect on the mutation ratio and cell division. However, for EACs no direct genotoxic-ity has yet been found, but xenoestrogens and phytoestrogen may work through many, if not all, of the other pathways mentioned above. Besides occupational exposures, the primary exposure route for humans to EACs is through the food chain, making both the biological pathways and time of exposure important considerations. Small disturbances for even a brief time may have profound and long lasting effects on the individual (e.g., *in utero* exposure to estrogen can change the differentiation process of the central nervous system, reproductive tract, and other organ systems [57, 58]).

4.3
Change in Risk Assessment Approach

The default approach to cancer risk assessment as described in the traditional EPA paradigm derives risk estimates based on observed response in animals or humans. Since reliable human dose-response studies are rare, theoretically con-servative methods such as the linearized multistage model (LMS) are used, along with the default assumptions, to take into account uncertainty when ex-trapolating from animal studies.

The LMS model is derived as the upper 95% confidence boundary on a poly-nomial function whose parameters are adjusted to fit the tumor data, predict-ing a non-zero response for all non-zero exposure levels [59]. Low-dose extrap-olations with this model are thought to be conservative and thus protective of the public health. As generally applied, the LMS model excludes mechanistic data that might be available on the compound being regulated. Use of mecha-nistic information with this model or others should improve risk estimates and avoid potential bias introduced by assumptions and the use of less data [60].

In 1996 the EPA revised its guidelines on cancer risk assessment in which a full characterization is being emphasized, using all the information available to de-sign dose-response approaches and expanding the role of mechanistic informa-tion. As a first step in risk estimation, tumor and mechanistic toxicology data are used to choose a biologically based model for evaluation in the range of observa-tion. And, as long as it makes biological sense, this model can be used for extrap-olation to exposures outside of the experimental/observational range (low-dose area). If biological models are not available, or there are insufficient data to be able to build a feasible model, curve fitting using standard statistical approaches becomes necessary [61]. Whether this statistical approach should consider lin-earity or non-linearity in the low-dose area will depend on the available data. The most common statistical method used is the "Benchmark Dose Modeling" ap-proach, defined as a lower statistical confidence limit for the dose corresponding to a specified increase in level of health effect over the background level [62]. The benchmark dose can be used for both carcinogenic and non-carcinogenic health effects; only the carcinogenic effects will be considered in this chapter, although one can discuss the importance of non-carcinogenic risk assessment for estro-gens. For agents believed to have a response proportional to dose, the extrapola-

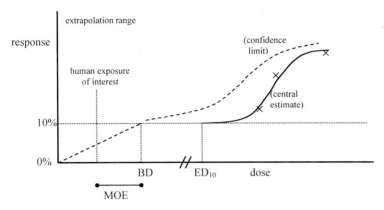

Fig. 5. Benchmark dose graph [87]

tion is linear below the benchmark dose (BD), and for others, a predetermined margin of exposure (MOE) is used to get an idea on how far exposed humans are from the effect actually observed in animals (see Fig. 5).

Incorporating biologically-based models for EACs into the risk assessment process for these compounds, can potentially reduce uncertainties in each step of this operation. These models reduce reliance upon assumptions for extrapolation issues, like high exposure to low exposure, animal to human, homogeneous populations to large heterogeneous populations and combining information from *in vivo* and *in vitro* systems, are addressed on a scientific basis. Furthermore, in many cases, exogenous and endogenous hormones work as agonists, antagonists, or as synergists. Even though it is a complicated, interactive system, the endocrine system is largely available for modeling because much is known about the regulation, effect, and interactions of natural hormones on various organ systems.

Biologically-based models also provide mechanistic understanding of pharmacokinetic and pharmacodynamic behavior, and provide an objective, quantitative structure useful for hypothesis testing. Furthermore, once a physiologically-based model has been established, it is possible to simply revise it whenever new toxicological and biological knowledge and data become available, and the same model structure can be used for other environmental agents with the same, or with similar mechanisms. Finally, developing mechanistic links between exposure, target tissue dose, short-term exposure and long-term exposure does not require *a priori* knowledge on whether a compound has carcinogenic capabilities. So when mechanisms are considered, risk-assessment for carcinogens and non-carcinogens can at least partially use the same knowledge base [63].

The requirement of an extensive, quantitative database, which is expensive and time-consuming to obtain, is one of the disadvantages of mechanistic modeling at this time. It is also time-consuming to create a reliable model as an iterative modeling approach is necessary. After a model has been developed, it needs to be tested by conducting experiments, evaluated against the data and modified as needed.

The diverse information available for the action of endogenous estrogen and EACs needs to be combined into a single model. Building these models uncovers knowledge gaps about such issues as chemical interactions, feedback systems, and biological connections. Nonetheless, carcinogen risk assessment needs to move forward, even in the absence of certain information. Thus, for now a trade-off between mechanism-based risk assessment and the use of mathematical methods, like the LMS model or BD approach seems inevitable, keeping in mind that risk assessment is only as good as the information upon which it is based.

5
Mechanistic Model Development for Estrogens

As described in the risk assessment part of this manuscript, over the last decade risk assessment in the United States shows a shift towards using physiologically realistic mathematical models. In order to get an understanding of the dimensions of the task that still lies ahead concerning mechanistic modeling of estrogens and estrogen-like compounds, an ideal situation is proposed. And an example of a physiologically-based endocrine system model is presented in the subsequent section.

5.1
Model Types

Even though many studies addressed the carcinogenic effects of estrogens and estrogen-like compounds in both humans and animals, hardly any mechanistic modeling has been done with regard to this subject. This is quite surprising as there is both a quite vast distribution of estrogenic agents in the external milieu, and a crucial role for endogenous estrogens in the endocrine system.

An ideal situation for EACs would be an overall mechanism based modeling structure (Fig. 6), in which they are able to interfere on every level, either direct or indirect. The change in effects or endpoints over the homeostatic background level, as a result of exogenous interference or even inter-individual variability, can subsequently be used for risk assessment.

This overall structure incorporates models that are currently being used in risk assessment, although they still do not displace the current default methods. Physiologically-based pharmacokinetic (PBPK) models combine detailed mechanisms by which chemicals are distributed from the external environment to the target tissues. Down at the cellular level of an organism, mechanisms by which target tissue doses are converted to adverse biological effects can be described by physiologically-based pharmacodynamic (PBPD) models, enclosing receptor interactions and cell signaling pathways. The third model type widely used nowadays to quantify risks, in this case tumor incidence rates, from exposure is a multistage model of carcinogenesis.

The task at hand is not only to merge available knowledge into a suitable physiological model for estrogenicity and associated risks, but also to link these

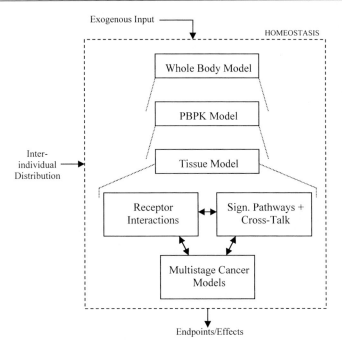

Fig. 6. Schematic overview of a complete mechanism-based mathematical model

models together in order to obtain an overall picture of the consequences of a possible perturbation. Predictions from this model can then be compared to data, making it possible to refine the model structure consistent with physiological and biological processes.

5.2
Estrus Cycle Model

The shortened textual description of a physiologically based model of the hormonal relationships during the rat estrus cycle, as developed by Willems et al. [64], is useful in visualizing the complexity of the endocrine system. This *estrus cycle model* (ECM) incorporates six physiological compartments belonging to the endocrine system; the hypothalamus (including the hypothalamus-pituitary portal system), the pituitary, the liver, the ovaries, the gastro-intestinal tract, and the liver (Fig. 7). The changes in hormone levels between these compartments are established with ordinary differential equations.

5.2.1
Hypothalamus System

GnRH is chosen as the starting point for describing the estrus cycle. After its synthesis and release GnRH is transported from the hypothalamus to the

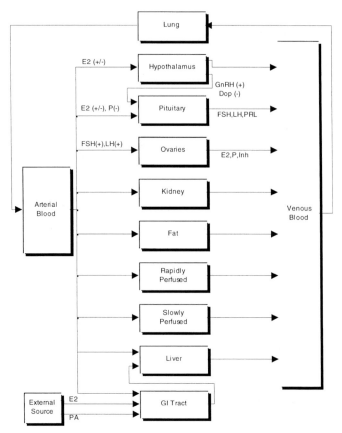

Fig. 7. Schematic overview of the estrus cycle model. Each compartment is characterized by physiological and biochemical parameters. There is a blood flow to each tissue (as a fraction of the total blood flow). The abbreviations of the hormone names (GnRH, Dop, E2, P, LH, FSH, PRL, and Inh) indicate in which tissues they are synthesized and where they exert their positive (+) or negative (−) feedback on the endocrine system

pituitary via the hypophysial portal veins where it stimulates the release of LH and FSH. Even though *in vivo* GnRH is excreted in a pulsatile way, this model assumes a constant signal produced by the intrinsic rhythm of the GnRH cells [65–67]. On the level of the hypothalamus, free E2 increases the synthesis of GnRH by receptor activation, and a partial agonist (PA) can compete with this free E2 for its receptor. The partial transcriptional activation of the estrogen receptor by the endocrine disrupter is described by the weighing term ω.

Cumulative exposure to E2 causes aging of the neuronal network of the hypothalamus, resulting in a decrease of its intrinsic signal [68] and modeled as a degradation of the synthesis rate. This can eventually lead to insufficient GnRH signaling to the pituitary. A second signal that originates in the hypothalamus is the neurotransmitter, annex hormone, dopamine (Dop), which is considered to be the predominant inhibiting factor for prolactin (PRL) synthesis [69].

5.2.2
Hypothalamus/Pituitary Portal System

GnRH and Dop are transported to their site of action via the hypophysial portal system, which runs directly from the hypothalamus to the pituitary gland over a distance of several micrometers. It is assumed here that no weakening of both signals will take place in this system.

5.2.3
Pituitary

Combined with feedback signals provided by E2 and progesterone (P), GnRH and Dop control synthesis and release of FSH, LH, and PRL. The production of the gonadotropins LH and FSH is stimulated by the GnRH signals and inhibited by the instantaneous levels of free E2 and P in the tissue. This model treats the LH and FSH signals to the ovaries as continuous serum levels.

The preovulatory gonadotropin surge, necessary for follicle growth and ovulation, is induced by an increase in circulating E2 concentrations that stimulates the secretion of GnRH from the hypothalamus as described before [70, 71] and enhances pituitary responsiveness to GnRH [72, 73]. This sudden rise in circulating LH and FSH levels is modeled by amplifying the GnRH signal.

The production of PRL by the lactotropic cells is inhibited by Dop, the inhibitory effects of which are blocked by E2 [74]. So when the free E2 concentration in the model exceeds a critical value, the inhibition of Dop is turned off for as long as the E2 level is elevated. The decreasing responsiveness of the pituitary gonadotropic membrane receptors [75] is modeled as a linear function of cumulative E2 exposure.

5.2.4
Ovaries

Recruitment and maturation of follicles, ovulation on the day of estrus and the differentiation of follicular-tissue into luteal tissue [76–78] is controlled by an interplay of neuronal and hormonal control mechanisms situated in the pituitary and central nervous system. Follicle cohorts need four cycles to grow to a size large enough to sustain the estrus cycle [79], and individual follicles (consisting of both theca and granulosa cells) are assumed to follow exponential growth kinetics under the influence of FSH. After ovulation, occurring as a result of a LH surge, the follicular tissue starts differentiating to luteal tissue and simultaneously degenerates.

As the follicles grow, they start producing amounts of E2 and inhibin (Inh). P is synthesized by the granulosa cells during proestrus and its peak is assumed to be controlled by the proestrus surge of LH [80]. The second P peak that occurs during the afternoon of metestrus arises from the newly formed corpora lutea (CL) [81]. Finally, simple first-order kinetics describe E2 and P degradation in the liver. GnRH, LH, FSH, PRL, Inh, and Dop are degraded in the blood using first order kinetics.

The next step would be trying to enhance the model, including possible signaling pathways, other tissues involved in endocrine and estrogen signaling, or linking it with a multistage cancer model. There seem to be numerous possibilities but also just as many data requirements.

6
Discussion

In this chapter we have given an overview of aspects that should be addressed when discussing the role of estrogens and estrogen-like agents in mechanism-based carcinogenic risk assessment. In the last decade, enormous amounts of information have become available with regard to the endocrine system and the role of estrogen within that system. It has become clear that estrogen, because of its extensive array of signaling pathways, wide variety of feedback signals, and large number of target tissues, is not a hormone that can be considered separately, apart from its interactive role with other hormones or endocrine active compounds. A broad toxicological approach is necessary to fully investigate its effect on human health, as opposed to the majority of toxicological studies conducted these days in which the effects of a single agent on a single target are examined. This, of course, complicates the assessment as new ways of dealing with endogenous and exogenous hormone interactions need to be explored. Furthermore, because the flow of information from biology (e.g., toxicology, molecular biology, and endocrinology) has grown exponentially, it is prudent and obvious that some new structure is necessary to comprehend and evaluate the available knowledge. These methods should include the recognition of the most significant data and the identification of knowledge gaps.

Even though a considerable amount of information is already available with regard to the action of endocrine active agents, it is essential that further research in this area be conducted. Methodologies for assessing human and wildlife health effects should focus on both endpoint effects and on unraveling mechanisms of action. Currently, there are inadequate *in vitro* and *in vivo* testing methodologies for several known potential targets such as the thyroid and androgen receptor systems. Furthermore, there is hardly any knowledge on the mechanisms of action of EACs on other endocrine targets, the effect on signal transduction, the role of new receptors like ER-β in diverse tissues and crosstalk between membrane-bound and nuclear receptors [82]. So action should be taken in the form of *in vitro* assays, short- and long-term *in vivo* assays, multigenerational studies and quantitative structure-activity relationship (QSAR) models, to gain more knowledge in these areas. Naturally, these protocols should be under close scrutiny to determine their advantages and shortcomings, and to be certain that the most critical gaps are closed first. The most prevalent areas of concern regarding the action of hormones are their carcinogenicity, the consequences for reproduction, their neurological effects, and their immunological effects [29], especially during development. Because these systems are closely intertwined, it is obvious that "crosstalk" will also take place at this higher level.

It is imperative that any new methods of analysis be more integrative. Animal data, *in vitro* assay results, clinical study outcomes, and physical and chemical characterizations need to be combined. A way of doing this is by means of bio-logically-based mathematical modeling, and even though it seems obvious to model these interactions, many hurdles still need to be overcome. First, one should know the size, length and time of exposure to estrogens or estrogen-act-ing compounds. This can be easily controlled in laboratory circumstances, and might even be predictable for human exposure to man-made chemicals. But lit-tle information is available describing the background levels caused by the up-take of, for example, phytoestrogens. Just as important is the question whether these background levels are the same throughout human lifetime or do peaks occur at particular moments? And do these moments coincide with more sen-sitive periods in human development? Keeping in mind that, even without this background interference, there is a distribution of homeostatic environments, no single number or set of numbers exists describing all effects and interactions for all humans. Furthermore, to create a full mechanism-based mathematical model, separate submodules need to be built first. Physiologically-based phar-macokinetic and dynamic models, and multistage cancer models are examples of these submodules. The next step links together these submodules, putting the emphasis on receptor interactions, signaling pathways, and crosstalk. It ampli-fies, once again, the importance of working on mechanisms of action rather than analyzing a single dataset on a single endpoint. A third problem, in work-ing towards an integrated model for endocrine action, would be the large num-ber of endogenous hormones and exogenous agents exerting control over an endocrine system. This is similar to the question on how to approach the analy-sis of toxicological interactions of chemical mixtures. It is difficult, possibly im-possible, to obtain and analyze data for every possible combination in, for in-stance, a 25 chemical mixture. The only realistic chance to deal with this level of complexity is the integration of PBPK and PBPD modeling with, e.g., Monte Carlo simulations or other, similar approaches [83].

Even though the issues described above seem complicated, one should keep in mind that even a small development is an improvement of the *status quo* in risk assessment. If done right, development of biologically-based dose-response models should logically lead to more precise and realistic estimates of risks than do default methods. In time, biologically-based modeling should also be able to improve extrapolation to low-dose areas of the dose-effect curve, pro-viding curve fitting in this area with a more solid scientific background. Nevertheless, empirical methods to estimate the risks in the lower area of the curve have also been improved.

The advantages of simpler techniques like the benchmark dose approach, in which the shape of a model is used to determine a point of departure for risk as-sessment (as opposed to the *no observed adverse effect level*, NOAEL), diminish when mechanistic information is ignored or used improperly. For example, for cancer, extrapolation below the benchmark dose is still assumed to be linear for genotoxins, regardless of the shape of the dose-response curve. This assumption is not based on data [84] and estimation of the risk in the low-dose area could thus be too conservative or too tolerant. Risk estimation (either at low doses or at benchmark responses) with biologically-based mechanistic models is more sen-sitive as subtle changes in an animal's response are measurable at low-dose exposure levels, like altered hormonal levels or change in proliferation rates. Assuming that physiological mechanisms function similarly at both high and low doses, these subtle changes can be predictive for all types of toxicity and the model makes it possible to identify potential non-linearity in low-dose response.

Endogenous levels of estrogen already explain some amount of carcinogenicity, as is the case for breast cancer. In this case, the expected dose-response turns from a possible non-linear shape into a shape governed by the addition to an ongoing process and the ability to saturate response on the upper end of the response curve [85]. The question is no longer whether people are at risk, because they are, but whether the exogenous exposure is acting by the same process (in which case response is likely to be proportionate to exposure) and whether that process has an upper or lower boundary of activity.

Finally, the success of mechanism-based risk assessment will be decided by the availability of both quantitative and qualitative information from researchers worldwide, and the willingness of researchers to work together. This need for cooperation might actually turn out to be one of the most challenging parts of the whole process.

7
References

1. Inano H, Suzuki K, Ishiiohba H, Ikeda K, Wakabayashi K (1991) Carcinogenesis 12: 1085
2. Kettles MA, Browning SR, Prince TS, Horstman SW (1997) Environ Health Perspect 105: 1222
3. Vom Saal F, Finch CE, Nelson JF (1994) Natural history and mechanisms of reproductive aging in humans, laboratory rodents, and other selected vertebrates. In: Knobil E, Neill JD (eds), The Physiology of Reproduction. Raven Press, New York, p 1213
4. Hammond G (1993) Extracellular steroid-binding proteins. In: Parker M (ed), Steroid Hormone Action. Oxford University Press, New York, p 1
5. Solomon S (1994) The primate placenta as an endocrine organ. In: Knobil E, Neill JD (eds), The Physiology of Reproduction. Raven Press, New York, p 863
6. Zhu B, Conney AH (1998) Carcinogenesis 19: 1
7. Hotchkiss J, Knobil E (1994) The menstrual cycle and its neuroendocrine control. In: Knobil E, Neill JD (eds), The Physiology of Reproduction. Raven Press, New York, p 711
8. McLachlan JA, Arnold SF (1996) Am Sci 84: 452
9. Clark J, Mani SK (1994) Actions of ovarian steroid hormones. In: Knobil E, Neill JD (eds), The Physiology of Reproduction. Raven Press, New York, p 1011
10. Green S, Chambon P (1991) The oestrogen receptor: from perception to mechanism. In: Parker M (ed), Nuclear Hormone Receptors. Academic Press, New York, p 15
11. Gronemeyer H, Laudet V (1995) Protein Profile 2: 1167
12. Katzenellenbogen BS (1996) Biol Reprod 54: 287
13. Barkhem T, Carlsson B, Nilsson Y, Enmark E, Gustafsson JA, Nilsson S (1998) Mol Pharmacol 54: 105
14. Paech K, Webb P, Kuiper GG, Nilsson S, Gustafsson JA, Kushner PJ, Scanlan TS (1997) Science 277: 1508
15. Moss R, Gu Q, Wong M (1997) Rec Prog Horm Res 52: 33
16. Fortunati N, Fissore F, Comba A, Becchis M, Catalano MG, Fazzari A, Berta L, Frairia R (1996) Horm Res 45: 202
17. Blaustein J, Lehman MN, Turcotte JC, Greene G (1992) Endocrinology 131: 281
18. Wehling M, Neylon CB, Fullerton M, Bobik A, Funder JW (1995) Circul Res 76: 973
19. Zakon H (1998) Trends Neurosci 21: 202

20. Smith C (1998) Biol Reprod 58: 627
21. Clark JH, Peck EJ Jr (1979) Female Sex Steroids: Receptors and Function. Springer Berlin Heidelberg New York
22. Simerly R, Young, BJ (1991) Mol Endocrinol 5: 424
23. Shupnik M, Gordon MS, Chin WW (1989) Mol Endocrinol 3: 660
24. Kuiper GG, Enmark E, Pelto-Huikko M, Nilsson S, Gustafsson JA (1996) Proc Natl Acad Sci USA 93: 5925
25. Perlmann T, Evans RM (1997) Cell 90: 391
26. Dauvois S, Parker MG (1993) Mechanism of action of hormone antagonists. In: Parker M (ed), Steroid Hormone Action. Oxford University Press Inc., New York, p 166
27. Fotherby K (1988) Ann NY Acad Sci 538: 313
28. Barton H, Andersen ME (1997) Reg Tox Pharmacol 25: 292
29. Kavlock R, Daston GP, De Rosa C, Fenner-Crisp P, Gray LE, Kaattari S, Lucier GW, Luster M, Mac MJ, Maczka C, Miller R, Moore J, Rolland R, Scott G, Sheehan DM, Sinks T, Tilson HA (1996) Environ Health Perspect 104: 715
30. Jensen E, Jacobson HI, Flesher JW (1966) Estrogen receptors in target tissues. In: Pincus G, Nakao T, Tait JF (eds), Steroid Dynamics. Academic Press, New York, p 133
31. Lippman M, Monaco ME, Bolan G (1977) Cancer Res 37: 1901
32. Bouton M, Raynaud JP (1979) Endocrinology 105: 509
33. Romano G, Krust A, Pfaff DW (1989) Mol Endocrinol 3: 1295
34. Anderson J, Peck EJ Jr, Clark JH (1975) Endocrinology 96: 160
35. Lan N, Katzenellenbogen BS (1976) Endocrinology 98: 220
36. Ruh T, Wassilak SG, Ruh MF (1975) Steroids 25: 257
37. Clark JH, Markaverich BM (1982) Pharmacol Ther 15: 467
38. Natrajan P, Greenblatt RB (1979) Clomiphene citrate. In: Greenblatt R (ed), Induction of Ovulation. Lea & Febiger, Philadelphia, p 35
39. Lyttle R, Damian-Matsumura P, Juul H, Butt TR (1992) J Steroid Biochem Mol Biol 42: 677
40. Guillette LJ Jr, Pickford DB, Crain DA, Rooney AA, Percival HF (1996) Gen Comp Endocrinol 101: 32
41. Klotz DM, Ladlie BL, Vonier PM, McLachlan JA, Arnold SF (1997) Mol Cell Endocrinol 129: 63
42. Spink D (1998) NIEHS/EPA Superfund Basic Research Program 'Research Brief' 28
43. Safe S, Krishnan V (1995) Toxicol Lett 82/83: 731
44. Arnold SF, Klotz DM, Collins BM, Vonier PM, Guillette LJ Jr, McLachlan JA (1996) Science 272: 1489
45. Ramamoorthy K, Wang F, Chen IC, Norris JD, McDonnell DP, Leonard LS, Gaido KW, Bocchinfuso WP, Korach KS, Safe S (1997) Endocrinology 138: 1520
46. McLachlan JA (1997) Science 277: 462
47. Hedden A, Muller V, Jensen EV (1995) Ann NY Acad Sci 761: 109
48. Hughes CJ (1988) Environ Health Perspect 78: 171
49. Murkies A, Wilcox G, Davis SR (1998) J Clin Endocrinol Metab 83: 297
50. Adlercreutz H, Hockerstedt K, Bannwart C, Bloigu S, Hamalainen E, Fotsis T, Ollus A (1987) J Steroid Biochem 27: 1135
51. Adlercreutz H, Hockerstedt K, Bannwart C, Hamalainen E, Fotsis T, Bloigu S (1988) Association between dietary fiber, urinary excretion of lignans and isoflavonic phyto-estrogens, and plasma non-protein bound sex hormones in relation to breast cancer. In: Bresciani F, King RTB, Lippman ME, Raynaud JP (eds), Progress in Cancer Research and Therapy. Raven Press, New York, p 409
52. Wang T, Sathyamoorthy N, Phang JM (1996) Carcinogenesis 17: 271
53. Sathyamoorthy N, Wang TTY, Phang JM (1994) Cancer Res 54: 957
54. Adlercreutz H (1995) Environ Health Perspect 103: 103
55. Ruh M, Zacharewski T, Connor K, Howell J, Chen I, Safe S (1995) Biochem Pharmacol 50: 1485
56. Lang L (1995) Environ Health Perspect 103: 334
57. Kavlock R, Ankley GT (1996) Risk Anal 16: 731

58. Sharpe R, Skakkebaek NE (1993) Lancet 341: 1392
59. Crump K (1982) J Environ Pathol Toxicol 5: 339
60. Conolly RB, Andersen ME (1993) Environ Health Perspect 101 (Supp 6): 169
61. Page N, Singh DV, Farland W, Goodman JI, Conolly RB, Andersen ME, Clewell HJ, Frederick CB, Yamasaki H, Lucier GW (1997) Fund Applied Toxicol 37: 16
62. Crump K (1984) Fund Applied Toxicol 4: 854
63. Conolly R (1995) Toxicology 102: 179
64. Willems BAT, Davis BJ, Portier CJ (2000) Comprehensive physiologically based model of the hormonal relationships in the rat estrus cycle I: description of the model. (in peer review)
65. Krsmanovic L, Stojilkovic SS, Mertz LM, Tomic M, Catt KJ (1993) Proc Natl Acad Sci USA 90: 3908
66. Weiner R, Wetsel W, Goldsmith P, Martinez de la Escalera G, Windle J, Padula C, Choi A, Negro-Vilar A, Mellon P (1992) Front Neuroendocrinol 13: 95
67. Wetsel W, Valenca MM, Merchenthaler I, Liposits Z, Lopez FJ, Weiner RI, Mellon PM, Negro-Vilar A (1992) Proc Natl Acad Sci USA 89: 4149
68. Keefe D, Garcia-Segura LM, Naftolin F (1994) Neurobiol Aging 15: 495
69. Kordon C, Drouva SV, Martinez de la Escalera G, Weiner RI (1994) Role of classic and peptide neuromediators in the neuroendocrine regulation of luteinizing hormone and prolactin. In: Knobil E, Neill JD (eds), The Physiology of Reproduction. Raven Press, New York, p 1621
70. Levine J, Ramirez VD (1982) Endocrinology 111: 1439
71. Sarkar D, Chiappa SA, Fink G, Sherwood NM (1976) Nature 264: 461
72. Baldwin D, Downs TR (1980) Biol Reprod 24: 581
73. Drouin J, Lagace L, Labrie L (1976) Endocrinology 99: 1477
74. Labrie F, Ferland L, DiPaolo T, Veilleux R (1980) Modulation of prolactin secretion by sex steroids and thyroid hormones. In: MacLeod R, Scapagnini U (eds), Central and Peripheral Regulation of Prolactin. Raven Press, New York, p 97
75. Wise P (1987) J Steroid Biochem 27: 713
76. Espey L, Lipner H (1994) Ovulation. In: Knobil E, Neill JD (eds), The Physiology of Reproduction. Raven Press, New York, p 725
77. Greenwald G, Roy SK (1994) Follicular development and its control. In: Knobil E, Neill JD (eds), The Physiology of Reproduction. Raven Press, New York, p 629
78. Niswender G, Nett TM (1994) Corpus luteum and its control in infraprimate species. In: Knobil E, Neill JD (eds), The Physiology of Reproduction. Raven Press, New York, p 781
79. Insler V, Kleinman D, Sod-Moriah U (1990) Gynecol Obstet Invest 30: 228
80. Freeman M (1994) The neuroendocrine control of the ovarian cycle of the rat. In: Knobil E, Neill JD (eds), The Physiology of Reproduction. Raven Press, New York, p 613
81. Smith M, Freeman ME, Neill JD (1975) Endocrinology 96: 219
82. Zacharewski T (1998) Environ Health Perspect 106: 577
83. El-Masri H, Reardon KF, Yang RSH (1997) Crit Rev Toxicol 27: 175
84. Hoel DG, Portier CJ (1994) Environ Health Perspect 102 (Suppl 1): 109
85. Portier CJ, Sherman CD, Kohn M, Edler L, Kopp-Schneider A, Maronpot RM, Lucier GW (1996) Tox Applied Pharmacol 138: 20
86. Portier CJ (1998) Personal communication
87. EPA (1996) Proposed Guidelines for Carcinogen Risk Assessment. EPA/600/P-92/003 C. U.S. Environmental Protection Agency, Washington, DC

Alterations in Male Reproductive Development: The Role of Endocrine Disrupting Chemicals

Shanna H. Swan[1], Frederick S. vom Saal[2]

[1] Family and Community Medicine, MA306 Medical Sciences Building, University of Missouri-Columbia, Columbia MO 65212, USA
E-mail: swans@health.missouri.edu
[2] Division of Biological Sciences, 114 Le Fevre Hall, University of Missouri-Columbia, Columbia MO 65212, USA
E-mail: vomsaal@biosci.mbp.missouri.edu

In this chapter we will address the following question: to what extent does current evidence from reproductive biology and epidemiology support a causal role for endocrine disrupting chemicals (EDCs) in the pathogenesis of altered male reproductive function? We have divided this discussion into two parts; epidemiology, presented first, followed by the relevant reproductive biology. Our discussion will focus primarily on semen quality, testicular cancer, hypospadias, and cryptorchidism, which we will refer to collectively as "adverse male endpoints". Other male reproductive parameters, including altered prostate development and prostate cancer, will also be discussed briefly. An historical overview of EDCs, beginning with the discovery of the first synthetic estrogen in 1933, is presented first.

The adverse male endpoints discussed here can result from perturbations of the hormonal environment during critical periods in fetal organogenesis. Such perturbations can result from multiple causes, including genetic defects and alterations in maternal physiology. These physiological changes are themselves related to a host of factors, which may include EDC exposure. For example, a number of aspects of pregnancy (e.g., birth order, birth weight, and multiplicity) or the pregnant woman (e.g., maternal age, ethnicity), referred to here as "pregnancy-related factors", may be directly related to prenatal hormone levels and, consequently, to adverse male endpoints. Additionally, these pregnancy-related factors may themselves modify effects of EDC exposures. The multiplicity of factors capable of perturbing the prenatal hormonal milieu, and the potential for their interactions, presents a challenge to scientists working in this field.

Hormonal exposures incurred by the fetus during normal pregnancy ("endogenous hormones") will be discussed as possible factors in the development of adverse male endpoints. We will also review studies on adverse male development in relation to pharmaceuticals with hormonal activity, particularly those to which the developing fetus may be exposed. These pharmaceuticals include oral contraceptives, hormones administered for pregnancy support including diethylstilbestrol (DES) and its congeners, and hormonal pregnancy tests.

For both endogenous and exogenous hormones, the most critical exposure period for adverse male endpoints appears to occur during organogenesis, during the embryonic and fetal stages of prenatal development. Changes induced by exposures at this time are typically irreversible. Moreover, as discussed below, exposure to extremely low doses during this time of heightened sensitivity may profoundly alter reproductive development. In contrast, reproductive changes induced during adulthood are usually reversible, and much larger doses are required to alter the reproductive system [1]. The impacts of such large doses on male reproduction can be seen in the occupational literature. Studies that report adverse male endpoints in association with occupational exposure to chemicals that are known or suspected to alter endocrine function will be summarized. There are also limited data from industrial accidents that have resulted in population exposure to EDCs and subsequent reproductive damage that will also be mentioned briefly.

The Handbook of Environmental Chemistry Vol. 3, Part M
Endocrine Disruptors, Part II
(ed. by M. Metzler)
© Springer-Verlag Berlin Heidelberg 2002

We then turn to a discussion of the relevant reproductive biology. We begin this section with a discussion of the biological plausibility that EDCs play a causal role in the development of adverse male endpoints. We review laboratory studies showing a causal relationship between EDCs and adverse male endpoints in animals. We note, however, that these adverse endpoints can also occur as the result of a variety of factors in addition to developmental exposure to EDCs. This complicates the assessment of causality in studies investigating the relationship of EDCs to adverse male endpoints in human populations. In this regard, we note the critical absence of literature directly relating adverse male endpoints in humans with measurements of EDCs at environmental levels. We will discuss reasons for this information gap and suggest steps to fill it.

Keywords. Development, Diethylstilbestrol, Endocrine disruption, Low-dose effects, Reproduction, Semen quality, Testicular function, Xenoestrogen

1	Introduction	133

2	Temporal Trends and Geographic Variability in Adverse Male Endpoints	136
2.1	Semen Quality	136
2.2	Incidence of Testicular and Prostate Cancer	140
2.3	Incidence of Hypospadias	141
2.4	Incidence of Cryptorchidism	142

3	Adverse Male Endpoints in Relation to Exogenous Hormone Exposure	142
3.1	Testicular Cancer	143
3.2	Male Genital Tract Abnormalities and Semen Quality	143

4	Adverse Male Endpoints in Relation to Endogenous Hormone Exposure	145
4.1	Testicular Cancer	146
4.2	Cryptorchidism	147
4.3	Hypospadias	148

5	Exposures in the Workplace and from Industrial Accidents	148

6	Plausibility of Adverse Effects of EDCs: Mechanisms of Gonadal and Accessory Reproductive Organ Differentiation	150

7	Effects of Estrogenic Chemicals on Accessory Reproductive Organs: Interaction with Endogenous Androgens	153

8	Hormonally Mediated Variability in Mammalian Development: Relationship to Bioactive Serum Concentration of Natural Hormones and Endocrine Disruptors	155

9   Issue of Dose in Toxicological Studies: Importance of Non-
Monotonic (Inverted-U) Dose-Response Curves 157

10   Development of the Prostate: Example of Opposite Effects of
High and Low Doses 158

11   Developmental Exposure to Diethylstilbestrol (DES) 159

12   Conclusions and Future Research Needs 161

13   References 163

List of Abbreviations

BPA   bisphenol A
DAS   4,4′-diaminostilbene-2,2′-disulfonic acid
DBCP dibromochloropropane
DDT   2,2-bis(p-chlorophenyl)-1,1,1-trichloroethane
DES   diethylstilbestrol
DHT   5α-dihydrotestosterone
EDC   endocrine disrupting chemical
EGE   ethylene glycol ether
OC   oral contraceptive
PCB   polychlorinated biphenyl
PSA   prostate specific antigen
PVC   polyvinylchloride
TTP   time to pregnancy

1
Introduction

A broad class of endocrine disrupting chemicals, the polychlorinated biphenyls (PCBs), were first manufactured in 1929 and widely used in electric transformers, capacitors, and hydraulic fluids [2]. In 1933, Cook and Dodds described the first synthetic estrogen and a test for estrogenicity [3]. Three years later Dodds (the discoverer of diethylstilbestrol, DES) synthesized a class of chemicals with bi-phenolic structures with full estrogenic activity in rats [4]. Notably this class in-cluded bisphenol-A (BPA), a chemical subsequently used in the manufacture of Bakelite and other plastics, and currently the subject of intensive study. DES was widely distributed soon after its discovery under a variety of trade names and for a range of indications [5]. Approval for use in pregnancy was obtained in 1947, despite considerable concern about its carcinogenic potential. Soon afterwards it was noted that not only pharmaceuticals, but also environmental chemicals, pos-sessed hormonal activity. In 1950 Burlington published the first observation in support of the estrogenicity of the widely used pesticide 2,2-bis(p-chlorophenyl)-

1,1,1-trichloroethane (DDT) [6]. Other reports of the estrogenicity of organo-chlorine pesticides appeared during the 1950s and 1960s when the use of these pesticides was at its height in the United States. For example, female-female pair-ings and significantly decreased male to female sex ratio in gull populations were observed [7, 8]. In the ensuing years, reproductive damage to wildlife, attributed to DDT, other pesticides and industrial chemicals with estrogenic, antiandro-genic, and antithyroid activity, was reported with growing frequency. DDT use was restricted in the United States in 1972 [9], and PCBs were banned shortly thereafter, following the enactment of the Toxic Substances Control Act in 1976.

In 1975, the first Conference on Estrogens in the Environment was held to ad-dress concerns that environmental chemicals with estrogenic properties could alter sexual development in the exposed offspring. The immediate focus of con-cern was a large cluster of premature breast development in Puerto Rico, which has continued until the present. This phenomenon remains largely unex-plained, though hormone contaminated food products and waste products from the manufacture of oral contraceptives were suspected [10]. The suspicion that an estrogenic compound was a likely explanation had been heightened by the discovery in 1971 that prenatal exposure to DES resulted in a rare vaginal cancer in a small proportion of the exposed offspring and also caused numer-ous other abnormalities in the reproductive system in a much greater propor-tion. DES was thus identified as a human transplacental carcinogen [11] and as a teratogen targeting the developing genital tract [12]. By the mid-1970s it was well established that exposure to DES during prenatal development was capable of profoundly altering reproductive development in both males and females. In the quarter century that followed, these initial DES findings were replicated and model systems developed to explicate mechanisms of action. These findings profoundly influenced the ensuing science. One consequence was the develop-ment of the field of teratology, a discipline devoted to the study of adverse ef-fects of such prenatal exposures.

At the same time as DES was becoming recognized as a transplacental car-cinogen, transplacental effects were reported in children exposed to potent en-docrine disruptors, not through pharmaceuticals, but in environmental set-tings, e.g., in Yusho [13] and Yu-Cheng [14]. In addition, sterility or subfertility was documented following high exposures in occupational settings, e.g., dibro-mochloropropane [15], ethylene dibromide [16], and kepone [17]. These episodes alerted the scientific community to the range of organic chemicals ca-pable of profoundly altering human reproductive development. Further, several widely used environmental chemicals had been shown to alter endocrine func-tion in the laboratory, and there was growing evidence that these chemicals were disrupting development in wildlife. However, it was not until the first Wingspread Conference in 1991 that the potential of these chemicals to impact profoundly human health at background levels present in the environment be-gan to be appreciated [18]. By then concern about the potential health effects of environmental estrogens had broadened to encompass a variety of chemicals that had the potential to alter endocrine function, agents that soon became known as Endocrine Disrupting Chemicals (EDCs). It has more recently been recognized that endocrine disruption refers not just to disruption of hormonal messengers transported in blood, but to disruption of any component of signaling systems that are involved in intercellular communication.

Concern about the health consequences of environmental EDCs has largely focused on male reproductive health, in contrast to the research on DES that was conducted primarily in prenatally exposed females. This may be a consequence, in part, of the difficulty in examining the reproductive tract of young females, an examination that requires invasive techniques such as those used to screen DES-exposed females at puberty or later. Since only the grossest of genital tract defects in females can be ascertained in infancy, no population-based surveillance data on these hidden reproductive endpoints are available. In contrast, a large literature describes the etiology and epidemiology of male genital tract anomalies, outcomes that can be ascertained at birth or by the end of the first year of life. Furthermore, population studies of biomarkers of female reproductive health (such as serum hormone levels) have not been conducted until recently. In contrast, the literature on semen quality dates from 1929 when the hemocytometer (designed originally to count white cells) was first used to count sperm [19].

As early as the 1970s, authors expressed concern that environmental factors may be contributing to a decline in male reproductive function. For example, Nelson and Bunge found sperm concentrations in 1970–1973 to be markedly lower than those reported in 1951 [20] and concluded: "The overall decrease in the sperm concentration and the semen volumes would tend to incriminate an environmental factor to which the entire population has been exposed" [21]. This study, and other analyses of historical data on semen quality [22, 23], which found declines in sperm density and suggested environmental causes, went relatively unnoticed. However, a 1992 analysis by Carlsen and colleagues, the most extensive such analysis published up until that time [24], was widely discussed, criticized, and reanalyzed, reflecting the considerable controversy concerning the safety of environmental chemicals in which the sperm decline issue had become imbedded. The link between a decline in sperm count and environmental factors was made more plausible by experimental evidence that developmental exposure to doses of estrogenic EDCs within the range of human exposure can permanently alter testicular function and decrease sperm production in laboratory animals [25, 26].

Sharpe and Skakkebaek noted in 1993 that decreasing sperm concentrations in Western countries were often paralleled by increases in the incidence of testicular cancer and male genital tract abnormalities. These authors suggested that since these reproductive endpoints share a common hormonal influence, they may result from a common cause and suggested prenatal disruption of Sertoli cell formation, a hormonally sensitive process [27]. Several authors [27–29] have hypothesized that these related trends might be the result of prenatal "over exposure" to endogenous estrogen, or possibly environmental estrogens. However, as these authors are careful to note, this is, as yet, a hypothesis to be tested in humans. Some authors have argued that it is unlikely that estrogenic compounds could have produced these trends. Other authors have questioned the validity of the trend analyses underlying this hypothesis [30, 31]. Here we review these trend analyses, beginning with semen quality and sperm density in particular, the measure of male reproductive function that has generated the greatest controversy to date.

2
Temporal Trends and Geographic Variability in Adverse Male Endpoints

2.1
Semen Quality

The question of a possible decline in semen quality was catalyzed in 1992 by the analysis of sperm concentration from 61 studies that had been conducted throughout the world and published between 1938 and 1990 [24]. This analysis, which used simple linear regression to model the changes in sperm concentration over time, concluded: "...reports published worldwide indicate clearly that sperm concentration has declined appreciably during 1938–1990".

On visual inspection these data [24] appear consistent with a model in which sperm density "levels off", or possibly increases slightly, after 1970 (See Fig. 1). This observation led several authors to question the use of a linear model [32, 33] and propose alternative models [34]. However, none of the models initially proposed accounted for the geographic distribution of these 61 studies, which varied considerably over the study period. For example, 11 of the 13 studies published before 1970 were conducted in the United States, while non-Western studies were few ($n=14$) and not published until 1978. Three analyses attempted to account for this unequal distribution of studies using different statistical models (see Table 1). Bahadur et al. [35] divided the studies into three strata, or "regions" (USA, Europe and elsewhere), and fitted separate linear regression models to each of these regions. The analysis by Becker and Berhane [36] was the first to reanalyze this data set using multiple regression methods. This latter analysis controlled for geographic region and type of study population, as defined by Carlsen (men of proven fertility or not). Becker's final model, which compared the slope for US studies to those from non-US studies, fitted the data somewhat better than did previous models.

Swan et al. [37] abstracted data from the studies analyzed by Carlsen in order to control for additional variables, such as age, abstinence time, method of spec-

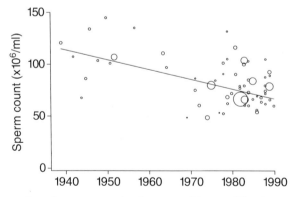

Fig. 1. Linear regression of mean sperm density reported in 61 publications 1938–1990 (data from [24])

Table 1. Global trends in sperm concentration: reanalyses of Carlsen et al. [24]

Author (year)	Model	Covariates	Adjusted R^2	Geographic region(s)	Results
Carlsen et al. [24]	Linear	None	0.36	All	Slope $= -0.93$ (<0.0001)
Olsen et al. [34]	Quadratic	None	0.42	All	Quadratic term stated to be significant (no probability given)
	Spline	None	0.44	All	Slopes pre-and post-1978: -1.40 and $+0.89$ respectively
	Step	None	0.45	All	Mean counts pre- and post-1964 109.7 and 72.7
Bahadur et al. [35]	Linear	None	0.71	USA only	$p<0.0001$ (no slope given)
	Quadratic	None	0.69	USA only	$p<0.0001$ (quadratic term not significant)
	Linear	None	0.04	Europe only	$p=0.32$
	Linear	None	0.00	Other countries only	$p=0.46$
Becker and Berhane [36]	Linear	Region, region $\times$ year, and population*	0.51	Single model; USA and non-USA	Significant slope for USA (-1.30); non-USA not significantly different from USA
	Linear	None	0.72	USA only	Slope -1.30 ($p<0.002$)
	Linear	None	0.02	Non-USA only	Slope -0.33 (not significant).
	Linear	None	0.07	Europe only	Slope -0.47 (not significant but not significantly different from USA)
Swan et al. [37]	Linear	Multiple**	0.80	Single model: USA, Europe, Other	Slope for USA (-1.50), Europe (-3.13) significant; slope for Other countries ($+1.56$) not significant
	Quadratic	Multiple**	0.78	Single model: USA, Europe, Other	Curvature (quadratic terms) not significant for any region
	Spline	Multiple**	0.79	Single model: USA, Europe, Other	USA slope pre-and post-1970similar (-1.52 and -1.47)
	Step	Multiple**	0.72	Single model: USA, Europe, Other	Post-1970 means (US, Europe, Other) significantly lower than pre-1970 mean in USA ($67.7, 75.0, 58.3$ vs 106.7)

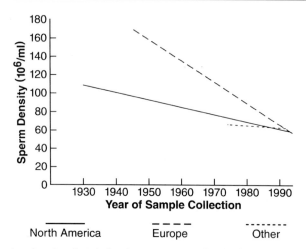

Fig. 2. Temporal and regional variation in mean sperm density (1938–1990) controlling for abstinence time, age, method of specimen collection, fertility status, study goal, and interaction of region and study year [37]

Table 2. Studies of local trends in semen quality: men from infertility clinics

Author (year)	Location	Years	Results
Bendvold et al. [237]	Stockholm, Sweden	1956–1986	Decrease in total count and normal morphology[a]
Adamopoulos et al. [238]	Athens, Greece	1977–1993	Decrease in total count and volume[a]
de Mouzon et al. [239]	France	1989–1994	Decrease in concentration by year of birth after 1950
Menchini-Fabris et al. [240]	Pisa, Italy	1975–1994	Decrease in concentration and motility
Vierula et al. [241]	Two cities, Finland	1967–1994	No change in concentration and count, decrease in volume
Berling and Wolner-Hanssen [242]	Lund, Sweden	1985–1995	Increase in concentration, motility, normal morphology; decrease in volume
Rasmussen et al. [243]	Odense, Denmark	1990–1996	No change in concentration or volume by year of birth
Zheng et al. [244]	Four centers, Denmark	1968–1992	Decrease in concentration by year of birth after 1950 (not before)
Younglai et al. [245]	11 centers, Canada	1984–1996	Decrease in concentration
Tortolero et al. [246]	Merida, Venezuela	1981–1995	Decrease in % men with high concentration ($>200 \times 10^6$/ml)

[a] Sperm concentration not reported.

Table 3. Studies of local trends in semen quality: unselected or fertile men

Author	Location	Years	Source	Results
Wittmaack and Shapiro [44]	Wisconsin, USA	1978–1987	Candidate donor	No change in concentration or motility
Auger et al. [39]	Paris, France	1973–1992	Candidate donor	Decrease in concentration, motility and normal morphology
Fisch et al. [42]	Three cities, USA	1970–1994	Pre-vasectomy	No change in concentration (increase for two cities)
Bujan et al. [247]	Toulouse, France	1977–1992	Candidate donor	No change in concentration
Paulsen et al. [43]	Washington, USA	1972–1993	Candidate donor	No change in any parameter
Van Waeleghem et al. [41]	Ghent, Belgium	1977–1995	Candidate donor	Decrease in concentration, motility and morphology
Irvine et al. [40]	Edinburgh, Scotland	1984–1995	Candidate donor	Decrease in all parameters by birth year
Bonde et al. [248]	Three cities, Denmark	1986–1995	Occupational cohort	Decrease in concentration and count with year of birth
Gyllenborg et al. [249]	Copenhagen, Denmark	1977–1995	Candidate donor	Increase in concentration, decrease in motility

imen collection, and percent of men with proven fertility, factors that might account for the observed decline, in whole or in part. In this analysis, studies were grouped into those from the US, Europe/Australia, and other (non-Western) countries. This analysis employed multiple regression analyses that included the entire data set in a single model, while allowing for different slopes in each region, a more powerful statistical technique than fitting separate models to each region. The result was a linear model that fitted the data better than those previously published and which demonstrated a significant decline, on average, in the United States (about 1.5% per year) and in Europe/Australia (about 3% per year) but not elsewhere (Fig. 2). Nonlinear models such as quadratic and spline did not improve this fit. Restricting the analysis to studies published between 1970 and 1990 demonstrated similar declines in mean sperm density in the US and Europe/Australia to those seen for the entire study period [38].

Variability between areas in the rates at which mean sperm density declined was demonstrated in these reanalyses. Heterogeneity is also demonstrated in the many local trend studies that have been published since 1992. Some of these have demonstrated declines in sperm density [39–41] within cities or countries, while others have found no decline or even slight increases in the last 20 years [42–44]. We have separated those studies conducted on men recruited through infertility clinics (Table 2) and those including candidate semen donors or other men unselected with respect to fertility (Table 3). These data demonstrate considerable temporal and geographic variation in sperm density and other semen parameters. Determining how much of this variation is attributable to environmental

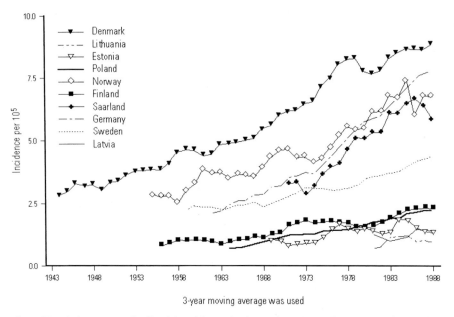

3-year moving average was used

Fig. 3. Trends in age-standardized (world standard population) incidence rates of testicular cancer [47]

factors and how much to methodological differences between studies, or true non-environmental differences (e.g., ethnicity or smoking), requires coordinated multinational studies of the kind we describe in our conclusions.

Have decreases in sperm concentration been accompanied by decreases in fertility? While one might expect sperm count to be directly correlated with fertility, this relationship is surprisingly complex. It is, first of all, complicated by the measure of fertility that is used. The simplest, and most uniformly recorded measure, number of births per population unit (e.g., among couples of reproductive age) reflects multiple factors, including voluntary delay of childbearing and contraception. For this reason, the reported decreases in fertility in many parts of the world that utilize this statistic [45] are unlikely to be informative about trends in true reproductive potential. On the other hand, a relationship between time to pregnancy (TTP), a more sensitive measure of reproductive capacity, and sperm density has been demonstrated in a recent Danish study [46]. This study found a significantly longer TTP among men with sperm density below 40×10^6/ml. Thus a trend towards decreasing sperm density, particularly one that implies a substantial increase in the proportion of the male population with sperm density below 40×10^6/ml, would be expected to impact fertility.

2.2
Incidence of Testicular and Prostate Cancer

Testicular cancer incidence has increased significantly for at least the past 20 years in most of the Caucasian populations that have been studied. For example, between 1943 and 1989 testicular cancer incidence increased by 2.3–3.4% annually in Nordic countries, where rates are high, particularly in men born after 1950 and in men aged 25–34 [29, 47, 48] (see Fig. 3).

This trend in testicular cancer rates has been particularly marked in Denmark, where the incidence of both seminomas and nonseminomas has been rising for several decades [49].

Data from the United States show an overall increased incidence between 1974 and 1994. For white males the increase was similar to that seen in the Nordic countries, about 2.4% per year (3.2 per 100,000 to 5.6 per 100,000), with little change seen among African-American males, in whom testicular cancer rates remain very low (<1 per 100,000) [50]. For example, in Alameda County, California, incidence in whites increased 71% between 1964 and 1987 (from 3.1 to 5.9 per 100,000) while only a slight increase was seen in African-Americans (0.0–0.7 per 100,000) [29]. Recently, increased risks of both seminomas and nonseminomas have been reported in Canada, particularly among cohorts of men born after 1950 [51]. Trends in Asian men, in whom testicular cancer rates are generally low, are inconsistent, with no change seen in China, India, or Chinese in Singapore, but increases among Japanese living in Japan and Hawaii [29]. Thus, both incidence rates and trends in testicular cancer rates appear to vary with both geographic region and ethnicity.

As with testicular cancer, prostate cancer rates showed a steady increase from the 1960s through the mid-1980s. However, once the use of the prostate specific antigen (PSA) test became widespread in the mid-1980s, trend data became difficult to interpret. Prostate cancer rates are highest in Northern Europe and North America. Unlike testicular cancer, incidence and mortality in the United States are highest among African-Americans [52]. Why African-Americans show a markedly higher incidence of prostate cancer and a markedly lower incidence of testicular cancer relative to Caucasians remains unknown.

2.3
Incidence of Hypospadias

An increase in the incidence of hypospadias was reported in 1975 by the newly instituted Norwegian birth defects monitoring system. While the number of cases in this report was small [53], a continuation of this initial increase was reported subsequently [54]. Czeizel [55] examined rates of isolated hypospadias in Hungary between 1971 and 1983 and found that the incidence had increased significantly ($p < 0.01$). Matlai and Beral [56] examined rates of malformations reported at birth and found a significant rise in cryptorchidism, hypospadias, and hydrocele between 1969 and 1983 in England and Wales (all p-values <0.001). A recent study in the United States reported a doubling of hypospadias rates between 1970 and 1993 (from 2 per 1000 to 4 per 1000) [57]. This increase was more marked among the most severe cases, which are least likely to be subject to diagnostic bias. Recent data from the International Clearinghouse for Birth Defects Monitoring Systems for 1960–1990 [54] show a leveling off, or a decline, in hypospadias rates in Hungary, Sweden, England, and Wales, beginning in the early 1980s. More recently Paulozzi [58] reviewed international trends through 1997 and found considerably variation within and between regions over time. He also noted that the incidence of hypospadias had leveled off, or declined somewhat, in the mid-1980s in many countries in which a rise had been seen previously. While these trends in hypospadias are less variable than those reported for sperm density and testicular cancer, the populations in which trends in hypospadias (and cryptorchidism) have been examined are quite homogeneous, limited, primarily, to Caucasian (European and North American) populations.

2.4
Incidence of Cryptorchidism

Rates of cryptorchidism are difficult to compare geographically or temporally because of their sensitivity to such factors as age at diagnosis, diagnostic practice, and age at orchidopexy. Nevertheless, it is notable that substantial increases have been reported in many countries in Northern Europe and the US since the mid-1950s. These trends are consistent with those for testicular cancer, for which cryptorchidism is a significant risk factor. In 1984 Chilvers et al. [59] reported an increase in the incidence of cryptorchidism in England and Wales, most marked among boys under ten years. Using hospital discharge data, the authors estimated that the cumulative incidence of cryptorchidism by age 15 had risen from 1.4% to 2.9% between 1952 and 1977. These authors suggested that a trend towards earlier (and possibly unnecessary) orchidopexy might have contributed to this increase. The incidence of cryptorchidism was assessed at three months of age among boys born in the Oxford area, using studies of similar design, at three points in time. The incidence among infants born 1984–1985 (1.58%) [60] was somewhat higher than that reported by Scorer [61] for births in the 1950s (0.96%). When this study was repeated for births in 1984–1988 the incidence was further increased (1.85%) [62], suggesting a doubling of risk between the 1950s and the late 1980s. In Scotland the number of hospital discharges for cryptorchidism among of boys less than 15 years of age increased substantially between 1961 and 1985. Data from the International Clearinghouse for Birth Defects Monitoring Systems [54] indicate that between 1974–1988 rates of cryptorchidism had increased throughout Scandinavia, with the exception of Sweden. In a recent review Paulozzi [58] examined more recent data, both published and unpublished, to update these analyses. He noted that in several countries in which incidence of cryptorchidism had been increasing, rates have remained constant (or in some cases declined) since the mid-1980s.

3
Adverse Male Endpoints in Relation to Exogenous Hormone Exposure

In this section we review the available literature on adverse male endpoints in relation to prenatal exposure to exogenous hormones, including DES. Often these exposures were ascertained more than 25 years after their use, making identification of specific hormones difficult. While some of these studies were

Table 4. Prenatal exposure to DES and other hormones; risk of testicular cancer

Author	N(cases)	N(controls)	RR
Henderson et al. [250]	78	78	5.0[a] $(p=0.11)$ 4.3[b] $(p=0.01)$
Schottenfeld et al. [251]	190[c]	166 (hospital) 143 (neighborhood)	1.8[c] $(p=0.20)$ 2.0[c] $(p=0.17)$
Depue et al. [63]	108	108	8.0[d] $(p=0.02)$
Brown et al. [64]	225	213	0.8[e] (NS)
Gershman and Stolley [65]	79	79	Two exposed cases vs 0 exposed controls[c] (NS)

[a] Hormone treatment not further specified.
[b] Hormone treatment or excessive nausea.
[c] Drug use for bleeding, spotting and/or threatened abortion (DES, other hormones or unknown).
[d] Exogenous hormones during first trimester of index pregnancy.
[e] Exogenous hormones during the index pregnancy.

able to estimate associations with DES specifically, others grouped exposures into such headings as "hormones to prevent miscarriages". Thus, estimates of these associations are often imprecise.

3.1
Testicular Cancer

Because of its low incidence, testicular cancer studies are most often of case-control design. In addition, most studies that examined risk associated with exogenous hormone use defined prenatal exogenous broadly. These studies are summarized in Table 4 where exposure definitions are footnoted. Depue et al. [63] found an eightfold increased risk associated with exogenous hormone use in the first two months of pregnancy $(p=0.02)$. History of a hormonal pregnancy test (estrogen-progestin) was the most frequent source of exogenous estrogen exposure. Overall, the associations between prenatal hormone exposure and testicular cancer tend to be positive, but studies are of limited statistical power and some studies found no association [64, 65].

3.2
Male Genital Tract Abnormalities and Semen Quality

Several DES-exposed cohorts and appropriate unexposed controls have been surveyed to examine the relationship between DES and both genital tract malformations and semen quality. Most of these studies demonstrate reproductive alterations in DES sons (Table 5).

The literature is less consistent with respect to exposure to oral contraceptives, or exogenous hormone exposure defined more generally. In an international study from eight birth defect monitoring programs, a significant associ-

Table 5. Prenatal exposure to DES; cryptorchidism and other genital abnormalities

Author	DES+	DES–	Cryptorchidism (DES+ vs DES–)	Other urogenital abnormalities (DES+ vs DES–)	Impaired semen quality
Henderson et al. [252]; Cosgrove et al. [253]	225	111	3 vs 1 (no p-value given)	Difficulty passing urine (p=0.0003) and penile abnormalities (p=0.017)	Not studied
Andonian and Kessler [254]	24	24	Not reported	13% vs 8% (NS)	17% vs 20% (NS)[a]
Gill et al. [255]	308	307	5.5% vs 0.3% (p<0.005)	31.5% vs 7.8% p<0.005[b]	18% vs 8%[a] (p<0.05)
Driscoll and Taylor [229][c]	31	Two control groups: 31 each	Not reported	p<0.001[d]	Not applicable
Leary et al. [256]	265	274	Not reported	No significant differences	No significant differences
Shy et al. [257]	51	29	8% vs 0% (p=0.07)	35% vs 4% (p=0.0006)[e]	21% vs 0% (p<0.025)[f]
Wilcox et al. [258]	253	241	Not reported	15% vs 5% (p<0.01)[g]	Not reported

[a] Severe pathological changes (Eliasson score >10).
[b] One or more of: epididymal cysts, hypoplastic testis, microphallus.
[c] Autopsy findings in male perinates exposed to estrogens with or without progestins.
[d] Prostate abnormalities; high ratio of Leydig cells to spermatogenic cells in the testis.
[e] Any genital abnormality except varicocele by physical exam.
[f] Only poor forward progression more frequent in exposed men.
[g] Significantly higher rate of abnormalities among men exposed before week 11 of gestation.

ation was seen between hormone therapy during pregnancy and hypospadias (OR=2.8, 95% CI 1.2, 6.9) [66]. In Hungary a significant correlation (R=0.85) between the sale of progestagens and isolated hypospadias was reported [55] but this was not confirmed in a later Hungarian study [67]. Rothman and Louik [68] found an increased risk for cryptorchidism for oral contraceptive (OC) use within one month of conception in comparison to pregnancies with no OC use within 36 months of conception (15.3 per 1000 vs 8.1 per 1000). These authors found no excess (or a possible decrease) in risk of hypospadias associated with recent OC use. In a matched case-control analysis Depue [69] found a relative risk of 2.8 (p=0.03) for cryptorchidism associated with exogenous estrogen use (most often DES), but no increased risk for use of progestins or oral contraceptives (OCs) during pregnancy. On the other hand, several authors have found little or no connection between any exogenous hormone use during pregnancy and risk of cryptorchidism [70, 71] or hypospadias [72]. In fact, the possible effects of early exogenous hormonal exposures have not been well studied and, given the large potential for exposure, this is an important research need.

4
Adverse Male Endpoints in Relation to Endogenous Hormone Exposure

While few data are available on pregnancy hormone levels in relation to adverse male endpoints, the literature supports relationships between these endpoints and several other pregnancy-related factors. There are also data suggesting that these factors, in turn, are reflective of the hormonal milieu of pregnancy. Birth order (and the correlated variable, maternal age) is quite consistently associated with adverse male reproductive outcomes, with risks of testicular cancer (particularly seminomas), cryptorchidism, and hypospadias greater among first birth. This elevated risk may be related to a higher level of estradiol [73–75], particularly unbound estradiol [76], in first compared to later pregnancies. The increase in risk was, in some studies, most marked among women experiencing their first birth at an older age. This finding may reflect a higher level of free estradiol in pregnancy among women delaying their first pregnancy until later in life. Bernstein et al. found free estradiol to be similar in young nulliparas and parous women, but to be higher among women having their first birth after age 34 [77]. Increased testosterone levels seen in the first trimester of first pregnancies (compared to later pregnancies) may also play a role in the increase in risk of these adverse male outcomes in first births [75].

Most studies also support an increased risk of testicular cancer, hypospadias, and cryptorchidism in association with low birth weight (and the correlated variable, prematurity). As discussed below, it is likely that this increase in risk reflects intrauterine growth retardation rather than preterm birth, per se.

We attempted to examine risks of these adverse male endpoints in relation to ethnicity. However, ethnic categories are themselves quite heterogeneous, and vary from one region to another, making it difficult to draw conclusions about ethnic variation. The strongest and most consistent association with ethnicity is seen for testicular cancer. The incidence of testicular cancer varies about ten-fold, worldwide, from 0.7 per 1000 in African Americans to nearly 1% (8.4 per 1000) in Denmark. This may be related to a lower serum testosterone level during the first trimester of pregnancy that has been reported in whites compared to African-Americans 78 which would suggest a role for reduced testosterone levels during early pregnancy in the etiology of testicular cancer and possibly cryptorchidism. The negative association for fetal serum testosterone and subsequent testicular cancer suggested by this comparison of testicular cancer rates in African-Americans and Caucasians appears inconsistent with the positive association suggested by the comparison of first with subsequent births described above. This suggests that any association between prenatal testosterone exposure and testicular cancer risk may be modified by ethnicity. In any case, the data presented here are inadequate to conclude that association between ethnicity and testicular cancer is, in whole or part, attributable to differences in endogenous hormone levels.

4.1
Testicular Cancer

First-born children have been found to be at increased risk of testicular cancer, though not consistently. The influences of birth order and maternal age appear to interact. The picture is further complicated by different patterns of association for seminomas and nonseminomas. Møller and Skakkebaek [79] examined this question and found that for seminomas the increased risk conveyed by older maternal age was most marked for first births (OR=4.1, 95% CI 1.1–14.6). However, this pattern was not seen for nonseminomas. Consistent with this finding, Akre et al. [80] found the associations between maternal age and the risk of seminomas and nonseminomas to be opposite; risk increased 1.13 per 5 years of age for seminomas but decreased 0.86 per 5 years for nonseminomas ($p=0.054$ for difference in trend). Prener et al. [81] found the risk of testicular cancer to be greater in first births; adjusted relative risk (ARR) for births of order four or higher relative to first births was 0.3 (95% CI 0.1–0.8). This association was particularly strong for seminomas (ARR=0.1, 95% CI 0.02–0.9). On the other hand, Westergaard et al. [82] found the association between testicular cancer and birth order to be similar for seminomas and nonseminomas, while Moss et al. [83] found no association between testicular cancer risk and either birth order or maternal age.

Most studies that have examined testicular cancer risk in relation to low birth weight and prematurity found one or the other of these correlated variables to be associated with increased risk. Brown et al. [64] examined the joint effect of these variables and found a 12-fold increased risk for testicular cancer among infants weighing 5 lb. or less (95% confidence interval (CI 2.8–78.1)), a risk that was greater for shorter gestation (RR=16.9, 95% CI 2.3–346.7). However, fetal age at birth (gestational age) was not associated with risk of testis cancer independent of birth weight. Similarly, Møller and Skakkebaek [79] found significantly increased risk for births less than 3000 and 2500 g (OR=1.5 and 2.6 respectively) but no association with gestational age. Depue [69] found a threefold increased risk of testicular cancer among infants weighing less than 6 lb. (95% CI 1.2–8.4). This study did not examine prematurity. Gershman and Stolley [65] found a significant tenfold increase in risk of testicular cancer for premature births, but did not examine birth weight. Two studies that examined seminomas and nonseminomas separately [79, 80] found the association of these variables to be stronger with nonseminomas than seminomas.

Risk of testicular cancer has also been studied in twins compared to singleton pregnancies. While little difference in testicular cancer incidence was seen relative to the general population for monozygotic twins, among dizygotic twins there was a 60% excess incidence ($p<0.05$). The ratio of observed to expected cases was greatest in men less than 35 years old (O/E ratio=2.6, 95% CI 1.1–4.2) [84]. This may be related to the elevation in maternal hormonal levels in twin pregnancies relative to singleton pregnancies, with the greatest differences seen in dizygotic pregnancies. These increases, summarized by Braun et al. [84], include a doubling of estrogen, human chorionic gonadotropin, and human placental lactogen.

With respect to prostate cancer, there is no evidence that either birth order or maternal age is related to risk [85]. In relation to low birth weight and prematurity, the pattern for prostate cancer appears to be opposite to that seen for testicular cancer. Ekbom et al. [85] found no premature births among 80 cases, compared to 6% of 196 controls ($p=0.02$) although mean birth weight was higher, though not significantly, among cases. Ross and Henderson [86] suggest that the dramatic (approximately 30-fold) difference in risk of prostate cancer between African-American and Japanese and Chinese men may be due, in part, to differences in androgen secretion and metabolism. For example, higher testosterone levels have been found in maternal serum of African-Americans [78], as well as in young adult males [87].

4.2
Cryptorchidism

Depue [69] suggested that elevated levels of free estradiol, particularly early in the first trimester, may produce hypoplastic testis, a risk factor for cryptorchidism [88]. Bernstein et al. [89] reported a significant 20% elevation in free estradiol in cryptorchid males compared to normal controls. Burton et al. [90], on the other hand, reported a 30% lower estradiol level in cases, though this was not statistically significant, and free estradiol was not examined, so these findings are not necessarily inconsistent. Key et al. [75] found serum testosterone in case mothers to be 25% lower in gestational weeks 6–14 ($n=18$, $p=0.06$), but 22% higher in weeks 15–20, though this latter group was small ($n=10$, $p=0.18$). This suggests that failure to control for gestational age at the time of serum sampling may account for some of the differences between study results.

Consistent with the findings on testicular cancer, a higher risk of cryptorchidism has been seen for first births [91, 92] though this too has not been seen consistently. Møller and Skakkebaek [79] reported a strongly decreased risk with higher birth order (AOR=0.5 for birth order 4+, p-value for trend=0.03), and Hjertqvist et al. [92] reported a significantly increased risk for first births ($p<0.001$). On the other hand, Key et al. [75] found no association between birth order and risk of cryptorchidism.

Low birth weight is consistently associated with an increased risk of cryptorchidism, independent of gestational age. Møller and Skakkebaek [79] found the risk of cryptorchidism to be inversely correlated with birth weight, with odds ratios ranging from 0.4 for births of 4500+ g to 2.3 for births less than 2500 g (p-value for trend, 0.001). These authors found little association between cryptorchidism and gestational age. Similarly, Jones and colleagues [93] found low birth weight to be significantly associated with cryptorchidism, with no effect of gestational age after controlling for birth weight.

In one study, geometric mean testosterone during gestational weeks 6–14 was reduced 25% ($p=0.06$) in whites compared to African-Americans [75]. A threefold greater risk of cryptorchidism in births to whites compared to African-Americans was reported in Los Angeles [78], a difference consistent with the far greater incidence of testicular cancer in whites in the United States. On the other hand, a cohort study in New York found no significant differences in ethnicity-specific prevalence rates of cryptorchidism at three months or one year of age [94].

We have focused this discussion on the maternal environment. However, there are two findings suggesting a possible role for paternal influence. Sweet and colleagues [95] reported a greater incidence of scrotal and testicular anomalies of the scrotum or testes in fathers of cases of hypospadias compared to normal controls (35% vs 3%, $p < 0.001$). Sperm concentration was also reduced among case fathers, but numbers were small. More recently, Fritz and Czeizel [96] reported poorer semen quality in fathers of cases of hypospadias compared to controls; $p < 0.02$ for differences in sperm concentration, percent motile, and several morphological defects.

4.3
Hypospadias

Hypospadias, which arises between weeks 6–14 of human embryonic development, can be regarded as a mild form of incomplete masculinization, a process that is androgen driven, and thus plausibly linked to endogenous hormone levels during pregnancy. Few studies on hypospadias have examined the association with birth order. Kallen et al. [97] found no association between birth order and hypospadias overall, but a higher risk among women over 40 at first birth, consistent with findings of Hay and Barbano [98], and consistent with the increase in risk of testicular cancer among women having their first birth at an older age.

Kallen et al. [97] reported an approximately doubled incidence of low birth weight (under 2500 g) in cases of hypospadias from seven national registries. An increased risk for premature birth was seen only among infants weighing less than 2500 g. An association between hypospadias and low birth weight is strengthened by the recent finding from a study of monozygotic twins discordant for hypospadias; in 16 out of 18 twins, the hypospadias occurred in the twin of lower birth weight (mean difference in birth weight 498 g; $p < 0.01$) [99].

In the Collaborative Perinatal Project, which recorded diagnoses of hypospadias at birth hospital across the United States, rates in whites were somewhat elevated compared to African-Americans (8.1 per 1000 vs 6.9 per 1000) and Puerto Rican (5.2 per 1000) births. Based on data from The International Clearinghouse for Birth Defects Monitoring Systems (ICBDMS) rates of hypospadias in Asian populations (Tokyo and Szechwan) were reduced relative to those in the United States. However, comparisons across registries are highly problematic because of the range of diagnostic definitions that may be utilized.

5
Exposure in the Workplace and from Industrial Accidents

Exposure to a variety of chlorinated and brominated pesticides, organic solvents, and other probable EDCs in occupational settings and in communities exposed as a result of industrial accidents have been shown to alter male reproductive function. Findings in occupationally exposed cohorts are summarized

Table 6. Adverse male endpoint in relation to occupational exposure to EDCs

Exposure	Endpoints	References
Dibromochloropropane (DBCP)	Azoospermia and oligospermia Decreased motility and morphology Elevated FSH and LH	[15, 101, 102, 259]
	Deficit of male births	[103, 104, 260]
Ethylene dibromide (EDB)	Decreased sperm counts	[16, 102, 105, 106]
Chlordecone (kepone)	Oligospermia, decreased sperm motility	[17, 109]
Perchloroethylene (PCE)	Dose-related morphological changes	[110]
Carbaryl	Impaired semen quality	[107, 108]
Ethylene glycol ethers (EGE)	Decreased sperm counts Decreased fertility	[111, 112, 113]
TCDD (dioxin)	Reduced serum testosterone, increased LH Deficit of male births	[261] [262]
p-Nitrophenol (PNP)	Decreased sperm concentration Decreased percent motile sperm Increased serum LH	[263] [264]
4,4'-Diaminostilbene-2,2'-disulfonic acid (DAS)	Reduced serum testosterone Impotence	[117] [116]
Ethinyl estradiol	Gynecomastia, impotence	[265]

in Table 6. A dramatic example of an occupational exposure to a male reproductive toxicant was reported in 1977. Whorton and Meyer [100] found a history of infertility in male workers strongly linked to the manufacture of the nematocide dibromochloropropane (DBCP). This finding was confirmed and exposure-dependent effects on sperm counts, LH and FSH, and impaired sperm motility and morphology documented [100]. Similar findings were reported in DBCP-exposed occupational cohorts in Israel [101] and Hawaii [102]. More recently, a decreased sex ratio has been reported in offspring of exposed workers who recovered sufficient sperm function to conceive [103, 104].

Other chlorinated pesticides, including ethylene dibromide [16, 102, 105, 106], carbaryl [107, 108], and chlordecone [17, 109] have been shown to impair semen quality among cohorts exposed occupationally. Organic solvents including perchlorethylene [110] and ethylene glycol ethers (EGE), a class of organic solvents widely used in semiconductor manufacture, shipyard paints, and other industrial uses, have been shown to be spermatotoxic in several occupational settings [111–113]. These solvents have also been shown to increase time to pregnancy [114].

Occupational exposure to polyvinyl chloride (PVC), but not other plastics, has been associated with increased risk of testicular cancer. In a case control study including 148 cases, the odds ratio for PVC exposure was 6.6 (95% CI 1.4–32). The association was strongest in men without a history of cryptorchidism, and was limited to seminomas [115].

The stilbene derivative 4,4'-diaminostilbene-2,2'disulfonic acid (DAS) is produced in the manufacture of whitening agents and laundry detergents and is structurally similar to diethylstilbestrol. A study in 1981–1983 of 39 DAS workers reported low testosterone in 37% of men tested, as well as impotence [116]. A follow-up study found significantly reduced testosterone (both total and free) in both current and former DAS-exposed workers compared to unexposed controls [117]. Thus, it is clear that at the high levels present in the work place, a variety of EDCs have the capacity to alter significantly male reproductive function.

6
Plausibility of Adverse Effects of EDCs: Mechanisms of Gonadal and Accessory Reproductive Organ Differentiation

The determination of gonadal sex in mammals is dependent on the presence of different sex chromosomes in males and females. Specifically, a gene on the Y chromosome in genetic males interacts with genes on other chromosomes to produce a signal that leads to the development of the testes. Gonads in embryos whose cells lack a Y chromosome, or whose Y chromosome lacks this gene, develop into ovaries [118]. The germ cells migrate into the nondifferentiated gonad, and differentiation commences by week 6 of gestation in human males and day 12 in male mice and rats [119, 120]. In males, the seminiferous cord mesenchyme differentiates into Sertoli cells, which form intimate contact with, and both support and regulate, the undifferentiated germ cells [1]. Sertoli cell proliferation begins on embryonic day 14 and ends between the second and third week of postnatal life in the mouse [121, 122].

The initial period of differentiation into a testis or ovary is not dependent on gonadal steroids in either males or females [123, 124]. Thus, mammals are unlike fish, amphibians, and reptiles, where gonadal differentiation (primary sex determination) is dependent on environmental factors, such as temperature, that appears to operate through altering gonadal steroid levels in some species [125, 126]. This fact makes it plausible that in lower vertebrates environmental EDCs can alter primary sex ratio in a population by directly influencing gonadal differentiation, and the resultant sex-reversed offspring are all fertile and "normal" in other respects [126, 127]. In contrast, in birds and mammals, high levels of hormones can interfere with normal gonadal differentiation, but sex reversal (for example, development of functional ovaries in genetic males) does not occur. The consequence of endocrine disturbance during gonadal development for birds and mammals is thus infertility and, with exposure to potent chemicals such as DES, quite often reproductive organ tumors [128–131].

While a switch in the course of differentiation of the gonads does not occur in mammals due to environmental factors, it is possible for EDCs to influence the early processes of testicular differentiation. There is a direct relationship between the number of Sertoli cells and the number of germ cells; the number of Sertoli cells is correlated with daily sperm production in rats [132]. This has led to the hypothesis that prenatal exposure to estrogen may permanently reduce Sertoli cell numbers, resulting in subsequent decreases in adult sperm output [27]. This could occur, for example, via an effect on FSH or thyroid hormones. It has been shown that, during testicular differentiation, FSH levels are positively related to Sertoli cell number and subsequent testicular sperm production [133, 134]. In addition, an experimental decrease in circulating thyroid hormone during testicular differentiation results in an increase in adult testicular sperm production [135]. There are thus multiple endocrine pathways via which testicular function can be influenced by developmental exposure to EDCs, and these developmental effects are permanent. While this hypothesis remains unexamined in humans, the animal data reviewed below suggest that such a relationship is plausible and worthy of study.

There are a number of experimental studies in laboratory animals relating developmental exposure to EDCs to changes in testicular function in adulthood. For example, a single administration of a low dose (50 ng/kg to 1 µg/kg of body weight) of dioxin to pregnant female rats on gestational day 15 resulted in altered sexual differentiation in male and female offspring [136]. The lowest dose of dioxin tested (50 ng/kg body weight) resulted in a concentration in fetal organs of approximately 5 parts per trillion (5 pg/g tissue). Since this dose is within the range of human exposure [137], these results are environmentally relevant. A single administration of dioxin to pregnant female rats produced a decrease in circulating testosterone and in anogenital distance in male offspring at birth. Effects on adult sex behavior in male offspring included changes in mounting, intromitting, number of ejaculations and latency to ejaculation, as well as the exhibition of the female sexually receptive posture (lordosis). Some of these effects were observed at prenatal doses as low as 50 ng/kg. In addition, monotonic dose-related decreases in testis and epididymis, as well as seminal vesicle and ventral prostate, weights were observed, as were decreases in daily sperm production and numbers of ejaculated sperm in males, although there was no effect on fertility [136, 138–142]. Changes in development of the reproductive system in female rat fetuses have also been observed following one-time maternal administration of dioxin on gestation day 15 of the same low doses (0.05–1.0 µg/kg) [143, 144].

Developmental exposure to low doses of estrogenic chemicals has been related to a permanent decrease in testicular function in rats and mice. Sharpe et al. [145] fed octylphenol (an alkylphenol surfactant used in detergents, plastics, and many other products) to pregnant and lactating rats and reported a decrease in daily sperm production in the adult male offspring. Specifically, developmental exposure to a dose of octylphenol estimated to be within the range of 100–400 µg/kg/day, while not influencing body weight of adult males, significantly reduced the ratio of testes/body weight, as well as weight of kidneys and ventral prostate, relative to untreated males. In addition, daily sperm production was significantly reduced. Moreover, administration of only 20 µg/kg/day dose of octylphenol to pregnant mice during days 11–17 resulted in a small (15%), but significant, decrease in daily sperm production in adult male offspring [146].

The effect of plasma steroid binding proteins on the bioactivity of octylphenol is unclear at this time. This is an important issue, since in vitro tests of potency of estrogenic EDCs are typically conducted using culture media that contains only a very small amount (or none) of these plasma proteins, which act in vivo to regulate the uptake into cells of steroids and lipophilic EDCs [147]. There are contradictory reports concerning the interaction of octylphenol with plasma binding proteins. There is a report that octylphenol shows low binding to plasma proteins and thus, when tested in vivo, an increase in bioactivity relative to estradiol [148]. In contrast, Nagel et al. [146] reported increased binding of octylphenol in human serum and thus decreased bioactivity relative to estradiol when tested in vivo.

Dodds and Lawson [4] synthesized the estrogenic chemical bisphenol A (BPA) in 1936. They showed by administration to ovariectomized rats that BPA and other chemicals with a structure similar to that of BPA had full estrogen agonistic activity (i.e., elicited a maximal response in the female reproductive system, similar to that elicited by estradiol). Subsequently, polymer chemists discovered how to create the polymer polycarbonate, and many other resin materials, by polymerizing BPA. Today, over two billion pounds of BPA are produced annually for use in lining metal food and beverage cans and to make many products, such as baby bottles, CDs, and food storage containers.

It is now recognized that exposure to BPA in the low ppb range can also occur due to eating canned foods [149] or drinking canned beverages [150]. It has been shown that BPA leaches out of plastic into food or beverages, particularly when heated [150, 151]. It has also been shown that leaching of BPA into water placed into the container occurs at room temperature [152] after polycarbonate is subjected to repeated washing and shows evidence of wear (scratching and discoloration). While new polycarbonate baby bottles release approximately 1 part per billion (ppb, which is a billionth of a gram per ml fluid) of BPA when heated, there is an increase in release of BPA (up to 6.5 ppb) from used polycarbonate baby bottles relative to new bottles [150, 153]. Human exposure resulting from use of a dental sealant made from BPA was approximately 13 ppb during the first hour after application of the sealant [154]. These findings provide the basis for determining the appropriate dose range to be used when conducting animal tests for BPA, a dose range that is relevant for human (environmental) exposure (expressed as dose per kg body weight). Whether exposures from typical use and disposal of environmental EDCs such as BPA are sufficient to alter human reproduction is a critical issue.

When doses of 2 and 20 µg per kg per day (2 and 20 ppb) of BPA were administered to pregnant mice for seven days, these doses altered development in the male and female offspring [26, 146]. The 20 ppb dose resulted in a 20% decrease in daily sperm production per g testes. The 2 ppb dose decreased weight of epididymides and seminal vesicles, and increased weight of prostate and preputial glands in the male offspring in adulthood [26, 146]. The 2 ppb dose also advanced the timing of puberty in female offspring, and the body weight at weaning was increased for both male and female offspring relative to controls [152]. Another study showed that prenatal exposure to a low (50 µg/kg/day) dose of BPA also permanently increased prostate size in mice and increased prostatic androgen receptors. In contrast, this low dose of BPA led to development of a significantly smaller epididymis. In addition, the developing urogenital sinus (from which the prostate differentiates) from gestation day 17 male mice was placed into culture medium with a physiological concentration of testosterone, and addition of 50 pg BPA/ml culture medium significantly increased the growth of the developing prostate and again increased prostatic androgen receptors, similar to effects of a 0.1 pg/ml culture medium dose of DES [155]. These findings, as well as others [156, 157], suggest that the dose of BPA of 50 µg/kg/day, previously estimated by the FDA to be "safe" for daily human consumption, may need to be lowered.

An important aspect of the rodent studies reviewed above is that estrogenic chemicals were administered to the mother, not directly to the offspring, and the offspring were seen to exhibit effects of maternal exposure. Thus, for example, during the initial period of testicular differentiation prior to birth, exposure to these chemicals would be across the placenta via transport through the maternal-fetal circulation. It is recognized that small lipophilic chemicals, such as DES, octylphenol, and BPA, readily cross the placenta, and DES remains in the fetal reproductive tract longer than it is retained in the pregnant female's organs [158]. BPA accumulates in pregnant mice when they are fed this chemical only once per day, so exposure of fetuses to such EDCs toward the end of pregnancy is being underestimated when estimates are based solely on daily dose.

Laboratory stocks of male mice subjected to selective breeding for high fecundity typically do not exhibit impaired fertility when exposed to low doses of estrogenic chemicals [159]. For example, when Newbold [130] examined the fertility of adult male CD-1 mice prenatally exposed to a range of doses of DES (0.1 – 100 µg/kg), fertility was only affected at the 100 µg/kg dose. Infertility can arise by multiple mechanisms and is a poor predictor of sperm count (particularly in these laboratory animals), as significant decreases in testicular or epididymal sperm reserves are not directly correlated with fertility [160].

7
Effects of Estrogenic Chemicals on Accessory Reproductive Organs: Interaction with Endogenous Androgens

The period of differentiation of the accessory reproductive organs varies from species to species. In long-gestation mammals, such as humans and pigs, differentiation of reproductive organs occurs primarily during prenatal life, although some aspects of sexual differentiation may occur after birth. In species with a very short gestation length, such as the mouse (18 – 19 days) and rat (21 – 23 days), gonadal and secondary sexual differentiation begins during the middle of gestation and continues into postnatal life for varying periods of time for different organs [161].

Between the seventh and eighth week of gestation in humans, stromal tissue in the developing testes derived from mesenchyme differentiates into the steroid secreting Leydig cells, which are located in the interstitial area between the seminiferous cords [162]. The Leydig cells rapidly begin secreting testosterone [163], the primary androgen secreted by testes in fetuses as well as adults. In addition, Müllerian inhibiting hormone (MIH), a glycoprotein, is secreted by the Sertoli cells and acts locally to suppress the development of the ipsilateral Müllerian duct [164, 165], and may also be involved in testicular development [166].

The secretion of androgen by the fetal testes is a necessary prerequisite to masculinization of the accessory reproductive organs, external genitalia, liver, kidney, and brain in males; in the liver, kidney, genitals, and brain, an important effect of androgen is the "imprinting" of enzyme systems that markedly influence tissue function and, more generally, homeostasis, throughout the remainder of life [1]. Without the secretion of androgen by the fetal testes, the secondary sexual characteristics are those typical of females. However, there is also evidence that estrogen plays an important role in the normal processes of masculinization of the brain and accessory reproductive system [161, 167, 168]. As a result, EDCs that interfere with the normal activity of either androgen or estrogen can disrupt the processes mediating the differentiation of accessory reproductive organs in males; these include the efferent ducts, epididymides, vas deferens, and seminal vesicles, which differentiate from the Wolffian (mesonephric) ducts under the local (not systemic) action of testosterone. In contrast, circulating testosterone mediates the differentiation of the external genitals and the urethra and associated glands, including the prostate and other periurethral glands; in rodents these include preputial glands which release pheromones involved in regulating social behaviors. These organs develop from tissues in the embryonic urogenital sinus and perineum and express the enzyme 5α-reductase during fetal (and in rodents also neonatal) life and differentiate under the control of 5α-dihydrotestosterone (DHT). Testosterone in the systemic circulation serves as the substrate for 5α-reductase rather than diffusion of testosterone from the ipsilateral testis, which controls differentiation of only the adjacent Wolffian duct.

The importance of the conversion of testosterone to DHT in selected target tissues is that DHT is as much as ten times more potent relative to testosterone [169]. Thus, while testosterone levels in the circulation are too low to induce differentiation of the Wolffian ducts without supplemental diffusion from the adjacent testis, 5α-reductase serves to amplify the action of testosterone in tissues by virtue of converting testosterone into the more potent DHT. Any chemical that interferes with the action of 5α-reductase will thus profoundly alter the differentiation of organs that express this enzyme during sexual differentiation [161].

This information is relevant to this discussion, since estrogen exerts an inhibitory effect on 5α-reductase and other steroid metabolizing enzymes, such as 17α-hydroxylase, which is involved in androgen biosynthesis [170]. When exposure occurs during fetal and neonatal life in rats, this effect is permanently imprinted in cells [171]. Both enzyme activity and steroid binding in male accessory reproductive organs are highly sensitive to the permanent, organizational effects of estrogen during fetal life [155, 167, 172].

In a number of studies involving BPA (e. g., see above), opposite effects of estrogenic chemicals have been observed on organs which differentiate from the urogenital sinus (prostate and preputial glands) relative to effects on organs which differentiate from the Wolffian ducts (epididymides and seminal vesicles). While such findings might appear confusing or contradictory, it is not unexpected that chemicals will exert different effects on embryonic tissues in which the hormonal mechanisms regulating differentiation differ.

Studying the (permanent) effects of fetal exposure to estrogenic chemicals on the brain (and behavior) and reproductive organs (preputial glands, prostate, seminal vesicles, epididymides, and testes) of mice increases insight into the multiple mechanisms by which EDCs act in different species. In mice, for example, preputial gland pheromones are involved in social communication between males and females [173] and influence aggressiveness between males [174, 175]. Preputial gland secretions pass through ducts which empty into the prepuce, which is specially adapted in mice for depositing urine marks [176]. The placing of these pheromones into a male mouse's environment via urine marking behavior is influenced by dominance status; dominant males mark at high rates, and subordination inhibits this behavior [177]. Fetal exposure to very low doses, within an environmentally relevant range, of estrogenic chemicals such as o,p'-DDT, increases preputial gland size and the rate at which male mice deposit urine marks in a novel environment [171], as well as inter-male aggressiveness [178]. This suggests that environmental estrogens may influence brain development and socio-sexual behaviors, as well as alter the functioning of organs involved in socio-sexual communication. In addition, these same chemicals can alter sperm development, sperm motility, and enzymes essential for fertilization. They may also alter the functioning of sperm after they are deposited in the female reproductive tract, for example by altering the functioning of the seminal vesicle and prostate glands that produce the components of seminal fluid [26, 179–181].

Differences and similarities in molecular systems across species are not well known, including species differences in the response to estrogens. However, at the receptor level, there is little evidence for significant differences between vertebrate species in the affinity of estradiol for estrogen receptors [182]. Therefore, if estradiol (and potentially other estrogenic chemicals) reaches estrogen receptors in target cells, the capacity to generate responses may not be very different among vertebrate species, although specific responses will vary not only between species, but within a species, as a function of age.

While the response of breast cancer cells and tissues in rodents and humans to estrogen appears to be very similar [183], comparisons of the sensitivity of different species to estrogenic chemicals are still needed. However, it is likely that individual differences in the response to estrogens as well as species differences are not mediated by differences in affinity of ligands for either the alpha or beta form of the estrogen receptor. Instead, such differences may be mediated by events that occur after a chemical has bound to the receptor, and these intermediate events are different in different target tissues [184].

8
Hormonally Mediated Variability in Mammalian Development: Relationship to Bioactive Serum Concentration of Natural Hormones and Endocrine Disruptors

Considerable individual variability in plasma steroid concentrations appears to be a normal part of pregnancy and is typically reported in studies of fetuses, e. g., humans [185], monkeys [186], rats [187], gerbils [188], and mice [189]. Since hormones regulate mammalian sexual differentiation (other than gonadal sex), it would be expected that sexual phenotype (morphology, physiology, and behavior) and their alterations by EDCs would also be highly variable. There is also considerable variability in the responsiveness of organs in different individuals to the same blood concentrations of hormones [159, 190]. This is not surprising, since, as described above, in addition to genetic factors, the response of tissues to hormones (including activity of metabolizing enzymes and hormone receptor numbers) is differentially "imprinted" in tissues as a consequence of variation in hormone levels during development.

Naturally occurring variation in the levels of testosterone and estradiol in both male and female mouse, rat, and gerbil fetuses (due to being positioned in utero between male or between female fetuses) leads to marked differences in a wide range of reproductive traits: morphology and functioning of internal and external genitals in males and females, and aggressive and sexual behavior in males and females [190–192]. The magnitude of the differences in hormone levels between male fetuses that develop in utero between two males (2M males) and male fetuses that develop between two females (2F males) in mice is significant. For example, on gestation day 18, one day after the initiation of prostate development, 2F males have significantly higher (101 pg/ml) total serum estradiol relative to levels in 2M males (78 pg/ml). 2F males average 0.2 pg/ml free serum estradiol, while 2M males average 0.16 pg/ml free serum estradiol. These findings suggest that a difference between 2M and 2F males in free serum estradiol of 0.04 pg/ml during sexual differentiation contributes to differences in the brain (and behavior), as well as virtually every aspect of reproductive function that has been examined. This hypothesis was verified in a study in which an experimental increase in free serum estradiol of 0.1 pg/ml during the last third of pregnancy in mice (via a maternal Silastic implant containing estradiol) resulted in enlargement of the prostate in male offspring, similar to the difference seen in comparisons of 2F and 2M mice [172]. Only males situated between a male and a female fetus (1MF males) were used in this study [167].

It is generally understood that the biologically active portion of total circulating steroid is the fraction that is not bound to plasma binding proteins [193]. These plasma-binding proteins are particularly important during pregnancy; in rodents, there is approximately a tenfold increase in circulating proteins that bind estradiol, relative to levels in adults. As a result, while the total amount of circulating estradiol is much higher in rat and mouse fetuses than in adults, the concentration of free estradiol, which is not bound to plasma proteins and thus can freely diffuse into cells and bind to receptors (the bioactive concentration), is actually very similar in fetuses and adults [194]. For example, in adult male mice, the total serum concentration of estradiol is approximately 5 pg/ml, with about 4% free, yielding a free estradiol concentration of approximately 0.2 pg/ml. In male fetuses about 0.2% of estradiol is free, due to the markedly higher levels of binding proteins relative to adults [195] leading to a free estradiol concentration of 0.2 pg/ml in male mouse fetuses [167].

The characterization of many estrogenic chemicals in the environment as "weak" and therefore not capable of causing effects at environmentally relevant doses does not take into consideration the very high potency and very low circulating free serum concentration of the reference hormone estradiol. The previously described experiment implies that the reference dose of an EDC sufficient to change the course of fetal development would be one that led to an increase in estrogenic activity in fetal serum equivalent in potency to an increase in free serum estradiol of between 0.04–0.1 pg/ml. To determine the biologically active concentration in fetal serum of an estrogenic chemical administered to a pregnant female would thus require knowledge of the affinity of that chemical for estrogen receptors expressed relative to estradiol, the total circulating concentration of the chemical, and the percent of the total concentration that is free in fetal serum. This information is available for a number of estrogenic endocrine disrupting chemicals and has proved very useful in predicting the bioactive dose in fetuses [146, 147]. The data presented above suggest that the disruption of normal development by low (ppb) concentrations, a common level of concentration for many estrogenic EDCs [147, 183, 196], is plausible.

9
Issue of Dose in Toxicological Studies: Importance of Non-Monotonic (Inverted-U) Dose-Response Curves

The issue of the sensitivity of fetuses to extremely small differences in circulating gonadal steroid levels is critical with regard to the very high doses traditionally used in toxicological studies. In the following two sections we will discuss data that demonstrate opposing effects of high and low doses of natural hormones, vitamins, and EDCs. These results suggest that it is inappropriate to assume that one can predict effects of low doses of EDCs from results obtained at high doses.

The dose level used in toxicological studies had not previously been considered a critical issue, since the dose-response curve had been assumed to be monotonic (e. g., assuming that response increases or remains constant with increasing dose). In sharp contrast, the endocrine literature provides abundant examples of hormones and hormone-mimicking chemicals with non-monotonic dose-response curves (e. g., the slope of the dose response curve changes with increasing dose). In this literature it is well known that dose is critical with regard to the effects that are observed. For example, there is experimental evidence from both in vitro and in vivo studies with natural and manmade estrogens and other EDCs (for example, dioxin, and aldicarb) that non-monotonic dose-response relationships can occur [167, 171, 197–201].

Normal development occurs within a limited physiological range for molecules that have a signaling function. For hormones and vitamins that operate via binding to receptors (such as thyroid hormone, vitamin A, and folic acid), either a deficiency or an excess (relative to normal physiological levels) can lead to adverse effects on development [202]. For example, retinoid (vitamin A) deficiency during pregnancy causes fetal death, or if fetuses survive, malformations in numerous organs (the vitamin A-deficiency syndrome) [203, 204]. Studies have also identified birth defects due to developmental exposure to high concentrations of a large number of natural and synthetic retinoids in numerous species [205, 206]. Thus, the dose-response curve for vitamin A forms a U-shaped function, with increased risk seen in response to vitamin A deficiency (below the normal range) and excess (above the normal range). The adverse effects of a developmental deficiency as well as a developmental excess of thyroid hormone are also well known [207]. The literature on deficiency and excess of hormones and vitamins has not yet been utilized in the design of developmental studies of EDCs that act as hormone-agonists or antagonists.

The issue of non-monotonicity of dose response curves and the use of very high doses to predict effects at low doses has only recently been raised in the context of the effects of EDCs [167, 202]. Therefore, the mechanisms that might give rise to inverted-U dose-response functions have not been well studied. There is, however, some understanding of the mechanisms that operate to reduce response to hormones at high doses. Factors such as down-regulation of receptors in the presence of high doses of a hormone may partially account for this phenomenon [208]. For example, different doses of estradiol administered to adult female rats via Silastic implants lead to up-regulation (an increase) of uterine estrogen receptors at a low dose and "down regulation" (a decrease) at a higher dose [209]. Other possibilities may involve the stimulation of response systems that are antagonistic to the initial response as saturation of that response occurs with the saturation of receptors [197]. There is considerable evidence for cross talk between estrogen and receptors for other steroids. These interactions do not normally occur when endogenous estrogens are present at physiological concentrations (due to low binding affinity for other receptors relative to the concentration of circulating estrogen), but are seen with addition of exogenous estrogens when total estrogenic activity in blood exceeds the physiological range [210]. The expectation is that cross talk between estrogen and receptors for other steroid hormones would result in inhibitory effects, rather than the responses typically seen in the binding of the appropriate steroid to its receptor.

The findings discussed above lead us to propose that there has been selection for hormones, vitamins, or other signaling molecules to operate within defined physiological ranges, with each individual exhibiting a unique "set point" within their range [190]. Disruption of normal development can occur either by increasing or decreasing the effective concentration of these signaling compounds possibly through the action of EDCs. As a result of developmental exposure to EDCs, the total hormonal activity (endogenous plus exogenous dose) would thus be different from that which would have occurred naturally. These chemicals may thus have the potential to disrupt the course of development.

10
Development of the Prostate:
Example of Opposite Effects of High and Low Doses

We focus here on the effects of estrogens on the prostate, since this is the most disease-prone organ in the human body. Prostatic buds form from the urogenital sinus just below the developing bladder. Buds begin forming from the urogenital sinus on gestation day 17 in mice and week 10 in humans; at birth, the mouse is similar to a human fetus at approximately week 17 of gestation [191]. The prostatic glands empty into the portion of the urethra surrounded by the prostate. Urogenital sinus mesenchyme, which shows estrogen as well as androgen binding [192, 211–213], regulates differentiation of the prostatic duct epithelium during fetal and neonatal life in mice [191, 214–217]. One mechanism through which estrogens affect epithelial cell differentiation in the prostate is through modulation of the action of androgen during prostate development.

In a recent study of developmental effects of natural and synthetic estrogens [167], evidence for an inverted-U dose response relationship in the prostate was obtained for both estradiol and DES. Specifically, as serum estradiol concentrations were increased in male mouse fetuses (via maternal Silastic implants) from 50–800% relative to controls, first an increase, and then a decrease in adult prostate weight was observed in male offspring. The 50% increase in free serum estradiol from 0.2 pg/ml (in controls) to 0.3 pg/ml. This 0.1 pg/ml (0.1 ppt) increase in free serum estradiol increased the number of developing prostate glands during fetal life and resulted in a permanent, sixfold increase in prostatic androgen receptors and a heavier (by 40%) prostate in adulthood relative to prenatally untreated males. In contrast, an eightfold increase in free serum estradiol in male fetuses significantly decreased adult prostate weight relative to the males exposed to the 50% increase in estradiol.

With maternal ingestion of 0.02, 0.2, or 2 ng of DES/g body weight/day, low dose stimulation of prostate growth in male mice occurred, while the high dose of 200 ng DES/g body weight/day (200 ppb) significantly decreased adult prostate weight [167]. This latter finding is consistent with numerous prior findings that exposure to a high dose of DES during development results in an abnormally small prostate in adulthood as well as a decrease in prostatic androgen receptors [167, 213, 218]. Stimulation of prostate development with a low dose (0.2 µg/kg/day) of DES and inhibition of development at a high dose (200 µg/kg/day) of DES, as well as low-dose stimulation and high-dose inhibition of androgen receptors has also been reported [155]. Damage to reproductive organs in females as well as males is seen with developmental exposure to high doses of synthetic estrogens [130, 219–222]. There was also a non-monotonic, inverted-U dose-response relationship between the maternal dose of DES and territorial behavior in male offspring [171]. Taken together, the above findings provide evidence that, at least in some cases, the effects at high doses of natural or manmade estrogens cannot be used to predict effects at low doses.

These findings demonstrate that the hypothesis of a monotonic relationship between dose and response for EDCs is not necessarily valid. This has particular significance with regard to assessing risks from exposure to low doses of EDCs. Thus, at least in some cases, the strategy currently used by the United States EPA and FDA of only testing chemicals at a few high doses and using a mathematical model to predict effects at much lower doses will lead to incorrect estimates of safe doses [202].

11
Developmental Exposure to Diethylstilbestrol (DES)

There is an extensive literature concerning the long-term effects of exposure to high doses of potent manmade estrogens, such as DES, during fetal/neonatal life in human [219, 222] and animal [130, 223, 224] studies, some of which is discussed above in Sect. 4. While the focus of most studies of DES has been the re-

Table 7. Abnormalities in males subsequent to prenatal exposure to diethylstilbestrol[a]

Abnormality	Organ	Outcome in humans (References)	Outcome in rodents (References)
Cancer	Testis	Positive [63, 250] or inconclusive [64, 65, 251]	Adenocarcinoma of the rete testes, interstitial cell carcinoma [266, 267]
	Prostate	No data available	Squamous cell cancer of dorsolateral prostate [268]
Other genital tract changes	Penis	Reduced size; hypospadias [252, 255, 258, 269]	Hypospadias [130, 270]
	Testis	Cryptorchidism, hypertrophy, capsular induration, epididymal cysts [68, 69, 255, 258, 269]	Cryptorchidism [128, 270, 271], epididymal cysts [128, 130]
	Prostate	Hyperplasia and metaplasia of prostatic ducts [229, 230]	Abnormal development, squamous metaplasia of prostatic and coagulating gland ductal epithelium [128, 167, 213, 231, 232, 268]
Adverse reproductive outcomes		Impaired semen quality and sperm concentration; impaired fertility inconsistent [254–258, 269]	Impaired semen quality and sperm concentration; impaired fertility [130, 270]

[a] Adapted from Newbold [130].

productive organs, effects on other tissues, such as bone, have been noted [225]. The overwhelming proportion of this research, at least in humans, has been conducted in the female offspring, with ongoing follow-up of thousands of DES daughters. Follow-up on the DES sons, on the other hand, has been limited to cohorts of only a few hundred exposed males. Therefore, while the consequences of prenatal DES exposure to females are now well established, the effects on human males are less certain. As seen in Table 7, the reproductive consequences of prenatal DES exposure are highly similar in mouse and man, and when the appropriate exposure window is considered, the effective doses are of the same order of magnitude. The high doses of DES used in animal studies have been selected to be similar to the doses that were administered to pregnant women during the 1950s and 1960s, and thus have clinical relevance. DES has a higher potency than estradiol during development [226, 227], due to limited binding by DES to plasma estrogen binding glycoproteins (sex hormone binding globulin in humans and alphafetoprotein in rodents) [147, 228].

Typically, estrogenic EDCs have a lower intrinsic estrogenic activity than DES [147, 171, 183, 196]. Since the very high doses of DES that have been used in prior animal studies were chosen for their clinical relevance, findings from these studies may not predict effects seen with environmentally relevant doses of estrogenic EDCs. However, the DES studies do provide valuable information linking effects in controlled animal studies with clinical outcomes of DES exposure in humans. Thus, information concerning exposure in early life to DES provides a link between animal and human responses to a man-made estrogen and identifies the types of clinical outcomes that should be examined when assessing effects from an EDC exposure.

Treatment of pregnant females with high doses of DES interferes with the action of Mullerian inhibiting hormone on Mullerian duct regression in male mice [128] and humans [229]. The utricular remnant within the prostate, which is of Mullerian origin, is enlarged and there is marked hyperplasia and metaplasia of prostatic ducts in the central zone of the adult prostate [229, 230]. As mentioned above, exposure of the neonatal mouse to high doses of DES has been shown to interfere with normal development of the prostate [213, 231, 232]. Squamous metaplasia of prostatic and coagulating gland (dorsocranial prostate) ductal epithelium in male mice and rats has also been reported after exposure to exogenous estrogen or estrogenic chemicals during early life [128]. This endpoint is also characteristic of the effect of estrogen on rat prostatic cells in culture [200]. Exposure of rats and mice to high doses of DES during development also alters testis development, leading to a decrease in testis size, sperm numbers, undescended testes, and epididymal cysts and infertility [130].

In some studies, the effects of exposure to DES during development were not noticeable prior to the animals reaching old age. For example, treatment of male rats with DES during the first month after birth did not result in observable malignancies at 6–9 months of age, but by 20 months (old age), squamous cell cancer was detected with involvement of the dorsolateral prostate, coagulating glands (dorsocranial prostate) and ejaculatory ducts [233]. There are not yet reports of prostate abnormalities in men exposed to DES during fetal life. However, DES was used in pregnancy during 1947–1971, with peak use in the 1950s and 1960s, so most DES sons are 40–50. Since few of these men have reached the age at which benign prostatic hyperplasia (BPH) or prostate cancer increases in frequency, prostate abnormalities in DES sons have not yet been studied and it is important that this cohort be monitored for these endpoints.

12
Conclusions and Future Research Needs

In this chapter we have provided indirect evidence from studies in laboratory animals that suggest that EDCs at environmental levels can induce adverse male endpoints in humans. However, the absence of data directly correlating low levels of EDC exposure with reproductive endpoints in humans is a critical gap that remains to be filled. We see several reasons for this gap.

First, the long latency until expression of an adverse reproductive outcome when exposure occurs prenatally (or during the early postnatal period) makes determining the consequences of endocrine disruption during development extremely difficult, particularly in humans. Scientists have only recently appreciated that endocrine factors related to decreased sperm count, testicular cancer, and genital tract defects should be sought in the man's own gestational period, which occurred decades before the diagnosis of his cancer or reproductive abnormality. Moreover, adult biomarkers of exposure are more likely to reflect childhood and adult exposures and are therefore not likely to be markers for his prenatal exposure; data regarding prenatal PCB exposure support this hypothesis [234]. For biomarkers of prenatal exposure, maternal serum, stored since a male's gestation, is required. Therefore, studies linking prenatal EDC levels to male reproductive outcomes in humans, either retrospectively using archived maternal serum, or prospectively, are essential. These studies must be of sufficient statistical power to detect effects in sensitive sub-populations and small, but biologically important changes in larger populations.

Second, assays for biomarkers of EDC exposure have, until very recently, been quite limited both in sensitivity and range, and very costly. Until the last few years, almost all studies that included biomarkers of exposure assayed only for DDT and its metabolites, PCBs, and, occasionally, dioxin. Furthermore, the high cost of these assays limited their usefulness for large cohort studies. An additional complicating factor is that exposure is typically to mixtures of chemicals with different mechanisms of action, making predictions of outcomes extremely difficult.

Third, results from animal studies suggest that changes induced by EDCs at environmental levels may be subtler than those usually studied by epidemiologists. Epidemiology must develop studies capable of detecting subtle changes and patterns of minor variations analogous to those that have been demonstrated as a consequence of low level EDCs in animal studies. We suggest, therefore, that epidemiologists seek "low level effects" from "low level exposures" and not limit their studies to the detection of isolated gross abnormalities such as hypospadias and sterility. While it is unlikely that additional evidence will be obtained linking historical trends in sperm density or other reproductive parameters to environmental factors, there is an indirect means of addressing this question. Factors influencing male reproduction, unless uniformly distributed geographically (or no longer present), would be expected to produce not only temporal variation in sperm density but also geographic variation in today's populations. Therefore, carefully controlled, cross-sectional studies conducted in several cities, with differing types and levels of environmental exposures, would be invaluable for assessing such geographic variation and its causes. If such a study identified significant differences between geographic areas in, for example, sperm density, differences that are not explained by host factors (such as age and ethnicity), or personal behaviors (such as smoking and drinking), it might be possible, using biomarkers of environmental exposures, to identify environmental factors that have contributed to these differences.

An international study of semen quality in partners of pregnant women, ongoing in the United States, Europe, and Japan, was designed to meet these objectives. Considerable effort is being made to standardize procedures and questionnaires so that results will be comparable across study centers. Within a few years more definitive information on the magnitude of geographic variation in sperm density will be available. Long-term follow-up of these study populations will provide invaluable information about trends in these parameters in the future. If differences in male reproductive parameters remain after controlling for non-environmental factors, environmental causes will be investigated. In addition to other sources of variation, possible chemical causes will be investigated. Candidate chemicals will include not only EDCs, but also chemicals (such as metals) that may alter reproduction through non-endocrine disrupting pathways.

Based on the information provided here, there is good reason to question the validity of currently established "safe" levels for specific chemicals. However, with the exception of pesticides and food additives, there are as yet no requirements for testing EDCs such as BPA, octylphenol, or nonylphenol. The U.S. Environmental Defense Fund reported that for over 75% of the highest volume chemicals (those for which over one million pounds are produced annually), no information on health effects is available [235]. This report notes that claims that the majority of the greater than 75,000 chemicals in use today are safe are thus based on the absence of any data. The assumption that chemicals for which there is very high human exposure are safe until "clear and convincing" evidence of human harm is demonstrated has been the subject of heated debate. It is challenged by those who prefer to assume that those chemicals that are persistent and bioaccumulate are harmful to humans until proven otherwise; e.g., to apply the "Precautionary Principle" [236].

13
References

1. vom Saal FS, Finch CE, Nelson JF (1994) Natural history and mechanisms of aging in humans, laboratory rodents and other selected vertebrates. In: Knobil E, Neill JD, Greenwald GS, Markert CL, Pfaff DW (eds) The physiology of reproduction, vol 2. Raven Press, New York, p 1213
2. Schettler T, Solomon G, Valenti M, Huddle A (1999) Generations at risk: reproductive health and the environment. MIT Press, Cambridge
3. Cook JW, Dodds EC, Howen CL (1933) Nature 131: 56
4. Dodds EC, Lawson W (1936) Nature 137: 996
5. Palmlund I (1996) J Psychosom Obstet Gynecol 17: 71
6. Burlington H, Lindeman VF (1950) Proc Soc Exp Biol Med 74: 48
7. Schreiber RW (1970) Condor 72: 133
8. Harper CA (1971) Condor 73: 337
9. Dunlop TR (1981) DDT: Scientists, citizens and public policy. Princeton Press, Princeton
10. Saenz CA, Toro-Sola M, Conde L, Bayonet-Rivera NP (1982) Bol Asoc Med Puerto Rico 74: 16
11. Herbst AL, Ulfelder H, Poskanzer DC (1971) N Engl J Med 284: 878
12. Herbst AL, Kurman RJ, Scully RE (1972) Obstet Gynecol 40: 287 098
13. Kuratsune M, Yoshimura T, Matsuzaka J, Yamagushi A (1972) Environ Health Perspect 1: 119
14. Hsu ST, Ma CI, Hsu SK, Wu SS, Hsu NH, Yeh CC, Wu SB (1985) Environ Health Perspect 59: 5
15. Whorton D, Krauss RM, Marshall S, Milby T (1977) Lancet 2: 1259
16. Wong O, Utidijian HMD, Karten VS (1979) J Occup Med 21: 98
17. Guzelian PS (1982) Drug Metab Rev 13: 663
18. Colborn T, Clement C (1992) Chemically induced alterations in sexual and functional development: the wildlife/human connection. Princeton Scientific Publishing, Princeton
19. Macomber D, Sanders MB (1929) N Engl J Med 200: 981
20. MacLeod MJ, Gold RZ (1951) J Urol 66: 436
21. Nelson CM, Bunge RG (1974) Fertil Steril 25: 503
22. Murature DA, Tang SY, Steinhardt G, Dougherty RC (1987) Biomed Environ Mass Spect 14: 473
23. James WH (1980) Andrologia 12: 381
24. Carlsen E, Giwercman A, Keiding N, Skakkebaek NE (1992) Brit Med J 305: 609
25. Sharpe RM (1995) Human Exp Toxicol 14: 463

26. vom Saal FS, Cooke PS, Buchanan DL, Palanza P, Thayer KA, Nagel SC, Parmiciani S, Welshons WV (1998) Toxicol Ind Health J 14: 239
27. Sharpe RM, Skakkebaek NE (1993) Lancet 341: 1392
28. Giwercman A, Carlsen E, Keiding N, Skakkebaek NE (1993) Environ Health Perspect 101(2): 65
29. Toppari J, Larsen JC, Christiansen P, Giwercman A, Grandjean P, Guillette LJ Jr, Jegou B, Jensen TK, Jouannet P, Keiding N, Leffers H, McLachlan JA, Meyer O, Muller J, Rajpert-De Meyts E, Scheike T, Sharpe R, Sumpter J, Skakkebaek NE (1996) Environ Health Perspect 104(4): 741
30. Farrow S (1994) Brit Med J 309: 1
31. Lerchi A, Nieschlag E (1996) Exp Clin Endocrinol Diab 104: 301
32. Brake A, Krause W (1992) Brit Med J 305: 1498
33. Lerchi A (1995) Brit Med J 311: 569
34. Olsen GW, Bodner KM, Ramlow JM, Ross CE, Lipshultz LI (1995) Fertil Steril 63: 887
35. Bahadur G, Ling KLE, Katz M (1996) Human Reprod 11: 2625
36. Becker S, Berhane K (1997) Fertil Steril 67: 1103
37. Swan SH, Elkin EP, Fenster L (1997) Environ Health Perspect 105: 1228
38. Swan SH, Elkin EP (1999) BioEssays 21: 614
39. Auger J, Kunstmann JM, Czyglik F, Jouannet P (1995) N Engl J Med 332: 281
40. Irvine S, Cawood E, Richardson D, MacDonald E, Aitken J (1996) Brit Med J 312: 467
41. Van Waeleghem K, De Clerq N, Vermeulen L, Schoonjans V, Comhaire F (1996) Human Reprod 11: 325
42. Fisch H, Goluboff ET, Olson JH, Feldshuh J, Broder SJ, Barad DH (1996) Fertil Steril 65: 1009
43. Paulsen CA, Berman NG, Wang C (1996) Fertil Steril 65: 1015
44. Wittmaack FM, Shapiro SS (1992) Wis Med J 91: 477
45. Shah I (1997) Eur J Contracept Reprod Health Care 2: 53
46. Bonde JP, Kold JT, Brixen LS, Abell A, Scheike T, Hjollund NH, Kolstad HA, Giwercman A, Skakkebaek NE, Keiding N, Olsen J (1998) Scand J Work Environ Health 24: 407
47. Adami HO, Bergstrom R, Mohner M, Zatonski W, Storm H, Ekbom A, Tretli S, Teppo L, Ziegler H, Rahu M (1998) Int J Cancer 59: 33
48. Bergstrom R, Adami HO, Mohner M, Zatonski W, Storm H, Ekbom A, Tretli S, Teppo L, Akre O, Hakulinen T (1996) J Natl Cancer Inst 88: 727
49. Forman D, Moller H (1994) Cancer Surv 19/20: 323
50. Ries LA, Miller BA, Hankey BF, Kosary CL, Harras A, Edwards BK (1994) SEER Cancer statistics review 1973–1991: tables and graphs. NIH Pub. No. 94–2789, National Cancer Institute, Bethesda
51. Weir HK, Marret LD, Maraven V (1999) Can Med Assoc J 160: 201
52. Parkin DM, Muir CS, Whelan SL, Gao YT, Ferlay J, Powell J (eds) (1992) Cancer incidence in five continents, vol VI. International Agency for Research on Cancer, Lyon
53. Bjerkedal T, Bakketeig LS (1975) Int J Epidemiol 4: 31
54. International Clearing House for Birth Defects Monitoring Systems (1991) Congenital malformations worldwide: a report from the International Clearing House for Birth Defects Monitoring Systems. Elsevier Science Publishers, New York
55. Czeizel A (1985) Lancet 1: 462
56. Matlai P, Beral V (1985) Lancet 1: 108
57. Paulozzi LJ, Erickson JD, Jackson RJ (1997) Pediatrics 100: 831
58. Paulozzi LJ (1999) Environ Health Perspect 107: 297
59. Chilvers C, Pike MC, Forman D, Fogelman K, Wadsworth ME (1984) Lancet 2: 330
60. John Radcliffe Hospital Cryptorchidism Study Group (1986) Cryptorchidism: an apparent substantial increase since 1960. Brit Med J 293: 1401
61. Scorer CG (1964) Arch Dis Child 39: 605
62. John Radcliffe Hospital Cryptorchidism Study Group (1992) Cryptorchidism, a prospective study of 7500 consecutive male births 1984–8. Arch Dis Child 67: 892
63. Depue RH, Pike MC, Henderson BE (1983) J Natl Cancer Inst 71: 115

64. Brown LM, Pottern LM, Hoover RN (1986) Cancer Res 46: 4812
65. Gershman ST, Stolley PD (1988) Int J Epidemiol 17: 738
66. Kallen BAJ, Martinez-Frias ML, Castilla EE, Robert E, Lancaster PAL, Kringelbach M, Mutchinick OM, Mastroiacovo P (1992) Int J Risk Safety Med 3: 183
67. Czeizel A, Toth J (1990) Teratology 41: 167
68. Rothman KJ, Louik C (1978) N Engl J Med 299: 522
69. Depue RH (1984) Int J Epidemiol 13: 311
70. Beard CM, Melton LJ, O'Fallon WM, Noller KL, Benson RC (1984) Am J Epidemiol 120: 707
71. McBride ML, Van den Steen N, Lamb CW, Gallagher RP (1991) Int J Epidemiol 20: 964
72. Polednak AP, Janerich DT (1983) Teratology 28: 67
73. Panagiotopoulou K, Katsouyanni K, Petridou E, Garas Y, Tzonou A, Trichopoulos D (1990) Cancer Causes Control 1: 119
74. Swerdlow AJ, Huttly SR, Smith PG (1987) Brit J Cancer 55: 571
75. Key TJ, Bull D, Ansell P, Brett AR, Clark GM, Moore JW, Chilvers CE, Mike MC (1996) Brit J Cancer 73: 698
76. Bernstein L, Depue RH, Ross RK, Judd HL, Pike MC, Henderson BE (1986) J Natl Cancer Inst 76: 1035
77. Bernstein L, Pike MC, Ross RK, Judd HL, Brown JB, Henderson BE (1985) J Natl Cancer Inst 74: 741.
78. Henderson BE, Bernstein L, Ross RK, Depue RH, Judd, HL (1988) Brit J Cancer 57: 216
79. Møller H, Skakkebaek NE (1997) Cancer Causes Control 8: 904
80. Akre O, Ekbom A, Hsieh CC, Trichopoulos D, Adami HO (1996) J Natl Cancer Inst 88: 883
81. Prener A, Hsieh CC, Engholm G, Trichopoulos D, Jensen OM (1992) Cancer Causes Control 3: 265
82. Westergaard T, Andersen PK, Pedersen JB, Frisch M, Olsen JH, Melbye M (1998) Br J Cancer 77: 1180
83. Moss AR, Osmond D, Baccetti P, Torti FM, Gurgin V (1986) Am J Epidemiol 124: 39
84. Braun MM, Ahlbom A, Floderus B, Brinton LA, Hoover N (1995) Cancer Causes Control 6: 519
85. Ekbom A, Hsieh CC, Lipworth L, Wolk A, Ponten J, Adami HO, Trichopoulos D (1996) Brit Med J 313: 337
86. Ross RD, Henderson BE (1994) J Natl Cancer Inst 86: 252
87. Ross RK, Bernstein L, Judd H (1986) J Natl Cancer Inst 76: 45
88. Berthelsen JG, Skakkebaek NE, Mogenson P, Sorenson BL (1979) Brit Med J 2: 363
89. Bernstein L, Pike MC, Depue RH, Ross RK, Moore JW, Henderson BE (1988) Br J Cancer 58: 379
90. Burton MH, Davies TW, Raggatt PR (1987) J Epidemiol Commun Health 41: 127
91. Swerdlow AJ, Wood KH, Smith PG (1983) J Epidemiol Commun Health 37: 238
92. Hjertqvist M, Damber JE, Bergh A (1989) J Epidemiol Commun Health 43: 324
93. Jones ME, Swerdlow AJ, Griffith M, Goldacre MJ (1998) Paediat Perinat Epidemiol 12: 383
94. Berkowitz GS, Lapinski RH, Dolgin SE, Gazella JG, Bodian CA, Holzman IR (1993) Pediatrics 92: 44
95. Sweet RA, Schrott HG, Kurland R, Culp OS (1974) Mayo Clinic Proc 49: 52
96. Fritz G, Czeizel AE (1996) J Reprod Fertil 106: 63
97. Kallen B, Bertollini R, Castilla E, Czeizel A, Knudsen LB, Martinez-Frias ML, Mastroiacovo P, Mutchinick O (1986) Acta Paediat Scand (Suppl) 324: 1
98. Hay S, Barbano H (1972) Teratology 6: 271
99. Fredell L, Lichtenstein P, Pedersen NL, Svensson J, Nordenskjold A (1998) J Urol 160: 2197
100. Whorton MD, Meyer CR (1984) Fertil Steril 42: 82
101. Potashnik G, Ben-Adaret N, Israeli R, Yanai-Inbar I, Sober I (1978) Fertil Steril 30: 444
102. Takahashi W, Wong L, Rogers BJ, Hale RW (1981) Bull Environ Contam Toxicol 27: 551
103. Potashnik G, Goldsmith J, Insler V (1983) Andrologia 16: 213

104. Goldsmith JR, Potashnik G, Israeli R (1984) Arch Environ Health 39: 85
105. Ratcliffe JM, Schrader SM, Steenland K, Clapp DE, Turner T, Hornung RW (1987) Br J Ind Med 44: 317
106. Schrader SM, Turner TW, Breitenstein MJ, Simon SD (1988) Reprod Toxicol 2: 183
107. Whorton MD, Milby TH, Krauss RM, Stubbs HA (1979) J Toxicol Environ Health 5: 929
108. Wyrobek AJ, Watchmaker G, Gordon L, Wong K, Moore D, Whorton D (1981) Environ Health Perspect 40: 255
109. Cannon SB, Veazey JM Jr, Jackson RS, Burse VW, Hayes C, Straub WE, Landrigan PJ, Liddle JA (1978) Am J Epidemiol 107: 529
110. Eskenazi B, Wyrobek AJ, Fenster L, Katz DF, Sadler M, Lee J, Hudes M, Rempel DM (1991) Am J Ind Med 20: 575
111. Welch LS, Schrader SM, Turner TW, Cullen MR (1988) Am J Ind Med 14: 509
112. Ratcliffe JM, Schrader SM, Clapp DE, Halperin WE, Turner TW, Hornung RW (1989) Br J Ind Med 46: 399
113. Veulemans H, Steeno O, Masschelein R, Groesneken D (1993) Brit J Ind Med 50: 71
114. Eskenazi B, Gold EB, Samuels SJ, Wight S, Lasley BL, Hammond SK, O'Neill RM, Schenker MB (1995) Am J Ind Med 28: 817
115. Hardell L, Ohlson CG, Fredrikson M (1997) Int J Cancer 73: 828
116. Quinn MM, Wegman DH, Greaves IA, Hammond SK, Ellenbecker MJ, Spark RF, Smith ER (1990) Am J Ind Med 18: 55
117. Grajewski B, Whelan EA, Schnorr TM, Mouradian R, Alderfer R, Wild DK (1996) Am J Ind Med 29: 49
118. McLaren A (1991) Biol Essays 13: 151
119. Schlegel RJ, Farias E, Russo NC, Moore JR, Gardner LI (1967) Endocrinology 81: 565
120. Moore KL (1982) The developing human: clinically oriented embryology. WB Saunders, Philadelphia
121. Kluin PM, Kramer MF, de Rooij DG (1984) Anat Embryol 169: 73
122. Vergouwen RP, Jacobs SG, Huiskamp R, Davids JA, de Rooij DG (1991) J Reprod Fertil 93: 233
123. Jost A (1972) Johns Hopkins Med J 130: 38
124. Lubahn DB, Moyer JS, Golding TS, Couse JF, Korach KS, Smithies O (1993) Proc Natl Acad Sci USA 90: 11,162
125. Bull JJ (1983) Evolution of sex determining mechanisms. Benjamin Cummings Pub, Menlo Park
126. Sheehan DM, Willingham E, Gaylor D, Bergeron JM, Crews D (1999) Environ Health Perspect 107: 155
127. Bergeron JM, Crews D, McLachlan JA (1994) Environ Health Perspect 102: 780
128. McLachlan JA, Newbold RR, Bullock B (1975) Science 190: 991
129. Newbold RR, Bullock BC, McLachlan JA (1990) Cancer Res 50: 7677
130. Newbold RR (1995) Environ Health Perspect 103: 83
131. Newbold RR, Hanson RB, Jefferson WN, Bullock BC, Haseman J, McLachlan JA (1998) Carcinogenesis 19: 1655
132. Berndtson WE, Thompson TL (1990) J Androl 11: 429
133. Meachem SJ, McLachlan RI, de Kretser DM, Robertson DM, Wreford NG (1996) Biol Reprod 54: 36
134. Orth JM (1984) Endocrinology 115: 1248
135. Cooke PS, Hess RA, Porcelli J, Meisami E (1991) Endocrinology 129: 244
136. Gray LE Jr, Ostby JS, Kelce WR (1997) Toxicol Appl Pharmacol 146: 11
137. DeVito M, Birnbaum L, Farland W, Gasiewicz T (1995) Environ Health Perspect 103: 820
138. Mably TA, Moore RW, Peterson RE (1992) Toxicol Appl Pharmacol 114: 97
139. Mably TA, Moore RW, Goy RW, Peterson RE (1992) Toxicol Appl Pharmacol 114: 108
140. Mably TA, Bjerke DL, Moore RW, Gendron-Fitzpatrick A, Peterson RE (1992) Toxicol Appl Pharmacol 114: 118
141. Peterson RE, Moore RW, Mably TA, Bjerke DL, Goy RW (1992) Male reproductive system ontogeny: effects of perinatal exposure to 2,3,7,8-tetrachlorodibenzo-p-dioxin. In: Colborn T, Clement C (eds) Chemically induced alterations in sexual and functional de-

velopment: the wildlife/human connection. Princeton Scientific Publishing, Princeton, p 175

142. Petersen RE, Theobald HM, Kimmel GL (1993) Crit Rev Toxicol 23: 293

143. Gray LE Jr, Ostby JS (1997) Toxicol Appl Pharmacol 133: 285

144. Gray LE Jr, Wolff C, Mann P, Ostby JS (1997) Toxicol Appl Pharmacol 146: 237

145. Sharpe RM, Fisher JS, Millar MM, Jobling S, Sumpter JP (1995) Environ Health Perspect 103: 1136

146. Nagel SC, vom Saal FS, Thayer KA, Dhar MG, Boechler M, Welshons WV (1997) Environ Health Perspect 105: 70

147. Nagel SC, vom Saal FS, Welshons WV (1998) Proc Soc Exp Biol Med 217: 300

148. Arnold SF, Collins BM, Robinson MK, Guillette, LJ Jr, McLachlan JA (1996) Steroids 61: 642

149. Brotons JA, Olea-Serrano MF, Villalobos M, Pedraza V, Olea N (1995) Environ Health Perspect 103: 608

150. Raloff J (1999) Science Service, Washington DC. http://www.sciencenews.org/sn_arc99/8_7_99/food.htm

151. Krishnan AV, Stathis P, Permuth SF, Tokes L, Feldman D (1993) Endocrinology 132: 2279

152. Howdeshell KL, vom Saal FS (2000) Am Zoologist 40: 429

153. Takao Y, Lee HC, Ishibashi Y, Kohra S, Tominaga N, Arizono K (1999) Jpn J Toxicol Environ Health 45: 39

154. Olea N, Pulgar R, Perez P, Olea-Serrano F, Rivas A, Novillo-Fertrell A, Pedraza V, Soto AM, Sonnenschein C (1996) Environ Health Perspect 104: 298

155. Gupta C (2000) Proc Soc Exp Biol Med 224: 61

156. Colerangle JB, Roy D (1997) J Steroid Biochem Mol Biol 60: 153

157. Steinmetz R, Brown NG, Allen DL, Bigsby RM, Ben-Jonathan N (1997) Endocrinology 138: 1780

158. Shah HC, McLachlan JA (1976) J Pharmacol Exp Therap 197: 687

159. Spearow JL, Doemeny P, Sera R, Leffler R, Barkley M (1999) Science 285: 1259

160. Gray LE Jr, Ostby JS, Ferrell J, Sigmond R, Cooper RL, Linder R, Rehnberg G, Goldman JM, Laskey J (1989) Correlation of sperm and endocrine measures with reproductive success in rodents. In: Burger EJ (ed) Sperm measures and reproductive success: Institute for Health Policy Analysis Forum on Science, Health, and Environment. Alan R Liss, New York, p 193

161. vom Saal FS, Montano MM, Wang MH (1992) Sexual differentiation in mammals. In: Colborn T, Clement C (eds) Chemically induced alterations in sexual and functional development: the wildlife/human connection. Princeton Scientific Publishing, Pinceton, p 17

162. Byskov AG, Hoyer PE (1994) Embryology of mammalian gonads and ducts. In: Knobil E, Neill J (eds) Physiology of reproduction, vol 1. Raven Press, New York, p 487

163. Huhtaniemi IT (1985) Functional and regulatory differences between the fetal and adult populations of rat Leydig cells. In: Jaffe RB, Dell'Acqua S (eds) The endocrine physiology of pregnancy and the peripartal period. Raven Press, New York, p 65

164. Donahoe PK, Hutson JM, MacLaughlin DT, Budzik GP (1985) Steroid interactions with Mullerian inhibiting substance. In: Jaffe RB, Dell'Acqua S (eds) The endocrine physiology of pregnancy and the peripartal period. Raven Press, New York, p 101

165. Josso N (1986) Endocrine Rev 7: 421

166. Hirobe S, He WW, Lee MM, Donahoe PK (1992) Endocrinology 131: 854

167. vom Saal FS, Timms BG, Montano MM, Palanza P, Thayer KA, Nagel SC, Dhar MD, Ganjam VK, Parmigiani S, Welshons WV (1997) Proc Natl Acad Sci USA 94: 2056

168. Hess RA, Bunick D, Lee KH, Bahr J, Taylor JA, Korach KS, Lubahn DB (1997) Nature 390: 509

169. Walvoord DJ, Resnick MJ, Grayhack JT (1976) Invest Urol 14: 60

170. Nozu K, Tamaoki B (1974) Acta Endocrinol 76: 608

171. vom Saal FS, Nagel SC, Palanza P, Boechler M, Parmigiani S, Welshons WV (1995) Toxicol Lett 77: 343

172. Nonneman D, Ganjam V, Welshons WV, vom Saal FS (1992) Biol Reprod 47: 723
173. Caroom D, Bronson FH (1971) Physiol Behav 7: 659
174. Mugford RA, Nowell NW (1972) Horm Behav 3: 39
175. Ingersoll DW, Morley KT, Benvenga M, Hands C (1986) Behav Neurosci 100: 187
176. Maruniak JA, Desjardins C, Bronson FH (1975) J Reprod Fertil 44: 567
177. Bronson FH (1979) Quart Rev Biol 54: 246
178. Palanza P, Morellini F, Parmigiani S, vom Saal FS (1999) Neurosci Biobehav Rev 23: 1011
179. Pang SF, Chow PH, Wong TM (1979) J Reprod Fertil 56: 129
180. Peitz B, Olds-Clarke P (1986) Biol Reprod 35: 608
181. Peitz B (1988) J Reprod Fertil 83: 169
182. Katzenellenbogen BS, Katzenellenbogen JA, Mordecai D (1979) Endocrinology 105: 33
183. Welshons WV, Nagel SC, Thayer KA, Judy BM, vom Saal FS (1999) Toxicol Ind Health 15: 12
184. Katzenellenbogen JA, O'Malley BW, Katzenellenbogen BS (1996) Mol Endocrinol 10: 191
185. Reyes FI, Boroditsky RS, Winter JSD, Faiman C (1974) J Clin Endocrinol Metab 38: 612
186. Resko JA, Ploem JG, Stadelman HL (1975) Endocrinology 97: 425
187. Weisz J, Ward IL (1980) Endocrinology 106: 306
188. Clark MM, Crews D, Galef BG (1991) Physiol Behav 49: 239
189. vom Saal FS, Bronson FH (1980) Science 208: 597
190. vom Saal FS (1989) J Animal Sci 67: 1824
191. vom Saal FS, Timms BG (1999) The role of natural and manmade estrogens in prostate development. In: Naz RK (ed) Endocrine disruptors: effects on male and female reproductive systems. CRC Press, Boca Raton, p 307
192. Timms BG, Petersen SL, vom Saal FS (1999) J Urol 161: 1694
193. Mendel C (1989) Endocrine Rev 10: 232
194. Montano MM, Welshons WV, vom Saal FS (1995) Biol Reprod 53: 1198
195. vom Saal F, Timms BG (1999) The role of natural and manmade estrogens in prostate development. In: Naz RK (ed) Endocrine disruptors: effects on male and female reproductive systems. CRC Press, Boca Raton, p 307
196. Soto AM (1995) Environ Health Perspect 103: 113
197. Amara JF, Dannies PS (1983) Endocrinology 112: 1141
198. Olson LJ, Erickson BJ, Hinsdill RD, Wyman JA, Porter WP, Binning LK, Bidgood RC, Nordheim EV (1987) Arch Environ Contam Toxicol 16: 433
199. Bigazzi M, Bradni ML, Bani G, Sacchi TB (1992) Cancer 70: 639
200. Martikainen PM, Makela SI, Santti RS, Harkonen PL, Suominen JJ (1987) Prostate 11: 291
201. Somjen D, Kohen F, Jaffe A, Amir-Zaltsman Y, Knoll E, Stern N (1998) Hypertension 32: 39
202. vom Saal FS, Sheehan DM (1998) Forum for Applied Research and Public Policy 13: 11
203. Wilson JG, Roth CB, Wabkany J (1953) Am J Anat 92: 189
204. Lohnes D, Kastner P, Dierich A, Mark M, LeMeur M, Chambon P (1993) Cell 73: 643
205. Cohlan SQ (1953) Science 117: 535
206. Wilhite CC, Wier PJ, Berry DL (1989) Crit Rev Toxicol 20: 113
207. Becks GP, Burrow GN (1991) Med Clin North Am 75: 121
208. Gorski J, Gannon F (1976) Ann Rev Physiol 38: 425
209. Medlock KL, Lyttle CR, Kelepouris N, Newman ED, Sheehan DM (1991) Proc Soc Exp Biol Med 196: 293
210. Fox TO (1975) Proc Natl Acad Sci USA 72: 4303
211. Cooke PS, Young P, Hess RA, Cunha GR (1991) Endocrinology 128: 2874
212. Cooke PS, Young P, Cunha GR (1991) Endocrinology 128: 2867
213. Prins GS (1992) Endocrinology 130: 2401
214. Cunha GR, Donjacour AA (1987) Stromal-epithelial interactions in normal and abnormal prostatic development. In: Coffey DS, Burchovsky N, Gardner WA Jr, Resnick MI, Karr JP (eds) Current concepts and approaches to the study of prostate cancer. Alan R Liss, New York, p 251
215. Small EJ, Prins GS (1996) Physiology and endocrinology of the prostate. In: Vogelzang NJ, Scardino PT, Shipley WU, Coffey DS (eds) Comprehensive textbook of genitourinary oncology. Williams and Wilkins, Baltimore, p 600

216. Prins GS (1997) Developmental estrogenization of the prostate gland. In: Naz RK (ed) Prostate: basic and clinical aspects. CRC Press, New York, p 247
217. Prins GS, Marmer M, Woodham C, Chang W, Kuiper G, Gustafsson JA, Birch L (1998) Endocrinology 139: 874
218. Santti R, Newbold RR, Makela S, Pylkkanen L, McLachlan JA (1994) Prostate 24: 67
219. Herbst AL, Bern HA (eds) (1981) Developmental effects of diethylstilbestrol (DES) in pregnancy. Thieme-Stratton, New York
220. Bern HA (1992) The fragile fetus. In: Colborn T, Clement C (eds) Chemically induced alterations in sexual and functional development: the wildlife/human connection. Princeton Scientific Publishing, Princeton, p 9
221. Greco TL, Duello TM, Gorski J (1993) Endocrine Rev 14: 59
222. Mittendorf R (1995) Teratology 51: 435
223. Vannier B, Raynaud JP (1980) J Reprod Fertil 59: 43
224. Bern HA, Edery M, Mills KT, Kohrman AF, Mori T, Larson L (1987) Cancer Res 47: 4165
225. Migliaccio S, Newbold RR, Bullock BC, McLachlan JA, Korach KS (1992) Endocrinology 130: 1756
226. Sheehan DM, Branham WS (1987) Teratogen Carcinogen Mutagen 7: 411
227. Harmon JR, Branham WS, Sheehan DM (1989) Teratology 39: 253
228. Sheehan DM, Young M (1979) Endocrinology 104: 1442
229. Driscoll SG, Taylor SH (1980) J Am Coll Obstet Gynecol 56: 537
230. Blacklock NJ (1983) The development and morphology of the prostate. In: Ghanadian R (ed) The endocrinology of prostate tumours. MTP Press, Lancaster, p 1
231. Turner T, Ederly M, Mills KT, Bern HA (1989) J Steroid Biochem 32: 559
232. Pylkkanen L, Makela S, Valve E, Harkonen P, Toikkanen S, Santti R (1993) J Urol 149: 1593
233. Arai Y, Chen CY, Nishizuka Y (1978) Jap J Cancer Res 69: 861
234. Jacobson JL, Jacobson SW (1996) N Eng J Med 335: 783
235. Roe D (1997) Toxic ignorance: the continued absence of basic health testing for top-selling chemicals in the United States. Environmental Defense Fund, New York
236. Raffensperger C, Tickner J (1999) Protecting public health and the environment: implementing the precautionary principle. Island Press, Washington DC
237. Bendvold, E, Gottlieb C, Bygdeman M, Eneroth P (1991) Arch Androl 26: 189
238. Adamopoulos DA, Pappa A, Nicopoulou S, Andreou E, Karamertzanis M, Michopoulos J, Deligianni V, Simou M (1996) Human Reprod 11: 1936
239. de Mouzon J, Thonneau P, Spira A, Multigner L (1996) Brit Med J 313: 43
240. Menchini-Fabris F, Rossi P, Palego P, Simi S, Torchi P (1996) Andrologia 28: 304
241. Vierula M, Niemi M, Keiski A, Saaranen M, Saarikoski S, Suominen J (1996) Int J Androl 19: 11
242. Berling S, Wolner-Hanssen P (1997) Human Reprod 12: 1002
243. Rasmussen PE, Erb K, Westergard LG, Laursen SB (1997) Fertil Steril 68: 1059
244. Zheng Y, Bonde JP, Ernst E, Mortensen JT, Egense J (1997) Int J Epidemiol 26: 1289
245. Younglai EV, Collins JA, Foster WG (1998) Fertil Steril 70: 76
246. Tortolero I, Bellabarba Arata G, Lozano R, Bellabarba C, Cruz I, Osuna JA (1999) Arch Androl 42: 29
247. Bujan L, Mansat A, Pontonnier F, Mieusset R (1996) Brit Med J 312: 471
248. Bonde JP, Ernst E, Jensen TK, Hjollund NH, Kolstad H, Henriksen TB, Scheike T, Giwercman A, Olsen J, Skakkebaek NE (1998) Lancet 352: 1172
249. Gyllenborg J, Skakkebaek NE, Nielsen NC, Keiding N, Giwercman A (1999) Int J Androl 22: 28
250. Henderson BE, Benton B, Jing J, Yu MC, Pike MC (1979) Int J Cancer 23: 598
251. Schottenfeld D, Warshauer ME, Sherlock S, Zauber AG, Leder M, Payne R (1980) Am J Epidemiol 112: 232
252. Henderson BE, Benton B, Cosgrove M, Baptista J, Aldrich J, Townsend D, Hart W, Mack TM (1976) Pediatrics 58: 505
253. Cosgrove MD, Benton B, Henderson BE (1977) J Urol 117: 220
254. Andonian RW, Kessler R (1979) Urology 13: 276

255. Gill WB, Schumacher GFB, Bibbo M, Straus FH, Schoenberg HW (1979) J Urol 122: 36
256. Leary FJ, Resseguie LJ, Kurland LT, O'Brien PC, Emslander RF, Noller KL (1984) J Am Med Assoc 252: 2984
257. Shy KK, Stenchever MA, Karp LE, Berger RE, Williamson RA, Leonard J (1984) Fertil Steril 42: 772
258. Wilcox AJ, Baird DD, Weinberg CR, Hornsby PP, Herbst AL (1995) N Eng J Med 332: 1411
259. Glass RI, Lyness RN, Mengle DC, Powell KE, Kahn E (1979) Am J Epidemiol 109: 346
260. Potashnik G, Porath A (1995) J Occupat Environ Med 37: 1287
261. Egeland GM, Sweeney MH, Fingerhut MA, Wille KK, Schnorr TM, Halperin WE (1994) Am J Epidemiol 139: 272
262. Mocarelli P, Brambilla P, Gerthoux PM, Patterson DG Jr, Needham LL (1996) Lancet 348: 409
263. Padungtod C, Savitz DA, Overstreet JW, Christiani DC, Ryan LM, Xu XP (2000) J Occupat Environ Med 42: 982
264. Padungtod C, Lasley B, Christinai DC, Ryan L, Xu X (1998) J Occupat Environ Med 40: 1038
265. Harrington JM, Rivera RO, de Morales AV (1978) Arch Environ Health 33: 12
266. Newbold RR, Bullock BC, McLachlan JA (1985) Cancer Res 45: 5145
267. Newbold RR, Bullock BC, McLachlan JA (1987) J Urol 138: 1446
268. Arai Y (1970) Endocrinology 86: 918
269. Gill WB, Schumacher GF, Bibbo M (1977) J Urol 117: 477
270. McLachlan JA (1981) Rodent models for perinatal exposure to diethylstilbestrol and their relation to human disease in the male. In: Herbst AL, Bern HA (eds) Developmental effects of diethylstilbestrol (DES) in pregnancy. Thieme-Stratton, New York, p 148
271. Bullock BC, Newbold RR, McLachlan JA (1988) Environ Health Perspect 77: 29

CHAPTER 6

Effects of Perinatal Estrogen Exposure on Fertility and Cancer in Mice

Retha R. Newbold

Developmental Endocrinology Section, Laboratory of Molecular Toxicology, National Institute of Environmental Health Sciences, Research Triangle Park, NC 27709, USA
E-mail: newbold1@niehs.nih.gov

Concerns have been raised regarding the reproductive and health hazards of chemicals in the environment that have potential endocrine disrupting effects. These concerns include increased incidences of breast, ovarian, and uterine cancer, endometriosis, fibroids, infertility, and early menopause in women; in men, alterations in sex differentiation, decreased sperm concentrations, benign prostatic hyperplasia, prostatic cancer, testicular cancer, and reproductive problems have been suggested. Studies with the potent synthetic estrogen diethylstilbestrol (DES) have shown that exogenous estrogen exposure during critical stages of development results in permanent cellular and molecular alterations in the exposed organism. These alterations manifest themselves in the female and male as structural, functional, or long-term pathological changes including neoplasia. Although DES is a potent environmental estrogen, studying its effects at low dose levels in an experimental animal model offers a unique opportunity to identify adverse effects that may be caused by weaker environmental estrogens which fit the category of endocrine disruptors.

Keywords. Diethylstilbestrol, DES, Estrogenic, Estrogen, Estrogen receptor, Embryogenesis, Perinatal exposure, Reproductive tract development, Genital tract, Sex hormones, Sex differentiation, Infertility

1	**Introduction**	172
2	**Prenatal DES Exposure as an Example**	173
3	**Effects on Males**	174
3.1	Fertility	175
3.2	Cancer	175
4	**Effects on Females**	176
4.1	Fertility	177
4.2	Cancer	178
5	**Mechanisms of Reproductive Tract Toxicity**	180
6	**Multigenerational Effects**	181
7	**Not Just DES – Other Examples**	182
8	**Summary and Conclusions**	184
9	**References**	185

The Handbook of Environmental Chemistry Vol. 3, Part M
Endocrine Disruptors, Part II
(ed. by M. Metzler)
© Springer-Verlag Berlin Heidelberg 2002

List of Abbreviations

DES   diethylstilbestrol
ER   estrogen receptor

1
Introduction

Recently there has been increasing concern among the scientific community, policy makers, and general public regarding the role of environmental endocrine disrupting chemicals as possible contributors to increased incidences of various reproductive and health hazards. Sharpe and Skakkebaek [1] have postulated that environmental chemicals with hormone-like activity may be the underlying cause of falling sperm counts and increases in testicular cancer, undescended testis, and malformations of the male genital tract that have been reported over the last half of this century. Furthermore, another report from Scotland revealed that men born after 1970 have sperm counts that are 25% lower than those born before 1959; this is an average decline of 2.1% per year [2]. They also reported that lower sperm counts were associated with poor semen quality [2]. On the contrary, Olsen and colleagues found that the average sperm count was not decreasing if they used alternative statistical models to analyze the data [3]. The debate continues; changes over time in semen quality and quantity remain a topic of controversy and have been recently reviewed by Swan et al. [4].

Extending the correlation of adverse health effects in men, Davis et al. [5] suggested that exposure to environmental endocrine disrupting chemicals was also responsible for the increased incidence of breast cancer in women that has been reported over the same time interval. Publication of the book "Our Stolen Future" [6] further raised concerns about endocrine disrupting chemicals not only for various human health consequences but also for damaging effects on wildlife. Thus, some environmentalists and researchers believe that wildlife and human species are approaching a fertility and reproductive health crisis, while others think that the available data are just too insufficient to support global concern. Although definitive data does not exist to prove that exposure to environmentally relevant levels of endocrine disrupting chemicals has a casual role in the development of adverse human health consequences, epidemiological studies and experimental animal models reveal a strong link between hormonal exposure and infertility/neoplasia.

It has been well documented that endogenous estrogens are a main risk factor in the development of breast cancer [7]. Further, estrogen replacement therapy has been shown to increase the incidence of endometrial cancer [8]. While these adverse effects have been mainly identified in individuals exposed to hormones after maturity, of greatest concern is the exposure of the fetus and/or neonate to chemicals with hormonal activity because of the increased susceptibility of this stage of development to environmental insult [9]. Considering what is currently known about chemicals that can disrupt the endocrine system, their effects (a) may be manifested differently, and with permanent conse-quences, in the embryo, fetus, and neonate as compared to effects resulting from exposure to adults, (b) can alter the course of development for the exposed or-ganism, with the outcome dependent on the specific developmental exposure periods, and (c) are often delayed and not recognized until the organism reaches maturity or perhaps even later in life, although the critical period of ex-posure occurred during embryonic, fetal, or neonatal life.

As an example, the profound effects of estrogens on the developing repro-ductive tract have been demonstrated by prenatal exposure to the synthetic es-trogen, diethylstilbestrol (DES) [10].

2
Prenatal DES Exposure as an Example

The scientific community has had a long history with teratogenic and carcino-genic effects of synthetic estrogenic compounds. Although it was recognized for centuries that the ovary controlled the estrous cycle, not until the early twen-tieth century were the biologically active substances produced by the ovary de-scribed. The term "estrogens" was coined for these substances because of their ability to induce estrus in animals. Many laboratories extensively studied the physiological, biochemical, and pharmacological properties of estrogens and the search for a synthetic substitute followed. DES, a nonsteroidal compound with properties similar to the natural female sex hormone, estradiol, was first synthesized in 1938. This potent synthetic estrogen was specifically designed for its estrogenic activity and ready solubility. Like many of today's environ-mental estrogens, DES was not structurally similar to the natural estrogens [11]. In fact, early research with DES demonstrated that compounds with widely di-verse structures could exhibit similar biological functions associated with es-trogens. A historical account of the development and use of DES and the early search for compounds with estrogenic activity has been previously summa-rized [12]. DES also demonstrated another significant point, the potential toxic effects of estrogens.

For almost 30 years, physicians prescribed DES to women with high-risk pregnancies to prevent miscarriages and other complications of pregnancy. In 1971, a report associated DES with a rare form of reproductive tract cancer termed "vaginal adenocarcinoma" which was detected in a small number of adolescent daughters of women who had taken the drug while pregnant. Subsequently, DES was also linked to more frequent benign reproductive tract problems in an estimated 90–95 % of the DES-exposed daughters; reproductive organ dysfunction, abnormal pregnancies, reduction in fertility, immune sys-tem disorders, and multiple other effects have been reported. Similarly, DES-ex-posed male offspring demonstrated structural, functional, and cellular abnor-malities following prenatal exposure; hypospadias, microphallus, retained testes, inflammation, and decreased fertility were reported [10]. DES became one of the first examples of an estrogenic toxicant in humans; it was shown to cross the placenta and induce direct effects on the developing fetus. Although DES is no longer used clinically to prevent miscarriage, a major concern remains that when DES-exposed women age, and reach the time at which the incidence of reproductive organ cancers normally increase, they will show a much higher incidence of cancer than unexposed individuals. Further, the possibility of second-generation effects has been suggested [13–15], which puts still another generation at risk for developing problems associated with DES treatment of their grandmothers. Thus, the DES episode continues to be a serious health consequence and is a reminder of the potential toxicities that can be caused by hormonally active chemicals. The well-documented effects in DES-exposed humans serve to justify the concern of developmental exposure to other environmental estrogens and endocrine disrupting chemicals.

Questions of the mechanisms involved in DES-induced teratogenic and carcinogenic effects prompted the development of experimental animal models to study the adverse effects of estrogens and other endocrine-disrupting chemicals on reproductive tract development and differentiation. A description follows of one murine animal model that has been used successfully to replicate and predict adverse effects in humans with similar DES exposure. First, some of the most significant findings are discussed in males.

3
Effects on Males

To study the effects of prenatal exposure to DES, outbred CD-1 mice were treated subcutaneously with DES on days 9–16 of gestation, the period of major reproductive tract organogenesis. The doses of DES ranged from 0.01 to 100 μg/kg during pregnancy, which is equal to or less than doses given therapeutically to pregnant women. In fact, the lower DES doses are comparable to weaker estrogenic compounds found in the environment. Mice were born on day 19 of gestation, and the offspring were followed as they aged. As seen in Table 1, the effects of prenatal exposure to DES on male mice and humans are comparable. Abnormalities such as undescended and hypoplastic testes, infertility, epididymal cysts, sperm abnormalities, hypospadias, microphallus, retained Mullerian duct remnants, prostatic inflammation, and squamous metaplasia of the prostatic ventricle have been reported in both species. The similarity of effects in mice and humans recommends the murine model as a good comparative study. Thus, treatment with DES during critical stages of differentiation permanently alters the developing male reproductive tract in both species.

Table 1. Similar developmental effects of prenatal exposure to DES in mice and humans – male offspring

Reproductive tract dysfunction
Subfertility/infertility
Undescended and hypoplastic testes
Epididymal cysts
Sperm abnormalities
Hypospadias and microphallus
Increased reproductive tract tumors

3.1
Fertility

Subfertility has been reported in both humans and mice following developmental exposure to DES. For mice, this parameter was assessed in the animal model by breeding prenatal DES-exposed male mice to control females of the same strain. In the DES-exposed mouse, a slight reduction in fertility was seen at the DES-10 µg/kg dose, but the two higher doses (50 and 100 µg/kg) were associated with noticeable decreases in reproduction; only 50% of DES-male offspring treated with 50 µg/kg, and 40% treated with 100 µg/kg, were fertile. It is interesting to note that rodent models have been reported to be insensitive to toxic insult with chemicals, requiring almost tenfold sperm count reduction to occur before fertility is affected [16]. Therefore, the decline in fertility that is seen in mice following prenatal DES exposure is probably biologically significant and relevant to humans. An extrapolation to humans, however, remains difficult. Even in unexposed humans, sperm counts differ widely; moreover, some individuals have counts that make them subfertile, and some have counts that are in a range at which any reduction would shift them into a subfertile category. Thus, these difficulties may help explain some of the inconsistencies in fertility reported in the prenatal DES-exposed human data [17, 18], as well as in humans exposed to environmental estrogenic contaminants [3, 19–22].

In summary, multiple factors appear to be related to the observed decreased fertility in the rodent model including (a) retained testes and Mullerian remnants, (b) abnormal sperm morphology and motility, (c) lesions in reproductive tract tissues, (d) abnormal reproductive tract secretions, and (e) inflammation. Cellular alterations in the testis, retained Mullerian ducts, and differentiation of Wolffian duct structures are further discussed.

3.2
Cancer

In men exposed prenatally to DES, several studies have reported an increased incidence of testicular seminomas [23, 24]. To examine the long-term effects of DES in mice and to determine if a similar increase in testicular tumors occurs, or if there is an increased incidence of tumors in other reproductive organs, male mice treated prenatally with DES were followed to 18 months of age. In these aged DES-treated mice, retained cryptorchid testes were a common finding; this abnormality was seen in the younger animals and persisted throughout their life. As expected, these retained gonads had a high degree of degeneration and some even showed mineralization in reminiscent seminiferous tubules. Sperm granulomas were also frequently seen. Interstitial cell tumors of the testes were seen in approximately 3% of the DES-exposed males but were not seen in the control CD-1 males that were used in this study. Interstitial cell hyperplasia and carcinoma have been produced experimentally in some strains of mice after prolonged treatment with estrogenic compounds [25], but in the prenatal DES-exposed males there were more interstitial cell carcinomas in proportion to interstitial cell tumors than those reported in other strains following prolonged estrogen treatment. Together these findings suggested that indeed DES had increased the incidence of testicular tumors in prenatal DES-exposed mice.

Further support for this statement can be seen with the increased incidence of another rare form of testicular cancer, rete adenocarcinoma. Rete testes hyperplasia (56%) and adenocarcinoma (5%) were also seen in aged prenatal DES-exposed mice [26]. This particular tumor has never been seen in any control males in our studies and is reported to be equally rare in humans. To date, rete adenocarcinoma in DES-exposed humans has not been reported, although it has been suggested that this lesion can be misdiagnosed as seminoma. Thus, the cases of seminoma that were reported in DES-exposed males [24] deserve careful re-examination in light of the rare finding of rete adenocarcinoma in the mouse model [26].

In addition to testicular cancer, increased incidence of tumors in retained Mullerian ducts and other reproductive tract tissues including prostate and seminal vesicle were also seen in the DES animal model. To date, an increased incidence of prostate cancer has not been reported in DES-exposed humans, but it may be due to their young age. Most of the DES-exposed human population is just reaching the fifth decade of life, a time when reproductive tract tumors, for example prostatic lesions, would be expected to increase. Since the mouse model has been predictive of other findings, continued close surveillance of DES-exposed men is warranted.

4
Effects on Females

As described for the prenatally DES-exposed males, outbred CD-1 mice were treated with subcutaneous injections of DES on days 9–16 of gestation at doses ranging from 0.01 to 100 µg/kg during pregnancy. Mice were born on day 19 and female offspring were followed for effects on fertility and incidence of reproductive tract neoplasia later in life. As seen in Table 2, the effects of prenatal DES exposure on female mice and humans are comparable. Abnormalities such as reproductive tract dysfunction, structural malformations, and increased tumors in the reproductive tract were observed in both species.

Table 2. Similar developmental effects of prenatal exposure to DES in mice and humans – female offspring

Reproductive tract dysfunction
Subfertility/infertility
Structural malformations
Oviduct, uterus, cervix, vagina; paraovarian cysts of mesonephric origin
Increased reproductive tract tumors
Vaginal adenosis and adenocarcinoma

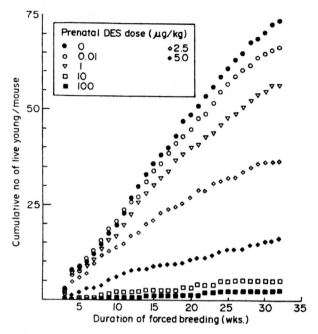

Fig. 1. Total reproductive capacity of mice exposed prenatally to DES. Mice were CD-1 female offspring exposed prenatally to DES on days 9–16 of gestation. Postnatal fertility was determined by a repetitive breeding technique and expressed as the total number of live young born per mouse over an 8 months (32 weeks) interval. The cumulative number of young per mouse is plotted on the x-axis. Note that there is a dose-related decrease in fertility at all DES doses and that even the mice exposed to low doses of DES exhibit decreased fertility as compared to corresponding control mice. This figure is reproduced from [27] with permission

4.1
Fertility

To assess the effects of prenatal DES exposure on postnatal reproductive tract function, the fertility of DES female mice was determined using a continuous breeding protocol [27]. The most striking effect observed was a dose-related decrease in reproductive capacity, ranging from minimal subfertility at the lower DES doses to total sterility at the highest DES doses (Fig. 1). It is interesting to note that over the course of the breeding schedule, even the low doses of DES (0.01 µg/kg/day) exhibited subfertility. The observed reduced reproductive capacity appeared to be a reflection of a decrease in the total number of litters and in litter sizes. A major component of the sterility seen in females that were given high doses of DES was in the oviduct and ovary; the number of ova recovered from the ampulla of the oviduct after induced ovulation was less than 30% that of the controls. Also, structural abnormalities were observed in the oviduct, uterus, cervix, and vagina, which contributed to subfertility. Together these data suggested that in utero exposure to DES, even at low doses, results in the permanent impairment of female mouse reproductive capacity. Numerous reports of altered pregnancy outcomes and decreased fertility in young women exposed in utero to DES [10], as well as accidental DES exposure to wildlife resulting in infertility, further point out the importance of these findings in mice, and demonstrate that environmental estrogens play an important role in decreased female fertility.

4.2
Cancer

To assess the long-term effects of prenatal DES exposure on the female, mice were sacrificed at 12 to 18 months of age, and reproductive tract tissues were studied for histological alterations. Histological examination revealed lesions throughout the reproductive tracts [28]. The vagina of DES-exposed mice was characterized by excessive keratinization and female hypospadias (urethra opens into the vagina rather than the vulva), and at the highest dose (100 µg/kg), 25% had epidermoid tumors of the vagina. Vaginal adenocarcinoma was seen in the 2.5, 5, and 10 µg/kg DES dose groups although its frequency was rare. The cervix of DES-exposed offspring was often enlarged, but the size of cervical lumen was not different from control untreated mice. Stromal stimulation was responsible for the enlargement in the cervical region. Further evidence of stromal alterations were documented by a low prevalence of benign (leiomyomas) and malignant (stromal cell sarcomas and leiomyosarcomas) tumors in the cervix. In the uterus, epithelial and stromal stimulation were also observed, and cystic endometrial hyperplasia was common even in the lower DES-dose animals; a low incidence of benign (leiomyomas) and malignant (stromal cell sarcomas and leiomyosarcomas) uterine tumors was also observed. The ovaries of prenatally DES-treated females were more cystic than controls; at the highest dose (100 µg/kg), ovarian tumors were noted and 100% of the oviducts were inflamed and congenitally malformed. Taken together, these data suggested that in utero exposure to DES results in not only subfertility/infertility and reproductive tract abnormalities including malformation, but also an increase in reproductive tract tumors. While this increase was low, it was still a significant increase as compared to untreated mice.

Since data from the literature suggested that developmental exposure to DES just during neonatal life resulted in a high incidence of vaginal abnormalities [29–31], we compared prenatal and neonatal DES exposure and resulting abnormalities. Outbred CD-1 mice were treated neonatally with DES (2 µg/pup/day) on days 1–5. This treatment resulted in a high incidence (90–95%) of uterine cancer when the mice aged to 18 months [32, 33]. Other species including rats and hamsters [34, 35] neonatally exposed to DES have also been reported to have a high incidence of reproductive tract abnormalities including uterine tumors. Thus, the neonatal mouse model replicates tumors seen in other experimental models and therefore may be predictive of the carcinogenic potential of estrogens in the adult human uterus as it ages. A representative histological picture of the murine uterine carcinoma associated with neonatal treatment with DES is shown in Fig. 2. These tumors rarely metastasized but, in aged animals (24 months of age or older), the lesions sometimes showed spread

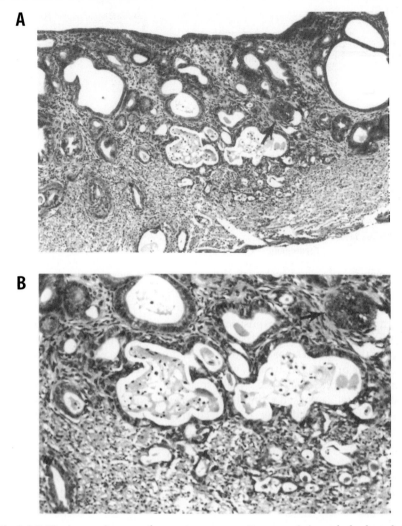

Fig. 2. A, B Uterine carcinoma – the most common pattern seen is irregularly shaped glands that vary from dilated structures to almost solid nests of cells (↑) (×10). **B** Uterine carcinoma – high power view of **A** showing solid nests of cells (↑) and columnar and cuboidal cells lining glandular elements of the tumor (×20)

to para-aortic lymph nodes or direct extension to contiguous organs. It is significant that the mouse tumors progress through the same morphological and biological continuum of hyperplasia to atypical hyperplasia to neoplasm as seen in women. Uterine carcinoma were not observed in the uterus of untreated control CD-1 mice at corresponding ages, nor after similar adult short-term exposure to estrogens, suggesting that the developmental stage of the uterus and the time of estrogen exposure were important factors in the development of the lesions.

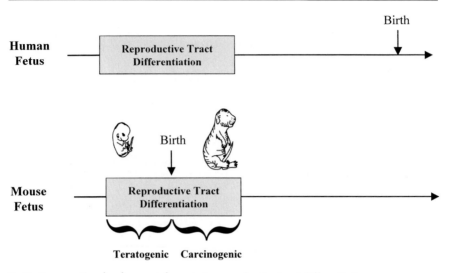

Fig. 3. Comparative developmental events in reproductive tract differentiation

For the female, perinatal exposure (prenatal or neonatal) to DES resulted in subfertility/infertility and increased incidences of reproductive tract tumors (benign and malignant). For the most part, tumor incidence was increased, but still low, after prenatal treatment, whereas neonatal exposure caused a higher incidence of tumors. Teratogenic (malformations) abnormalities were higher in animals prenatally exposed rather than neonatally exposed. Since developmental events in the reproductive tract occurring in early neonatal life for the mouse occur entirely prenatally in humans, the perinatal mouse (prenatal and neonatal) is useful to model human fetal development. This is schematically explained in Fig. 3. This figure illustrates, as previously stated, that the timing of exposure or the stage of tissue differentiation determines the subsequent resulting abnormalities.

5
Mechanisms of Reproductive Tract Toxicity

Many studies have demonstrated that perinatal exposure to DES interferes with the normal differentiation of both female and male reproductive tract. Although the mechanisms are not completely understood, epigenetic and/or genetic components may be involved. Recent studies [36, 37] have suggested a molecular mechanism responsible for the structural alterations observed in oviduct, uterus, cervix, and vagina. Cellular changes may also be closely linked to these structural alterations. Furthermore, we have described permanent abnormal gene imprinting in which neonatal exposure to DES causes demethylation of an estrogen-responsive gene in the mouse uterus [38]. We are currently investigating the relationship of this finding to tumor induction.

The role of the estrogen receptor (ER) in the induction of abnormalities and tumors following developmental exposure to DES has also been studied using transgenic mice which overexpress ERα (MT-mER). Transgenic ER mice were treated with DES during neonatal life as described for the CD-1 mice and followed as they aged. We hypothesized that because of the abnormal expression of the ER, the reproductive tract tissues of the MT-mER mice would be more susceptible to tumors after neonatal exposure to DES. In fact, it is interesting to note that mice overexpressing ER were indeed at a higher risk of developing abnormalities including uterine adenocarcinoma in response to neonatal DES as compared to DES-treated wild type mice; at 8 months, 73% of the DES-treated MT-mER mice compared to 46% of the DES-treated wild type mice had uterine adenocarcinoma. Further, these abnormalities occurred at an earlier age as compared to wild type DES mice [39]. These transgenic mouse studies suggested that the level of ERα present in a tissue may be a determining factor in the development of estrogen-related tumors. However, the specific role of ER in the induction and progression of the lesions requires additional study. Other transgenic mouse models which express variant forms of ER or the ER knockout, as well as experimental models constructed with ERβ, will also aid in determining the role of the ER in the development of these reproductive lesions. Since various estrogenic compounds have been reported to bind preferentially to either ERα or ERβ, these murine models will be essential in extrapolating human health risks to various environmental estrogen exposures.

6
Multigenerational Effects

As described, prenatal exposure of humans and perinatal exposure of experimental animals to DES have been associated with the subsequent development of reproductive tract abnormalities, including poor reproductive outcome and neoplasia in both males and females. Additional concerns with DES and other endocrine disrupting chemicals, however, hypothesize that adverse effects may be transmitted to subsequent generations [40–42]. In fact, experimental animal studies with various chemical carcinogens have suggested that this may indeed be a valid concern [43–45]. With this as background, we evaluated whether DES adverse effects could be passed on to succeeding generations. Outbred CD-1 mice were treated with DES either prenatally or neonatally: the prenatal group was treated (2.5, 5 or 10 µg/kg maternal body weight) on days 9–16 of gestation, which is the time of major organogenesis; the neonatal group was treated on days 1–5 of neonatal life (0.002 µg/pup/day). The doses that were chosen for these multi-generation studies were in the low dose range of the perinatal studies described earlier in the text because higher doses were not compatible with fertility. Fertility (the ability to produce a second-generation study) actually set the doses used in the multigeneration experiment. Female mice (F1) in each DES group were raised to sexual maturity and bred to control males. As described earlier, fertility of the F1 DES-exposed females was decreased in all groups. Female offspring (DES lineage or F2) of these matings were raised to maturity and housed with control males for 20 weeks. The fertility of these DES lineage female mice was not affected by DES exposure of their "grandmothers". DES lineage mice were sacrificed at 17–24 months of age. An increased incidence of malignant reproductive tract tumors, including uterine adenocarcinoma, was seen in DES-lineage mice; the range and prevalence of tumors increased with age. Because uterine adenocarcinomas were seen in both prenatal and neonatal groups, both developmental exposure periods were considered susceptible to the adverse effects of DES. These data suggested that the reduced fertility observed in the DES F1 female mice was not transmitted to their descendants; however, increased susceptibility to tumor formation was apparently transmitted to subsequent generations. More details of the multi-generation effects of DES can be found in another publication [15]. Additional studies with the F2 male siblings showed that there is also an increased incidence of reproductive tract tumors in these mice as they age [46]. Furthermore, studies from other laboratories have shown similar findings of DES lineage effects [13, 14, 47]. While our studies have only focused on adverse effects that are carried through the female (F1), studies from Turusov et al. [13] have reported similar increased tumor incidences being passed through DES males (F1).

The mechanisms involved in these transgenerational events remain unknown. Because DES has been reported to have genetic/epigenetic effects [36–38, 48–53], damage or altered imprinting of the germ cell is a possibility.

It is important to point out that increased occurrence of reproductive tract tumors in DES-lineage humans has not been reported; however, these people have not reached a comparable age at which the mice tumors were observed. Since the DES mouse studies have been predictive of other human abnormalities, continued close surveillance of the prenatally DES-exposed cohort and their offspring (grandchildren) is certainly advised. With regards to other endocrine disrupting chemicals, DES demonstrated that multigenerational effects are possible if exposure to estrogenic compounds occurs at critical stages of development.

7
Not Just DES – Other Examples

To determine if the adverse effects observed with perinatal exposure to DES were unique to DES or were common to other environmental compounds, we looked at other chemicals with estrogenic activity. Structural diversity of environmental estrogens has been previously reported [41]. Note the diversity of the compounds that we studied, which are illustrated in Table 3. We focused on only treatment during the neonatal period and only on uterine tumor induction; no multigenerational studies were included. In brief, mice were treated neonatally on days 1–5 with varying compounds and sacrificed from 12–18 months of age. A wide dose range was used for most compounds. Uterine tumor induction is summarized in Table 3. All compounds tested thus far have been carcinogenic to the neonatal mouse except methoxychlor. This is probably due to the use of the pure compound which is only weakly estrogenic unless metabolized to more active estrogenic substances. Since the neonatal mouse is not capable of significant metabolism to convert the pure compound to active estrogenic metabolites,

Table 3. Uterine tumor induction following neonatal treatment with various estrogens

Chemicals	Structure	Tumor
DES		+++
Estradiol		±
Hexestrol		+++
Tetrafluoro-DES		+++
Tamoxifen		++
Genistein		+
2-OH-Estradiol		±
4-OH-Estradiol		++++
Ethinyl Estradiol		++

Table 3. (continued)

Chemicals	Structure	Tumor
Bisphenol A		?
Nonylphenol		+
Methoxychlor		−

methoxychlor had little effect. Data from bisphenol A is still under study and evaluations are not complete. It is interesting to note that 17β estradiol was also carcinogenic in this experimental model, but only at high doses (three times greater than DES) [33]. Thus far the general trend for all of these compounds is that carcinogenesis is related to estrogenicity.

8
Summary and Conclusions

Sufficient evidence has now been accumulated by many laboratories that shows perinatal exposure of the developing fetus or neonate to exogenous estrogens like DES adversely affects reproductive tract function. Further, it is increasingly clear that long-term changes, including benign and malignant tumors, also occur. Fertility was decreased in mice after perinatal DES exposure in a dose-dependent manner, but it was adversely affected in low dose estrogen-exposed animals only later in life. Therefore fertility may not be a sensitive marker of reproductive toxicants in rodents until they age. Reproductive tract tumors were also seen in the animals as they aged, even animals that were exposed to low doses of environmental estrogens. Of particular significance is the increased prevalence of tumors that is apparently passed on to subsequent generations.

Although DES is a potent estrogenic compound, it provides markers of the adverse effects of exposure to estrogenic and other endocrine-disrupting substances if exposure occurs during development. These exposures may come from naturally occurring chemicals, synthetic or environmental contaminants, and/or pharmaceutical agents. While the mature organism may suffer adverse effects to these chemicals, the developing animal is particularly sensitive to perturbation by these compounds and experience may permanent, long-lasting consequences. Ongoing mechanistic studies will help us further assess the risks of exposure to various endocrine-disrupting chemicals in the environment.

Acknowledgements. The author would like to thank Ms. Wendy Jefferson and Ms. Jennifer Hagelbarger for the preparation of this manuscript.

9
References

1. Sharpe RM, Skakkebaek NE (1993) Lancet 341: 1392
2. Brake A, Krause W (1992) Brit Med J 305: 1498
3. Olsen GW, Bodner KM, Ramlow JM, Ross CE, Lipshultz LI (1995) Fertil Steril 63: 887
4. Swan SH, Elkin EP, Fenster L (1997) Environ Health Perspect 105: 1228
5. Davis DL, Bradlow HL, Wolff M, Woodruff T, Hoel DG, Anton-Culver H (1993) Environ Health Perspect 101: 372
6. Colborn T, Dumanski D, Myers JP (1996) Our stolen future. Penguin Books, New York
7. Snedeker SM, Diaugustine RP (1996) Prog Clin Biol Res 394: 211
8. Siiteri PK (1978) Cancer Res 38: 4360
9. Bern H (1992) The fragile fetus. In: Colborn T, Clement C (eds) Chemically-induced alterations in sexual and functional development: the wildlife/human connection. Princeton Scientific Publishing Company, Princeton, p 9
10. Herbst AL, Bern HA (1981) Developmental effects of diethylstilbestrol (DES) in pregnancy. Thieme-Stratton, New York
11. McLachlan JA (1985) Estrogens in the environment. Elsevier Science Publishing, New York
12. Newbold RR, McLachlan JA (1996) Transplacental hormonal carcinogenesis: diethylstilbestrol as an example. In: Huff J, Boyd J, Barrett JC (eds) Cellular and molecular mechanisms of hormonal carcinogenesis. Wiley-Liss, New York, p 31
13. Turusov VS, Trukhanova LS, Parfenov YD, Tomatis L (1992) Int J Cancer 50: 131
14. Walker BE, Haven MI (1997) Carcinogenesis 18: 791
15. Newbold RR, Hanson RB, Jefferson WN, Bullock BC, Haseman J, McLachlan JA (1998) Carcinogenesis 19: 1655
16. Meistich ML (1989) Interspecies comparison and quantitative extrapolation of toxicity to the human male reproductive system. In: Working P (ed) Toxicology of the male and female reproductive system. Hemisphere Publishing, New York, p 303
17. Gill WB, Schumacher GF, Bibbo M (1976) J Reprod Med 16: 147
18. Wilcox AJ, Baird DD, Weinberg CR, Hornsby PP, Herbst AL (1995) N Engl J Med 332: 1411
19. Carlsen E, Giwercman A, Keiding N, Skakkebaek NE (1992) Brit Med J 305: 609
20. Bromwich P, Cohen J, Stewart I, Walker A (1994) Brit Med J 309: 19
21. Irvine DS (1994) Brit Med J 309: 476
22. Auger J, Kunstmann JM, Czyglik F, Jouannet P (1995) N Engl J Med 332: 281
23. Gill WB, Schumacher GFB, Hubby MM, Blough RR (1981) Male genital tract changes in humans following intrauterine exposure to diethylstilbestrol. In: Herbst AL, Bern HA (eds) Developmental effects of diethylstilbestrol (DES) in pregnancy. Thieme-Stratton, New York, p 103
24. Conley GR, Sant GR, Ucci AA, Mitcheson HD (1983) JAMA. 249: 1325
25. Huseby RA (1976) J Toxicol Environ Health Suppl 1: 177
26. Newbold RR, Bullock BC, McLachlan JA (1985) Cancer Res 45: 5145
27. McLachlan JA, Newbold RR, Shah HC, Hogan MD, Dixon RL (1982) Fertil Steril 38: 364
28. McLachlan JA, Newbold RR, Bullock BC (1980) Cancer Res 40: 3988

29. Forsberg JG, Kalland T (1981) Cancer Res 41: 721
30. Plapinger L, Bern HA (1979) J Natl Cancer Inst 63: 507
31. Iguchi T, Takase M, Takasugi N (1986) Proc Soc Exp Biol Med 181: 59
32. Newbold RR, McLachlan JA (1982) Cancer Res 42: 2003
33. Newbold RR, Bullock BC, McLachlan JA (1990) Cancer Res 50: 7677
34. Leavitt WW, Evans RW, Hendry WJ III (1982) Adv Exp Med Biol 138: 63
35. Hendry WJ III, Leavitt WW (1993) Differentiation 52: 221
36. Ma L, Benson GV, Lim H, Dey SK, Maas RL (1998) Develop Biol 197: 141
37. Miller C, Degenhardt K, Sassoon DA (1998) Nat Genet 20: 228
38. Li S, Washburn KA, Moore R, Uno T, Teng C, Newbold RR, McLachlan JA, Negishi M (1997) Cancer Res 57: 4356
39. Couse JF, Davis VL, Hanson RB, Jefferson WN, McLachlan JA, Bullock BC, Newbold RR, Korach KS (1997) Mol Carcinogen 19: 236
40. Colborn T, Clement C (1992) Chemically-induced alterations in sexual and functional development: the wildlife/human connection. Princeton Scientific Publishing Company, Princeton
41. McLachlan JA (1985) Estrogens in the environment II: influences on development. Elsevier, New York
42. Colborn T, vom Saal FS, Soto AM (1993) Environ Health Perspect 101: 378
43. Mohr U, Emura M, Aufderheide M, Riebe M, Ernst H (1987) Transplacental carcinogenesis: rat, mouse, hamster. In: Jones TC, Mohr U, Hunt RD (eds) Genital system. Springer, Berlin Heidelberg New York, p 148
44. Napalkov NP, Rice JM, Tomatis L, Yamasaki H (1989) Perinatal and multigenerational carcinogenesis. IARC Scientific Publications, Lyon
45. Tomatis L (1994) Jpn J Cancer Res 85: 443–454
46. Newbold RR, Hanson RB, Jefferson WN, Bullock BC, Haseman J, McLachlan JA (2000) Carcinogenesis 21: 1355
47. Walker BE (1984) J Natl Cancer Inst 73: 133
48. Tsutsui T, Barrett JC (1997) Environ Health Perspect 105(Suppl. 3): 619
49. Endo S, Kodama S, Newbold RR, McLachlan JA, Barrett JC (1994) Cancer Genet Cytogenet 74: 99
50. Boyd J, Takahashi H, Waggoner SE, Jones LA, Hajek RA, Wharton JT, Liu FS, Fujino T, Barrett JC, McLachlan JA (1996) Cancer 77: 507
51. Gladek A, Liehr JG (1991) Carcinogenesis 12: 773
52. Moorthy B, Liehr J, Randerath E, Randerath K (1995) Carcinogenesis 16: 2643
53. Forsberg JG (1991) Terat Carcin Mutagen 11: 135

CHAPTER 7

Genotoxic Potential of Natural and Synthetic Endocrine Active Compounds

Manfred Metzler

Institute of Food Chemistry and Toxicology, University of Karlsruhe, P.O. Box 6980, 76128 Karlsruhe, Germany
E-mail: *manfred.metzler@chemie.uni-karlsruhe.de*

In addition to their hormonal activity, endocrine active compounds may have the potential to cause DNA adducts, oxidative DNA damage, and disturbances of the mitotic apparatus of the cell. They can therefore give rise to gene mutations as well as structural and numerical chromosomal aberrations in the same way as do chemical carcinogens. This genotoxic potential is independent of the hormonal activity and often related to metabolic activation. For 17β-estradiol and the equine estrogens equilin and equilenin, the formation of catechols, in particular by hydroxylation at C-4, and subsequent oxidation to ortho-quinones appears to be an important pathway to the ultimate carcinogenic metabolite. Some phytoestrogens, e. g., genistein and coumestrol, exhibit clastogenic and mutagenic effects without metabolic activation. The synthetic antiestrogen tamoxifen is a potent liver carcinogen in rats and forms DNA adducts after allylic hydroxylation followed by sulfation. Diethylstilbestrol, an established human and animal carcinogen, can also be activated to DNA binding metabolites, but is a powerful aneuploidogenic agent by disturbing the assembly of microtubules even without metabolism. Likewise, bisphenol A and 17β-estradiol are able to interfere with the mitotic apparatus and cause near-diploid aneuploidy. Thus, certain endocrine active compounds of both natural and anthropogenic origin are capable of causing various kinds of genotoxic effects, which may act in concert with the hormonal activity in the initiation of estrogen-induced carcinogenesis.

Keywords. Bisphenol A, Diethylstilbestrol, Estradiol, Equilenin, Equine estrogens, Genistein, Genotoxicity, Lignans, Phytoestrogens, Steroidal estrogens, Tamoxifen

1   Introduction 188

2   **Natural Compounds** 189
2.1   Steroidal Estrogens 189
2.1.1 Estradiol and Estrone 189
2.1.2 Equine Estrogens 193
2.2   Phytoestrogens 194

3   **Synthetic Compounds** 197
3.1   Tamoxifen 197
3.2   Diethylstilbestrol 200
3.3   Bisphenol A 202

4   **Conclusion** 204

5   **References** 205

The Handbook of Environmental Chemistry Vol. 3, Part M
Endocrine Disruptors, Part II
(ed. by M. Metzler)
© Springer-Verlag Berlin Heidelberg 2002

List of Abbreviations

BPA   bisphenol A
CMTC   cytoplasmic microtubule complex
COM   coumestrol
CREST   calcinosis, Raynaud's phenomenon, oesophagal motility
        abnormalities, sclerodactyly, and telangiectasia
CYP   cytochrome P450
DAI   daidzein
DES   diethylstilbestrol
E   estrogen
E1   estrone
E2 17β-estradiol
EAC   endocrine active compound
EN   equilenin
END   enterodiol
ENL   enterolactone
EQ   equilin
GEN   genistein
GSH   glutathione
hprt   hypoxanthine phosphoribosyl transferase
MAT   matairesinol
MT   microtubule
MTP   microtubule proteins
ROS   reactive oxygen species
SEC   secoisolariciresinol
SHE   Syrian hamster embryo
TAM   tamoxifen

1
Introduction

Whereas there is no doubt that endogenous steroidal estrogens are indispensable for normal growth and development of mammalian organisms, there is increasing concern about possible adverse effects of the very same hormones, in particular when an individual is exposed at the wrong time of development or to unphysiological concentrations. Estrogen-associated diseases include cancer of the breast and uterus in women and of the prostate and testes in men, as well as developmental disorders, e.g., premature thelarche in girls and cryptorchidism and hypospadias in boys. The concern has considerably increased with the observation (i) that numerous environmental agents, termed endocrine active compounds (EACs), of both natural and anthropogenic origin can mimic the hormonal activity of the endogenous steroid estrogens, (ii) that the incidence of estrogen-associated human diseases is on the rise in Western industrialized countries, and (iii) that adverse developmental effects typical of sex hormones are observable in wildlife inhabiting areas of heavy chemical contamination.

Of particular concern is the association of endogenous estrogens and exogenous estrogen-like compounds with cancer in humans. For example, cumulative and excessive exposure to endogenous steroidal estrogen, the most active of which is 17β-estradiol (E2), is considered a risk factor for mammary and uterine tumors. An example of a carcinogenic exogenous hormone-like agent is the estrogenic drug diethylstilbestrol (DES), prescribed in the USA and other countries between 1952 and 1970 to pregnant women to prevent miscarriage. DES is now known to be the cause of vaginal and cervical tumors in the female offspring and of testicular cancer in the male progeny of the treated women.

The increasing evidence for the carcinogenicity of at least certain estrogens from epidemiological studies and also from experimental studies with laboratory animals has raised interest in the molecular mechanisms of hormonal carcinogenesis. One important question is whether the carcinogenic effect of an estrogenic agent is exclusively due to its hormonal activity or whether it involves the induction of genetic damage, which is generally considered a hallmark in the mechanism of carcinogenesis induced by chemicals. There is no dispute that the hormonal activity, e.g., the stimulation of cell proliferation, plays an important role in estrogen-mediated neoplasia. In contrast, the role of genetic damage in the process of estrogen carcinogenesis is still controversial. This chapter briefly reviews the present evidence for the genotoxic potential of several major EACs. In addition to E2 and DES, the equine estrogens equilin (EQ) and equilenin (EN), the phytoestrogens genistein (GEN) and coumestrol (COM), and the synthetic compounds tamoxifen (TAM) and bisphenol A (BPA) will be discussed. The chemical structures of these compounds are depicted in Fig. 1.

2
Natural Compounds

2.1
Steroidal Estrogens

2.1.1
Estradiol and Estrone

The mammalian estrogens E2 and estrone (E1), which are interconvertible in the cell by the enzyme 17β-hydroxysteroid dehydrogenase, are human and animal carcinogens [1]. An established animal model is the male Syrian golden hamster, in which carcinogenic estrogens lead to a 100% incidence of kidney tumors. The observation that several potent steroidal estrogens, e.g., 17α-ethinyl-E2 and 2-fluoro-E2, are virtually noncarcinogenic for the hamster kidney [2] is not consistent with the paradigm that the stimulation of cell proliferation and the eventual accumulation of spontaneous mutations [3] is the cause of hormonal cancer. This was one of the reasons that prompted several laboratories to study other mechanisms, in particular the role of genotoxic metabolites. Several recent reviews have detailed the current state of knowledge in this field [4–10]. As depicted in Fig. 2, the major pathways in the oxidative metabolism of E2 and E1 comprise hydroxylation at C-2, C-4, and C-16. The catechol es-

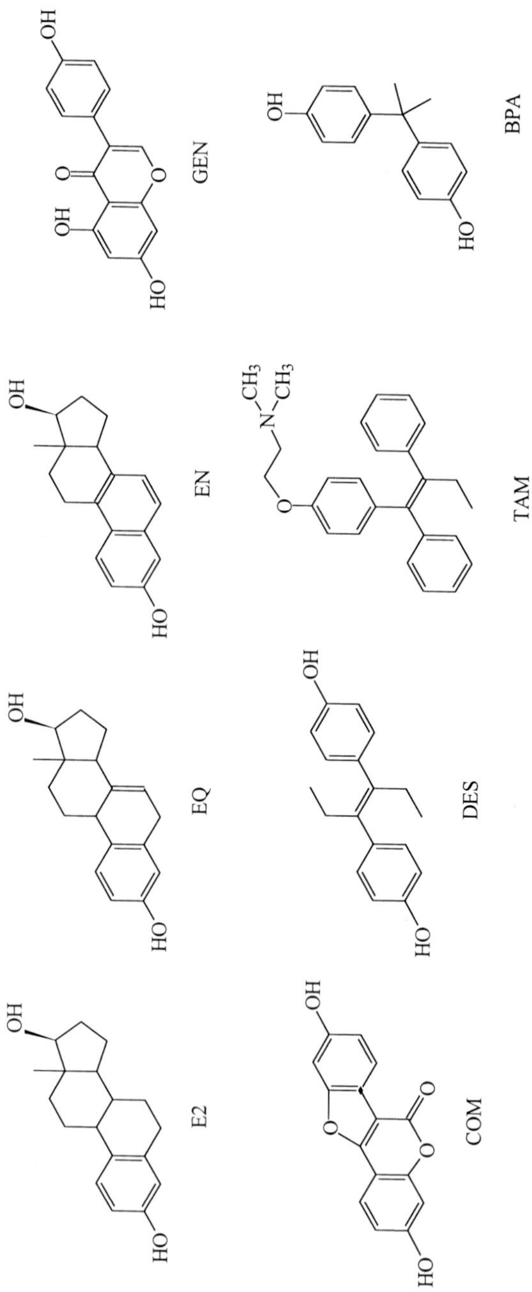

Fig. 1. Chemical structures of various EACs

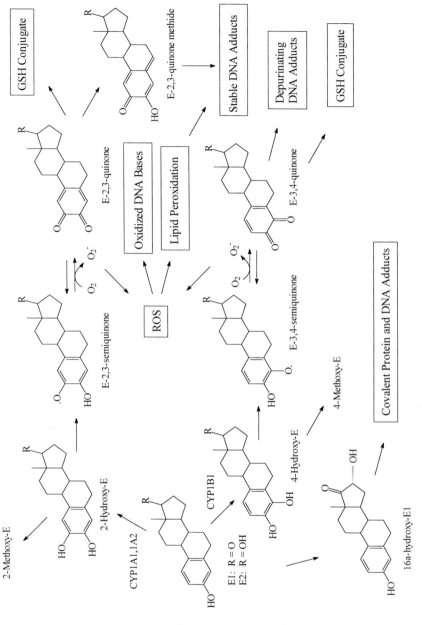

Fig. 2. Pathways proposed for the metabolic activation of E2 and E1

trogens 2-hydroxy-E and 4-hydroxy E can either be conjugated, e.g., methylated, or further oxidized to semiquinones and quinones. The latter are reactive metabolites, covalently binding to glutathione (GSH), proteins, and DNA. Several adducts of the catechol estrogens with the DNA bases guanine and adenine have been identified in in vitro systems and, to a limited extent, also in vivo. The chemical structures of the adducts show that the 3,4-quinone reacts primarily through C-1 with the N7-position of guanine, whereas the 2,3-quinone tautomerizes to a quinone methide prior to reacting through C-6 with the exocyclic amino groups of adenine and guanine. The DNA adducts of the 2,3-quinone are chemically stable, in contrast to the adduct of the 3,4-quinone which is released from the DNA by spontaneous depurination.

It is believed that 4-hydroxylation is the critical pathway for the carcinogenicity of E2. The major reasons for this proposition are (i) 4-hydroxy-E2 but not 2-hydroxy-E2 is carcinogenic in the Syrian hamster kidney and also in another animal model, the neonatal CD-1 mouse uterus [11], (ii) 4-hydroxylation is predominating over 2-hydroxylation in the target tissues of E2 carcinogenicity, e.g., the hamster kidney and the human breast and uterus, due to the activity of the hydroxylating enzyme cytochrome P4501B1 (CYP1B1), and (iii) the depurinating DNA adducts are considered a more critical lesion for the initiation of cancer than the stable DNA adducts. Moreover, 4-hydroxy-E are less rapidly methylated by catechol-O-methyltransferase and thus less efficiently inactivated than their 2-hydroxylated isomers.

In addition to the formation of direct DNA adducts, catechol estrogens can give rise to indirect DNA adducts through redox cycling (Fig. 2). The generation of reactive oxygen species (ROS) can lead to oxidative DNA damage and lipid peroxidation, the latter generating reactive aldehydes which adduct DNA. Both types of ROS-mediated DNA modifications have been observed with E2 in target tissues in vivo.

Due to their potential to cause DNA damage, E2 and its catechol metabolites should give rise to gene mutations. Whereas earlier studies in standard mutagenicity test systems such as the Ames test gave negative results, a weak mutagenic response was found in more recent studies [12, 13]. Interestingly, E2 mutagenicity was only observed at very low (10–100 pmol/l) and relatively high (0.1–1 µmol/l) concentrations in one of the studies [12]. This may be due to the fact that catechol estrogens have both prooxidant and antioxidant properties, and the prooxidant effects may only be detected at certain concentrations [14]. Moreover, it may be assumed that E2 exhibits only very weak mutagenicity, which is not readily detected by the conventional short-term assays designed for strong mutagens.

The third major metabolite of E2, namely 16α-hydroxy-E1 (Fig. 2), has also been implicated in the mechanism of E2 carcinogenesis [15, 16]. Due to the ability of the 17-keto group to form a Schiff base with amino groups and subsequently undergo an Amadori rearrangement with the 16α-hydroxyl group, this metabolite can form a stable protein or DNA adduct. It has been postulated that such a covalent modification of the estrogen receptor could lead to a persistent stimulation of cell proliferation in the mammary epithelium and thus favor the neoplastic progression of breast cancer.

To increase further the complexity of possible mechanisms of genetic toxicity of E2, numerical chromosomal aberrations (aneuploidy) have been observed with E2 and its catechol metabolites in vitro and in vivo. E2 and E1 caused aneuploidy and neoplastic transformation in cultured Syrian hamster embryo (SHE) fibroblasts even in the absence of detectable gene mutations [13]. The biochemical mechanisms underlying the aneuploidogenic effects are still poorly understood: E2 does not interfere with the mitotic spindle and may act through other components of the mitotic machinery, whereas the catechol metabolites, after oxidation to the quinones, bind covalently to critical sulfhydryl groups of tubulin and inhibit microtubule assembly and spindle formation [17].

In view of the plethora of effects of E2 and some of its metabolites on the genetic material, future in-depth mechanistic studies are required to clarify further the mechanism of E2-mediated carcinogenesis.

2.1.2
Equine Estrogens

Long-term estrogen replacement therapy of postmenopausal women is associated with a clear risk for endometrial cancer and a possible risk for mammary cancer [18, 19]. The most widely prescribed estrogen replacement formulations

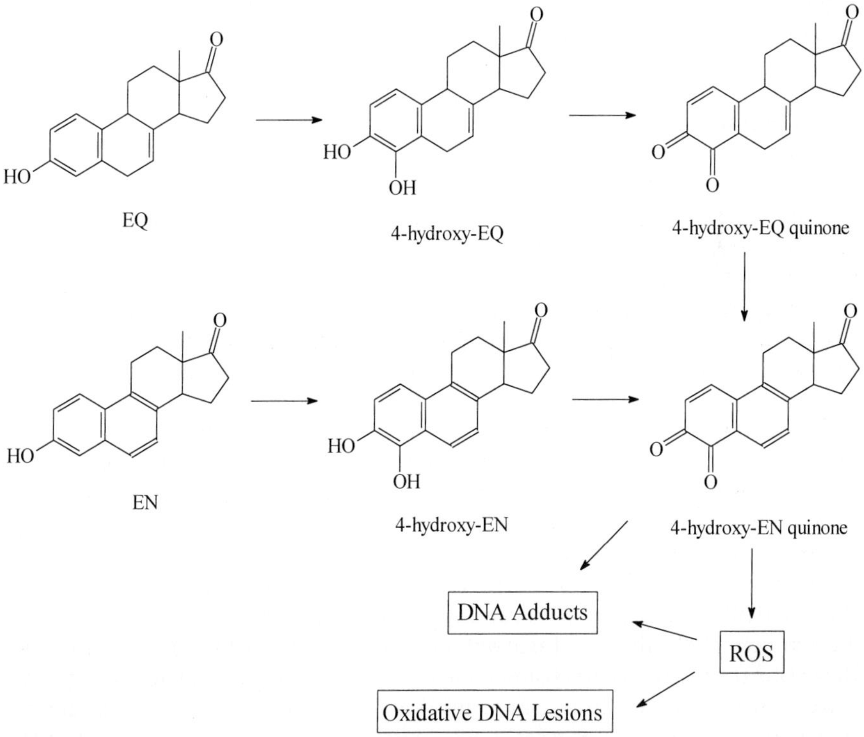

Fig. 3. Pathways proposed for the metabolic activation of equine estrogens

in the USA are derived from the urine of pregnant mares (e.g., Premarin) and contain, in addition to E2 and E1, significant amounts of steroidal estrogens with an unsaturated B ring, i.e., equilin (EQ) and equilenin (EN). Both EQ and EN are carcinogenic in the Syrian hamster kidney model [20]. Studies on the metabolism and genotoxicity of EQ and EN have recently been summarized by Bolton et al. [21]. In analogy to E2 and E1 (see above), EQ and EN undergo metabolic hydroxylation at C-2 and C-4 to yield catechols (Fig. 3). Interestingly, increasing the number of double bonds in the B ring favors hydroxylation at C-4 over C-2. As discussed above for the 4-hydroxylated E1 and E2, this pathway may lead to carcinogenic metabolites. It has been observed that both 4-hydroxy-EQ and 4-hydroxy-EN readily autoxidize to the respective *ortho*-quinones, and the quinone of 4-hydroxy-EQ isomerizes and autoxidizes to the quinone of 4-hydroxy-EN (Fig. 3). 4-Hydroxy-EN appears to undergo redox cycling and to generate ROS, as it causes DNA single strand breaks and oxidative damage to DNA bases in cell-free systems and also in human breast cancer cell lines [22, 23]. Reaction of 4-hydroxy-EN and also 4-hydroxy-EQ with DNA or deoxynucleosides in vitro gave rise to several different types of lesions including bulky stable adducts and apurinic sites [24]. More recently, 4-hydroxy-EN was shown to induce ROS, DNA single strands breaks, and oxidized DNA bases in two murine fibroblast cell lines to a much greater extent than 4-hydroxy-E1 [25]. The occurrence of DNA lesions in whole animals after administration of EQ and EN and the mutagenicity of these equine estrogens or their metabolites remain to be demonstrated.

2.2
Phytoestrogens

More than 300 plants are known to contain compounds exhibiting estrogenic activity [26]. Typical examples of such phytoestrogens are the isoflavones GEN and daidzein (DAI, Fig. 4), prominent in soy beans, the coumestan COM, present in sprouts of soy beans and alfalfa, and lignans, present in oilseeds, vegetables, and fruit. The two major plant lignans, secoisolariciresinol (SEC) and matairesinol (MAT, Fig. 4), are metabolized by intestinal bacteria to the mammalian lignans enterodiol (END) and enterolactone (ENL). Depending on the diet, some phytoestrogens may be ingested in considerable amounts, e.g., GEN and DAI with soy-based food and SEC and MAT with food containing flaxseed, and lead to plasma levels in the micromolar range [27]. Recently, isoflavones and lignans have gained interest due to their putative beneficial health effects. Consequently, there is an increasing tendency to consume phytoestrogens as dietary supplements.

Despite their widespread occurrence and dietary consumption, as well as their structural similarity with carcinogenic estrogens like DES and E2, very little is known about the genotoxic potential of phytoestrogens and their metabolites. The results of recent studies on the effects of several prominent phytoestrogens in cultured Chinese hamster V79 cells at various endpoints of genetic damage, e.g., induction of micronuclei, interference with cytoplasmic and mitotic microtubules, and mutations at the hprt gene locus [28, 29], are summa-

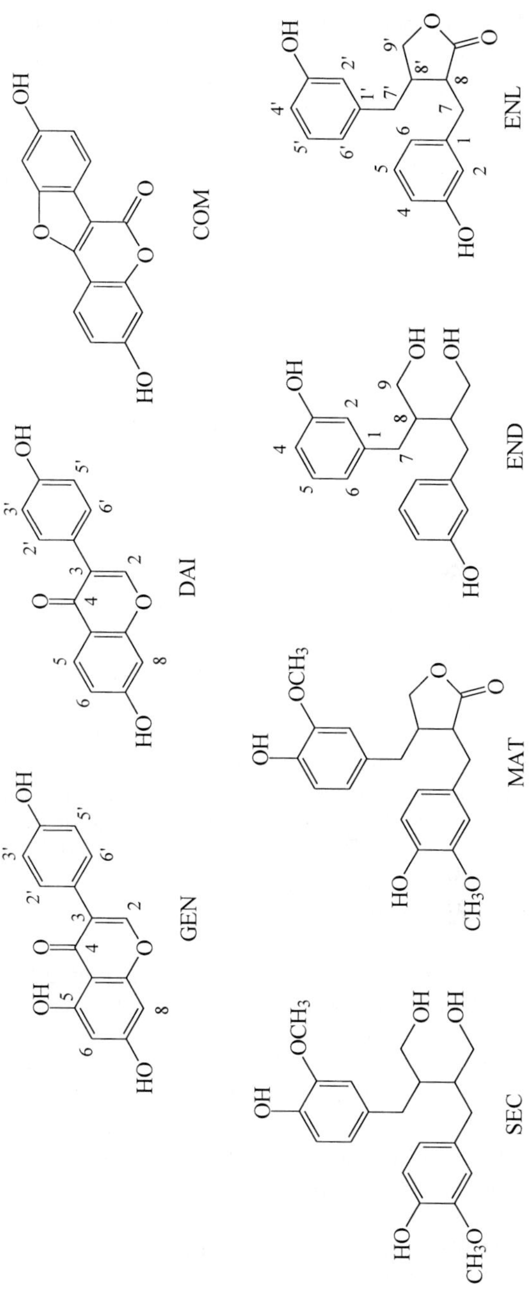

Fig. 4. Chemical structures of prominent phytoestrogens

Table 1. Genotoxic effects of various phytoestrogens in cultured V79 cells, according to [28] and [29]

Effect	GEN	DAI	COM	END	ENL	SEC	MAT
Induction of micronuclei							
CREST-positive	–	–	–	–	–	–	–
CREST-negative	+++	–	++	–	–	–	–
Interference with MT							
Disruption of CMTC	–	–	–	–	–	–	–
Mitotic spindle	–	–	–	–	–	–	–
Cell-free MT assembly	–	–	–	–	–	–	–
Induction of mitotic arrest	–	–	–	–	–	–	–
Mutations at hprt locus	(+)	–	++	–	–	–	–

rized in Table 1. A clearly distinct genotoxic profile was obtained for the seven phytoestrogens: whereas the four lignans and DAI did not show an effect at any of the endpoints, the induction of micronuclei containing acentric chromosomal fragments (as tested with the antikinetochore CREST antibody) and gene mutations at the hprt locus were observed for GEN and COM. GEN also proved mutagenic in human lymphoblastoid cells [30] and caused DNA strand breaks in the human colon tumor cell line HT29 [31]. The clastogenicity of GEN and COM, but not DAI, was also clearly demonstrated in cultured human peripheral blood lymphocytes [32]. It should be noted that GEN and DAI differ by just one hydroxyl group (Fig. 4), indicating that the genotoxicity is closely associated with the chemical structure. The specificity of the genotoxic effect is also illustrated by comparing COM with the structurally similar DES (Fig. 1): whereas COM is a clastogen and gene mutagen in V79 cells without any aneuploidogenic potential, DES is a clear aneugen in the same cells without any clastogenicity (see Sect. 3.2).

The biological significance of the clastogenicity and mutagenicity of GEN remains to be studied. With the exception of a recent report associating leukemia in Japanese infants with a soy diet [33], no epidemiological studies appear to exist linking isoflavones to increased cancer incidence in humans. A long-term carcinogenicity study in rodents is currently being conducted by the National Toxicology Program. In neonatal CD-1 mice, an animal model used to demonstrate the carcinogenicity of E2 (see Sect. 2.1.1) and DES (see Sect. 3.2), GEN induced the same incidence (about 30%) of uterine adenocarcinoma as did DES when administered at an equiestrogenic dose [34]. However, it is still unclear whether genotoxicity plays a role in the mechanism of estrogen carcinogenesis in this animal model.

In addition to the parent phytoestrogens, their metabolites, in particular hydroxylation products, may contribute to a possible genotoxic risk. At present, however, no data exist on the genotoxic potential of phase I or phase II metabolites of isoflavones and lignans. Surprisingly little was known until recently about the phase I metabolism of prominent phytoestrogens in humans and ex-

Table 2. Oxidative metabolites of GEN, DAI, END and ENL identified in human urine after ingestion of soy and flaxseed, according to [36] and [40]. HO, hydroxy; position of HO- see Fig. 4. The sequence of listing does not reflect the amount of the respective metabolite

Phyto-estrogen	Metabolites
GEN	6-HO-GEN, 8-HO-GEN, 3'-HO-GEN, 3',6-diHO-GEN, 3',8-diHO-GEN
DAI	6-HO-DAI, 8-HO-DAI, 3'-HO-DAI, 6,8-diHO-DAI, 3',6-diHO-DAI, 3',8-diHO-DAI
END	2-HO-END, 4-HO-END, 6-HO-END
ENL	2-HO-ENL, 4-HO-ENL, 6-HO-ENL, 2'-HO-ENL, 4'-HO-ENL, 6'-HO-ENL

perimental animals other than the biotransformation by intestinal bacteria. However, both isoflavones and lignans appear to be substrates for mammalian cytochrome P450, as rat and human hepatic microsomes are capable of generating several hydroxylated metabolites of DAI and GEN [35–38], and of END and ENL [39]. Most of these oxidative in vitro metabolites have also been detected in the urine of humans after ingestion of a diet containing soy or flaxseed (Table 2). All of the GEN and DAI metabolites are catechols, whereas the major metabolites of END and ENL are hydroquinones. It should prove interesting to examine the genotoxic potential of these metabolites.

3
Synthetic Compounds

3.1
Tamoxifen

The anti-estrogen TAM has been widely used since 1971 for adjuvant therapy in the treatment of women with breast cancer. It is currently tested on a large scale as a chemopreventive agent in healthy women with a high familial risk for mammary tumors. In both breast cancer patients and healthy women treated with TAM, a small increase in the incidence of endometrial cancers was observed, as reviewed in [41]. In male and female rats, but not mice, TAM induced a high incidence of hepatocellular carcinomas. No uterine tumors were observed in adult rats or mice after treatment with TAM [41]. However, exposure of newborn CD-1 mice [42] and Wistar rats [43] to TAM during the first five postnatal days led to a significant increase in uterine tumors late in life.

In order to assess better the carcinogenic risk of TAM for humans, several laboratories are working on the mechanisms underlying the observed carcinogenic effects in rodents. There is general agreement that TAM acts as a genotoxic agent in the rat liver due to the formation of one or more reactive metabolites causing DNA adducts. TAM is metabolized via N-desmethylation, 4-hydroxylation, α-hydroxylation, and N-oxidation in rat hepatic microsomes, rat hepatocytes, and in vivo [44]. α-Hydroxylation and 4-hydroxylation of TAM may represent the initial steps of metabolic activation (Fig. 5). For the genotoxicity of TAM in the rat liver, α-hydroxylation followed by sulfation appears to be

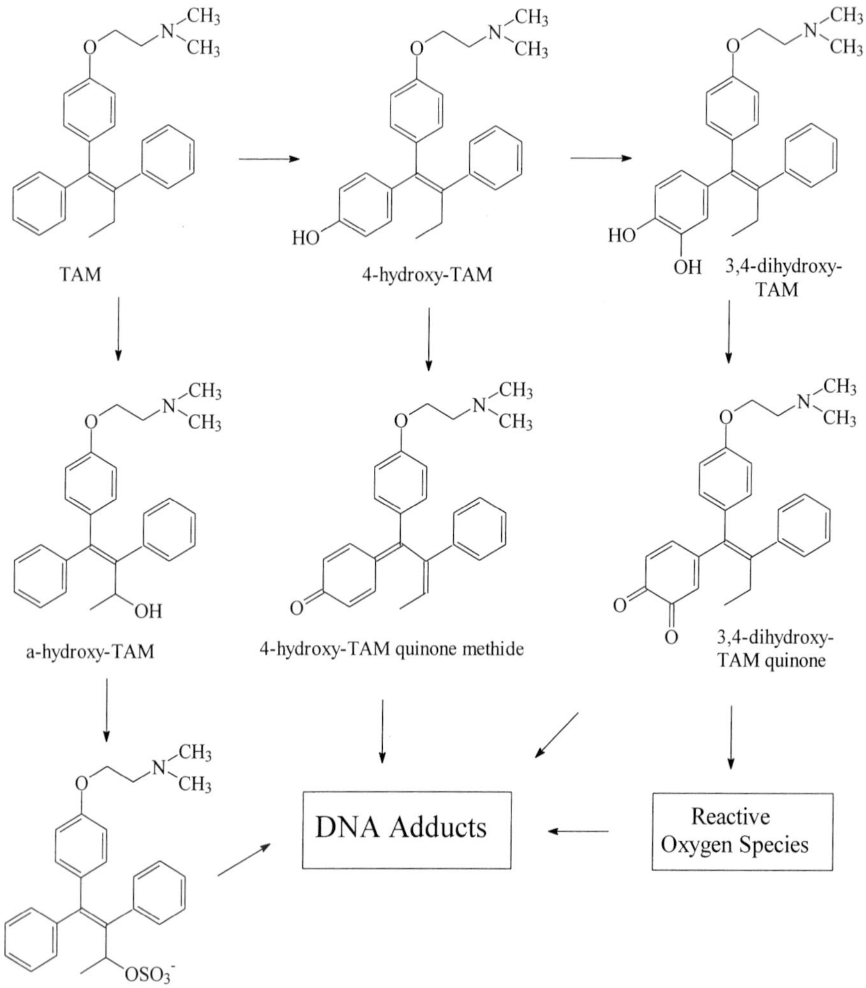

Fig. 5. Pathways proposed for the metabolic activation of tamoxifen

the critical pathway leading to DNA adduction. The two major DNA adducts from the liver of TAM-treated rats were identified by mass spectrometry and comparison with authentic reference compounds [45, 46] and are derived from the reaction of the exocyclic amino group of guanine with the α-position of TAM and N-demethyl-TAM (Fig. 6). 4-Hydroxy-TAM is not DNA reactive per se but can be oxidized to a quinone methide (Fig. 5), which reacts in vitro with DNA by a 1,8-Michael addition to yield (*E*)- and (*Z*)-α-(deoxyguanosin-*N2*-yl)-4-hydroxy-TAM (Fig. 6) as the major adducts. However, when 4-hydroxy-TAM was administered to rats or incubated with rat hepatocytes, no adducts could be detected in liver cell DNA by the 32P-postlabeling technique [47, 48]. This sug-

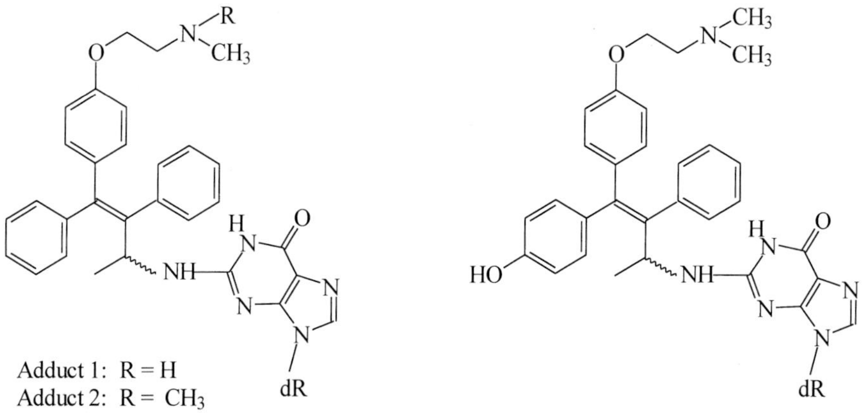

Adduct 1: R = H
Adduct 2: R = CH3

Fig. 6. DNA adducts of tamoxifen detected in rat liver (*left*) and obtained from 4-hydroxy-TAM in vitro (*right*)

gests that neither 4-hydroxy-TAM nor the product of subsequent hydroxylation, 3,4-dihydroxy-TAM (Fig. 5), play a significant role in the hepatocarcinogenicity of TAM in the rat.

Presently there are no epidemiological data suggesting that TAM is carcinogenic in the human liver. No TAM-specific DNA adducts were detected by 32P-postlabeling in liver biopsy samples of TAM-treated breast cancer patients [49], and incubation of human hepatocytes with TAM or α-hydroxy-TAM failed to yield significant levels of characteristic DNA adducts [50].

The etiology of the aforementioned endometrial cancer observed at low incidence in TAM-treated women is much less clear than that of TAM-induced rat liver cancer. Since previous attempts to detect DNA adducts in endometrial tissue failed or gave ambiguous results, a hormonal mechanism due to the partial estrogenic effect of TAM on the human uterus has been assumed. However, by using an ultrasensitive modification of the 32P-postlabeling technique, DNA adducts were recently detected in the endometrium of 6 out of 13 breast cancer patients but not in control subjects, and were identified as E- and Z-isomers of α-(N2-deoxyguanosinyl)-TAM [51], the same type of adduct as identified in rat liver (Fig. 6). The level of the endometrial adducts was in the range of 1–10 per 10^8 nucleotides, which is two orders of magnitude below the adduct level in rat liver at doses giving rise to tumors.

Why is TAM a potent genotoxic hepatocarcinogen in the rat and a poor genotoxin, if at all, in humans, in spite of the fact that the critical metabolite, α-hydroxy-TAM, is formed in both species? Studies by Glatt et al. [52] provide an explanation. Genetically engineered *Salmonella typhimurium* and Chinese hamster V79 cells, expressing various forms of rat and human sulfotransferases, were used to study the ability of α-hydroxy-TAM to induce DNA adduct formation and gene mutations in these cells. α-Hydroxy-TAM was mutagenic and gave rise to the same pattern of DNA adducts as observed in rat liver in vivo only in cells expressing rat hydroxysteroid sulfotransferase a, a liver-specific enzyme. Cells expressing the corresponding human sulfotransferase or six other human xenobiotic-metabolizing sulfotransferases, including the two sulfotransferases identified in endometrium, were at least 20 times less active in metabolizing α-hydroxy-TAM to a genotoxin. This strongly suggests that the species difference between rat and mouse with respect to the genotoxicity of TAM resides in the different activities of a phase II enzyme, sulfotransferase, necessary to generate the ultimate carcinogen, i.e., the sulfate conjugate of α-hydroxy-TAM. The fact that the human liver and the human endometrium sulfotransferases are not capable of activating α-hydroxy-TAM should provide some degree of protection.

3.2
Diethylstilbestrol

DES is a diphenolic compound synthesized in 1938 [53] in the search for an orally effective estrogen. It matches E2 in terms of its affinity to the estrogen receptors and is superior in terms of oral efficacy. DES contributed much to the present concern about carcinogenic effects of estrogens, as it proved to be the culprit of a rare form of vaginal and cervical tumors discovered in 1970 by Herbst et al. [54] and was shown to be a potent carcinogen in several animal models including the Syrian hamster kidney [55] and the neonatal mouse uterus [56]. Despite the efforts of several laboratories, its mechanism of tumorigenesis is still poorly understood. The formation of reactive metabolites has been demonstrated and their involvement in DES-mediated carcinogenesis proposed as early as 1975 [57, 58]. Today, several pathways for the metabolic activation of DES are discussed; for review see [6, 9, 10]. The best studied route for the generation of reactive metabolites is depicted in Fig. 7 together with the associated biochemical effects. Briefly, DES can be oxidized by cytochrome P450 or peroxidases to its 4′,4″-semiquinone and 4′,4″-quinone, which tautomerizes spontaneously to Z,Z-dienestrol. This metabolic sequence has been shown with microsomes from various organs and is believed to occur in target tissues of DES carcinogenicity. The semiquinone intermediate may reduce oxygen to superoxide radicals and thus give rise to ROS, which in turn causes lipid peroxidation and oxidative DNA damage. Moreover, DNA may be adducted by the electrophilic DES-4′,4″-quinone and by reactive products of lipid peroxidation. Both types of adducts have been demonstrated by the 32P-postlabeling technique in the Syrian hamster kidney, together with oxidized DNA bases. Other pathways of DES metabolism proposed for metabolic activation are the formation of the catechol 3-hydroxy-DES with subsequent oxidation to the respective ortho-semiquinone and -quinone in analogy to the activation of E2 and EN discussed earlier (see Sect. 2.1), and α-hydroxylation at C-2 or C-5 of the ethyl groups to yield an allylic alcohol which becomes reactive after sulfate conjugation [59], as described for tamoxifen (see Sect. 3.1). By the same token, it can be speculated that 1-hydroxy-Z,Z-dienestrol, which is also an allylic alcohol, can be activated by sulfation. Thus, DES can cause DNA lesions by various mechanisms. It remains to be demonstrated which are relevant for its carcinogenic effects.

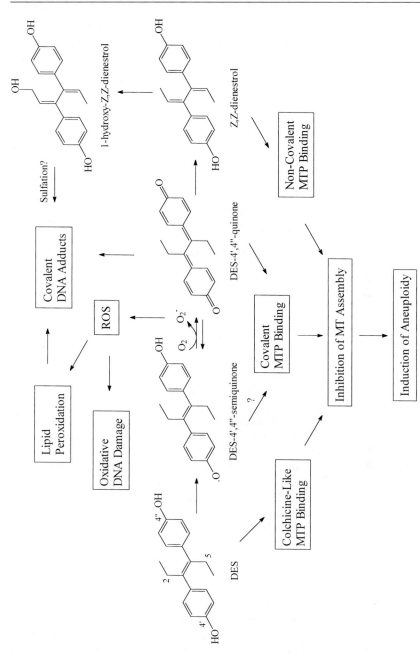

Fig. 7. Proposed metabolic activation and genotoxic effects of DES

In addition to damaging DNA, DES has been shown to give rise to numerical and structural chromosomal aberrations in cultured cells [60] and in target organs of DES carcinogenesis, e.g., the Syrian hamster kidney [61]. As depicted in Fig. 7, DES and several of its oxidative metabolites affect microtubule proteins (MTP) through different mechanisms [17]. Binding to MTP inhibits the assembly of microtubules and disrupts the mitotic spindle, thereby causing aneuploidy. DES itself binds to the major MTP, the heterodimer of α- and β-tubulin, in a colchicine-like manner. The binding of DES-4′,4″-quinone and Z,Z-dienestrol to MTP has been less well characterized, but both agents inhibit cell-free MT assembly [62].

Induction of aneuploidy represents an important mechanism of neoplastic cell transformation [63, 64]. DES was shown to induce near-diploid aneuploidy as well as morphological and neoplastic transformation of primary Syrian hamster embryo (SHE) fibroblasts in the absence of detectable gene mutations or DNA damage [60, 65, 66]. However, gene mutations were observed in the same cells with DES in the presence of exogenous metabolic activation [67]. Both aneuploidy induction and DNA adduct formation correlated with DES-induced cell transformation and may be important in the mechanism of DES carcinogenesis [65].

3.3
Bisphenol A

BPA is a diphenolic compound with a chemical structure similar to that of DES (Fig. 1) but a much lower estrogenicity. It is widely used, e.g., for the production of polycarbonates and epoxy resins, and produced in large amounts. Some of the materials are used for containing food and potable water and provide a source for human exposure, since BPA has been reported to leach from such containers.

Conventional short-term assays have shown no evidence for genotoxicity (as summarized by Tsutsui et al. [68], and a carcinogenicity bioassay in male and female Fischer 344 rats and B6C3F1 mice was negative with the exception of a marginally significant increase in leukemia in male rats only [69].

More recently, the potential of BPA to form DNA adducts and to induce aneuploidy has been studied. BPA was chemically (with potassium nitrosodisulfonate) and enzymatically (with peroxidase/hydrogen peroxide) oxidized to a catechol and subsequently to the respective *ortho*-quinone (Fig. 8), which gave rise to several DNA adducts under cell-free conditions [70]. After oral or intraperitoneal administration of a very high dose (200 mg/kg body weight) of BPA to male CD-1 rats, two major and several minor DNA adducts were detected in the liver at a level of about 10 total adducts per 10^9 nucleotides using the 32P-postlabeling technique [71]. The major in vivo liver DNA adducts of BPA matched two of four products of the cell-free reaction of BPA-quinone with DNA in thin layer chromatography. This led the authors to conclude that the quinone of a BPA-catechol is the DNA-binding metabolite (Fig. 8). However, the exact chemical structures of the adducts have not yet been reported, nor has the unequivocal formation of BPA catechol in vivo. An in vitro study with hepatic

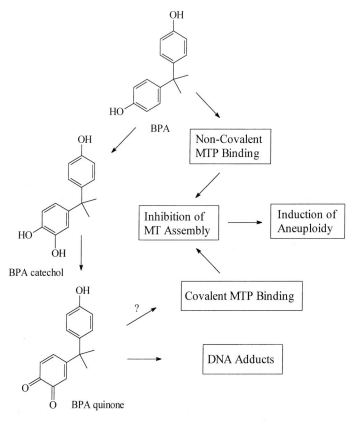

Fig. 8. Proposed metabolic activation and genotoxic effects of BPA

microsomes from aroclor-induced male rats showed a very poor oxidative metabolism of BPA with the formation of only trace amounts of BPA catechol, and no type I binding spectrum to cytochrome P450 with BPA concentration up to 20 nM; higher concentrations caused denaturation of microsomal proteins [72]. Recent reports have studied the inhibitory effects of BPA on various hepatic cytochrome P450 activities of rats [73] and humans [74] in detail. A recent study on the metabolism of 14C-labeled BPA in Fischer 344 rats showed that orally administered BPA is rapidly glucuronidated [75]. These data imply that BPA, in contrast to other phenolic compounds discussed above, like E2, EN, GEN, and DES, is a poor substrate for catechol formation. This fact and the rapid glucuronidation of BPA render it unlikely that significant DNA adduction will occur at low levels of exposure.

Because of the structural similarity of BPA to DES, the potential of BPA to interfere with microtubules (MT) and to induce aneuploidy has been studied. Indeed, BPA inhibited cell-free MT assembly and induced mitotic arrest and micronuclei containing whole chromosomes in cultured V79 cells in a DES-like manner, although with a lower activity [76, 77]. Treatment of SHE cells with BPA induced morphological transformation and near-diploid aneuploidy but failed to induce gene mutations [78], also behaving like DES in this respect (see Sect. 3.1). When four other bisphenols structurally related to BPA were studied, a good correlation of cell-free MT inhibition and micronucleus induction in V79 cells [76] with aneuploidy induction and cell transformation in SHE cells was observed [68]. As discussed above for DES, aneuploidy induction may also play an important role in the genotoxicity of BPA and other bisphenols.

4
Conclusion

This paper summarizes the current evidence suggesting that several prominent EACs not only exert hormonal effects on cells but also have the potential to inflict lesions of the genetic material which are well established in the field of chemical carcinogenesis. Such lesions include DNA adducts and oxidative DNA damage as well as structural and numerical chromosomal aberrations. With most of the EACs, these genotoxic effects and their sequelae, i.e., gene mutations and near-diploid aneuploidy, are subtle and not easily detectable in conventional short-term assays or in whole-animal studies. Systematic studies of a large number of steroidal estrogens and their metabolites as well as synthetic EACs have been carried out in SHE cells over the past years [13, 60, 65, 67, 68, 78]. SHE cells are primary fibroblasts that allow one to measure cellular transformation and various genetic effects in the same target cells. Moreover, SHE cells do not express measurable levels of estrogen receptors, and estrogen treatment does not stimulate cell growth. Thus, any observed alterations of the cells cannot be due to receptor-mediated hormonal effects of the tested EAC. On the other hand, SHE cells do have endogenous enzymes with cytochrome P450 and peroxidase activity and are therefore able to metabolize EACs. It has been observed that many, but not all, tested EACs lead to a concentration-dependent morphological and neoplastic transformation of SHE cells, and that the transforming activity corresponds to at least one of the genotoxic effects of the EAC tested, i.e., aneuploidy, structural chromosome aberration, or gene mutation (reviewed by [65]). This and the observation from whole-animal studies that not every powerful estrogen is a carcinogen [2] strongly suggest that genotoxic effects are involved in the initiation of EAC-induced carcinogenesis. It appears that estrogen-mediated carcinogenesis is the combined effect of hormonal stimulation of cell proliferation, heritable reprogramming of cellular differentiation, and genotoxicity [65, 79].

The proposition that EACs are not alike in terms of their carcinogenic potential has important implications for their risk assessment. First, it means that testing of the hormonal activity of a presumed EAC is not sufficient to assess its potential carcinogenic risk. Second, it may allow a better understanding of possible interactions of EACs with other environmental or dietary factors. For example, it has been reported that catechol metabolites of polychlorinated biphenyls inhibit the catechol-O-methyltransferase-mediated metabolism of catechol estrogens [80], which may modify the carcinogenicity of E2 and other EACs activated via catechol formation (see Sect. 2.1.). Similarly, the carcinogenic activity of phenolphthalein, used in over-the-counter laxatives, may in part be explained by the observation that its catechol metabolite also inhibits catechol-*O*-methyltransferase, thereby potentiating the genotoxicity of endogenous catechol estrogens [81].

Finally, a better understanding of the role of genotoxicity in the carcinogenic effects of estrogens may eventually help risk assessment regulatory agencies to establish tolerable concentrations of EACs. For compounds causing exclusively or predominantly aneuploidy, the linear dose-response relationship used for classical gene mutagens to extrapolate from high to low concentrations may not apply, and a threshold of action may exist. Indeed, a recent study on the induction of hyperdiploidy and polyploidy by DES and E2 in cultured human lymphocytes has shown sublinear dose-response relationships with likely threshold concentrations [82].

Acknowledgement. Studies from our laboratory cited in this paper were supported by the Deutsche Forschungsgemeinschaft (Grants Me 574/9 – 1 and Me 574/9 – 2).

5
References

1. International Agency for Research on Cancer (1979) IARC monographs on the evaluation of the carcinogenic risk of chemicals to humans, vol 21: sex hormones. International Agency for Research on Cancer, Lyon, p 173
2. Liehr JG (1983) Mol Pharmacol 23: 278
3. Feigelson HS, Henderson BE (1996) Carcinogenesis 17: 2279
4. Yager JD (2000) J Natl Cancer Inst Monogr 27: 67
5. Liehr JG (2000) Endocrine Rev 21: 40
6. Roy D, Liehr JG (1999) Mutat Res 424: 107
7. Bolton JL, Pisha E, Zhang F, Qui S (1998) Chem Res Toxicol 11: 1113
8. Zhu BT, Conney AH (1998) Carcinogenesis 19: 1
9. Metzler M, Kulling SE, Pfeiffer E, Jacobs E (1998) Z Lebensm Unters Forsch A 206: 367
10. Yager JD, Liehr JG (1996) Ann Rev Pharmacol Toxicol 36: 203
11. Newbold RR, Liehr JG (2000) Cancer Res 60: 235
12. Kong LY, Szaniszlo P, Albrecht T, Liehr JG (2000) Int J Oncol 17: 1141
13. Tsutsui T, Tamara Y, Yagi E, Barrett JC (2000) Int J Cancer 86: 8
14. Markides CSA, Roy D, Liehr JG (1998) Arch Biochem Biophys 360: 105
15. Swaneck GE, Fishman J (1988) Proc Natl Acad Sci USA 85: 7831
16. Davis DL, Telang NT, Osborne MP, Bradlow (1997) Environ Health Perspect 105(Suppl 3): 571
17. Metzler M, Pfeiffer E, Schuler M, Rosenberg B (1996) In: Li JJ, Li SA, Gustafsson JA, Nandi S, Sekely LI (eds) Hormonal carcinogenesis II. Springer, Berlin Heidelberg New York, p 193
18. Grady D, Gebretsadik T, Kerlikowske K, Emster V, Petitti D (1995) Obstet Gynecol 85: 304
19. Colditz GA, Hankinson SE, Hunter DJ, Willett WC, Manson JE, Stampfer MJ, Hennekens C, Rosner B, Speizer FE (1995) N Engl J Med 332: 1589
20. Li JJ, Li SA, Oberley TD, Parsons JA (1995) Cancer Res 55: 4347
21. Bolton JL, Trush MA, Penning TM, Dryhurst G, Monks TK (2000) Chem Res Toxicol 13: 135
22. Chen Y, Shen L, Zhang F, Lau SS, van Breemen RB, Nikolic D, Bolton JL (1998) Chem Res Toxicol 11: 1105

23. Chen Y, Liu X, Pisha E, Constantinou AI, Hua Y, Shen L, van Breemen RB, Elguindi EC, Blond SY, Zhang F, Bolton JL (2000) Chem Res Toxicol 13: 342
24. Bolton JL, Pisha E, Zhang F, Qui S (1998) Chem Res Toxicol 11: 1113
25. Pisha E, Lui X, Constantinou AI, Bolton JL (2001) Chem Res Toxicol 14: 82
26. Farnsworth NR, Bingel AS, Cordell GA, Crane FA, Fong HHS (1975) J Pharmaceut Sci 64: 717
27. Barnes S, Sfakianos J, Coward L, Kirk M (1996) In: American Institute for Cancer Research (ed) Dietary phytochemicals in cancer prevention and treatment. Plenum Press, New York, p 87
28. Kulling SE, Metzler M (1997) Food Chem Toxicol 35: 605
29. Kulling SE, Jacobs E, Pfeiffer E, Metzler M (1998) Mutat Res 416: 115
30. Morris SM, Chen JJ, Domon OE, McGarrity LJ, Bishop ME, Manjanatha MG, Casciano DA (1998) Mutat Res 405: 41
31. Pool-Zobel BL, Adlercreutz H, Glei M, Liegibel UM, Sittlingon J, Rowland I, Wähälä K, Rechkemmer G (2000) Carcinogenesis 21: 1247
32. Kulling SE, Rosenberg B, Jacobs E, Metzler M (1999) Arch Toxicol 73: 50
33. Abe T (1999) Leukemia 13: 317
34. Newbold RR, Banks EP, Bullock B, Jefferson WN (2001) Cancer Res 61: 4325
35. Kulling SE, Honig DM, Simat TJ, Metzler M (2000) J Agric Food Chem 48: 4963
36. Kulling SE, Honig DM, Metzler M (2001) J Agric Food Chem 49: 3024
37. Roberts-Kirchhoff ES, Crowley JR, Hollenberg PF, Kim H (1999) Chem Res Toxicol 12: 610
38. Coldham NG, Howells LC, Santi A, Montesissa C, Langlais C, King LJ, Macpherson DD, Sauer MJ (1999) J Steroid Biochem Mol Biol 70: 169
39. Jacobs E, Metzler M (1999) J Agric Food Chem 47: 1071
40. Jacobs E, Kulling SE, Metzler M (1999) J Steroid Biochem Mol Biol 68: 211
41. White INH (1999) Carcinogenesis 20: 1153
42. Newbold RR, Jefferson WN, Padilla-Burgos E, Bullock BC (1997) Carcinogenesis 18: 2293
43. Carthew P, Edwards RE, Nolan BM, Martin EA, Heydon RT, White INH, Tucker MJ (2000) Carcinogenesis 21: 793
44. Boocock DJ, Maggs JL, White INH, Park BK (1999) Carcinogenesis 20: 153
45. Rajaniemi H, Rasanen I, Koivisto P, Peltonen K, Hemminki K (1999) Carcinogenesis 20: 305
46. da Costa GG, Hamilton LP, Beland FA, Marques MM (2000) Chem Res Toxicol 13: 200
47. Osborne MR, Davis W, Hewer AJ, Hardcastle IR, Phillips DH (1999) Chem Res Toxicol 12: 151
48. Beland FA, McDaniel LP, Marques MM (1999) Carcinogenesis 20: 471
49. Martin EA, Rich K, White INH, Woods KL, Powles TJ, Smith LL (1995) Carcinogenesis 16: 1651
50. Phillips DH, Carmichael PL, Hewer A, Cole KJ, Hardcastle IR, Poon GK, Keogh A, Strain AJ (1996) Carcinogenesis 17: 89
51. Shibutani S, Suzuki N, Terashima I, Sugarman SM, Grollman AP, Pearl ML (1999) Chem Res Toxicol 12: 646
52. Glatt HR, Davis W, Meinl W, Hermersdorfer H, Venitt S, Phillips DH (1998) Carcinogenesis 19: 1709
53. Dodds EC, Golberg L, Lawson W, Robinson R (1938) Nature 141: 247
54. Herbst AL, Ulfelder H, Poskanzer DC (1971) N Engl J Med 284: 878
55. Kirkman H, Bacon RL (1950) Cancer Res 10: 122
56. Newbold RR, Bullock BC, McLachlan JA (1990) Cancer Res 50: 7677
57. Metzler M (1975) Biochem Pharmacol 24: 1449
58. Metzler M, McLachlan JA (1978) Biochem Biophys Res Comm (1978) 85: 874
59. Moorthy B, Liehr JG, Randerath E, Randerath K (1995) Carcinogenesis 16: 2643
60. Tsutsui T, Maizumi H, McLachlan JA, Barrett JC (1983) Cancer Res 43: 3814
61. Banerjee SK, Banerjee S, Li SA, Li JJ (1994) Mutat Res 311: 191
62. Pfeiffer E, Metzler M (1991) In: Li JJ, Li SA, Nandi S (eds) Hormonal carcinogenesis. Springer, Berlin Heidelberg New York, p 313

63. Duesberg P, Rasnick D, Li R, Winters L, Rausch C, Hehlmann R (2000) Anticancer Res 19: 4887
64. Oshimura M, Barrett JC (1986) Environ Mutagen 8: 129
65. Tsutsui T, Barrett JC (1997) Environ Health Perspect 105(Suppl 3): 619
66. Barrett JC, Wong A, McLachlan JA (1991) Science 212: 1402
67. Tsutsui T, Suzuki N, Maizumi H, McLachlan JA, Barrett JC (1986) Carcinogenesis 7: 1415
68. Tsutsui T, Tamura Y, Suzuki A, Hirose Y, Kobayashi M, Nishimura H, Metzler M, Barrett JC (2000) Int J Cancer 86: 151
69. National Toxicology Program (1982) National Toxicology Program Report Series No 215, US Department of Health and Human Services
70. Atkinson A, Roy D (1995) Biochem Biophys Res Commun 210: 424
71. Atkinson A, Roy D (1995) Environ Mol Mutagen 26: 60
72. Pfeiffer E, Metzler M (1998) Toxicologist 42: 279
73. Hanioka N, Jinno H, Tanaka-Kagawa T, Nishimura T, Ando M (2000) Chemosphere 41: 973
74. Niwa T, Tsutsui M, Kishimoto K, Yabusaki Y, Ishibashi F, Katagiri M (2000) Biol Pharmaceut Bull 23: 498
75. Pottenger LH, Domoradzki JY, Markham DA, Hansen SC, Cagen SZ, Waechter JM Jr (2000) Toxicol Sci 54: 3
76. Pfeiffer E, Rosenberg B, Deuschel S, Metzler M (1997) Mutat Res 390: 21
77. Pfeiffer E, Rosenberg B, Metzler M (1996) In: Li JJ, Li SA, Gustafsson JA, Nandi S, Sekely LI (eds.) Hormonal carcinogenesis II. Springer, Berlin Heidelberg New York, p 450
78. Tsutsui T, Tamura Y, Yagi E, Hasegawe K, Takahashi M, Maizumi N, Yamaguchi F, Barrett JC (1998) Int J Cancer 75: 290
79. Barrett JC, Tsutsui T (1996) Mechanisms of estrogen-associated carcinogenesis. In: Huff J, Boyd J, Barrett JC (eds) Cellular and molecular mechanisms of hormonal carcinogenesis. Wiley-Liss, New York, p 105
80. Garner CE, Burka LT, Etheridge AE, Mathews HB (2000) Toxicol Appl Pharmacol 162: 115
81. Garner CE, Mathews HB, Burka LT (2000) Toxicol Appl Pharmacol 162: 124
82. Schuler M, Hasegawa L, Parks R, Metzler, M, Eastmond DA (1998) Environ Mol Mutagen 31: 263

Emerging Issues Related to Endocrine Disrupting Chemicals and Environmental Androgens and Antiandrogens

L. Earl Gray Jr.[1], Christy Lambright[2], Louise Parks[1], Rochelle W. Tyl[2],
Edward F. Orlando[3], Louis J. Guillette Jr.[3], Cynthia J. Wolf[1], John C. Seely[4],
Tsai-Yin Chang[2], Vickie Wilson[5], Andrew Hotchkiss[5], Joseph S. Ostby[1]

[1] Endocrinology Branch, Reproductive Toxicology Division, US Environmental Protection
   Agency, Research Triangle Park, North Carolina 27711, USA
   E-mail: *emgray@mindspring.com, gray.earl@epa.gov*
[2] Research Triangle Institute, Research Triangle Park, North Carolina, USA
[3] University of Florida, Zoology Dept, Gainesville, Florida, USA
[4] Pathco, Research Triangle Park, North Carolina, USA
[5] North Carolina State University/US Environmental Protection Agency Cooperative
   Research Program

Disclaimer. The research described in this article has been reviewed by the National Health and Environmental Effects Research Laboratory, U. S. Environmental Protection Agency, and approved for publication. Approval does not signify that the contents necessarily reflect the views and policies of the Agency nor does mention of trade names or commercial products constitute endorsement or recommendation for use.

Wildlife populations from contaminated ecosystems display a variety of reproductive alterations including cryptorchidism in the Florida panther, small baculum in young male otters, small penises in alligators, sex reversal in fish, and altered social behavior in birds. In some cases, clear cause and effect relationships exist between exposure to endocrine disrupting chemicals (EDCs) and adverse effects in fish, wildlife, and domestic animals. The formation of biologically plausible hypotheses regarding toxicant-induced disruption of reproduction can be facilitated by definitive mechanistic rodent studies. To this end, we have investigated the in vivo and in vitro effects of suspect antiandrogenic and dioxin-like substances. In vivo studies examined short-term effects, effects on puberty, and toxicant-induced alterations of rat sexual differentiation. We utilize short-term in vivo and in vitro assays (receptor binding, transfected cell, and steroidogenesis assays) in order to confirm the suspected mechanism of action. To date, in vitro and in vivo studies have identified several antiandrogenic toxicants including vinclozolin, procymidone, linuron, several phthalate esters, and p,p'-DDE, all of which alter male rat sex differentiation. 2,3,7,8-Tetrachlorodibenzo-p-dioxin (TCDD) and the TCDD-like congener PCB 169 affect both male and female offspring, inducing dramatic reductions in ejaculated sperm numbers at low dosage levels. In utero exposure to antiandrogenic and TCDD-like chemicals result in profiles of effects in the offspring that are pathognomonic for each mechanism of action. Mechanistic information from rodent studies using dosing regimes that produce relevant toxicant tissue levels, coupled with an understanding of the endocrine factors regulating reproductive development in the species of concern, can enhance our ability to predict the effects of EDCs on human and wildlife reproduction. For some EDCs (i. e., PCBs, PCDDs, PCDFs, or p,p'-DDE), developmental effects are seen in rats using dosing regimes that produce fetal exposure levels that appear to be within the reported range of some segments of the human population.

Keywords. Sexual differentiation, Androgens, Dioxins, PCBs Vinclozolin, Procymidone, DDT and DDE, Linuron, Pesticides, Toxic substances, Pulp mill effluent, Intersex

The Handbook of Environmental Chemistry Vol. 3, Part M
Endocrine Disruptors, Part II
(ed. by M. Metzler)
© Springer-Verlag Berlin Heidelberg 2002

1	**Introduction** 212

2	**Effects of Endocrine Disrupting Chemicals (EDCs)** 215
2.1	Cause-and-Effect Relationships for EDCs 216
2.2	Known Effects of Drugs on Human Sexual Differentiation 217
2.3	Known Effects of Plant and Fungal Estrogens in Animals and Humans 217
2.4	Known Effects of Plant Antiandrogens in Humans and Rodents . . 218
2.5	Known Effects of Pesticides and Toxic Substances on Endocrine Function and/or Reproduction in Humans 219
2.5.1	Known Effects of PCBs, PCDDs, and PCDFs in Humans 219
2.5.2	Known Effects of DDTs in Humans 219
2.5.3	Occupational Exposure to Chlordecone and Dibromochloropropane 219
2.5.4	Occupational Exposure to Amsonic Acid 220

3	**Environmental Androgens Revealed** 220

4	**Potential Effects of Toxicants on Sex Ratio in Humans and Animals** 221

5	**Vinclozolin** 225
5.1	In Vitro and In Vivo Effects 225
5.2	Dose-Response of Developmental Effects 225
5.3	Effects During Puberty 226
5.4	Effects on the Liver and Adrenal 227
5.5	Effects in Lower Vertebrates 228

6	**Procymidone** 228
6.1	In Vitro Effects 228
6.2	Dose-Response of Developmental Effects 229

7	**Linuron** 229
7.1	In Vitro and Short-Term In Vivo Effects 229
7.2	Developmental Effects 230

8	**DDT and DDE** 231
8.1	Background 231
8.2	In Vitro and Developmental/Reproductive Effects of p,p'-DDE . . . 231
8.3	Potential Antiandrogenic Effects of p,p'-DDE in Wildlife 232
8.4	Endocrine Mechanisms of Eggshell Thinning 232
8.5	p,p'-DDE and Lake Apopka Alligators and Birds 233

9	**Antiandrogenic Effects of Phthalate Esters During Development** . . 235

10   Polychlorinated Aromatic Compounds 237

10.1   TCDD: A Putative Antiandrogen? 237
10.2   PCB 169: A Putative Antiandrogen? 238

11   Inhibitors of Steroid Hormone Synthesis 239

11.1   Ketoconazole 239
11.2   Fenarimol 240

12   Conclusions 240

13   References 241

List of Abbreviations

AGD	anogenital distance
AR	androgen receptor
ATPase	adenosine triphosphatase
BBP	benzyl butyl phthalate
CHO	Chinese hamster ovary
CNS	central nervous system
DAS	4,4′-diaminostilbene-2,2′-disulfonic acid
DBCP	dibromochloropropane
DBP	dibutyl phthalate
p,p'-DDD	2,2-bis(p-chlorophenyl)-1,1-dichloroethane
p,p'-DDE	2,2-bis(p-chlorophenyl)-1,1-dichloroethylene
o,p'-DDT	2-(p-chlorophenyl)-2-(o-chlorophenyl)-1,1,1-trichloroethane
p,p'-DDT	2,2-bis(p-chlorophenyl)-1,1,1-trichloroethane
DEHP	di-2-ethylhexyl phthalate
DEP	diethyl phthalate
DES	diethylstilbestrol
DHT	5α-dihydrotestosterone
DOP	dioctyl phthalate
DOTP	dioctyl terephthalate
DRE	dioxin response elements
EC50	effective concentration for 50% effect
ED50	effective dose for 50% effect
EDC	endocrine disrupting chemical
EE	ethinyl estradiol
EPA	Environmetal Protection Agency
ER	estrogen receptor
ERKO	estrogen receptor knock-out
ET	eggshell thinning
FQPA	Food Quality Protection Act
FSH	follicle stimulating hormone
GD	gestational day

IQ	intelligence quotient
KME	kraft mill effluent
LE	Long Evans
LH	luteinizing hormone
LOAEL	lowest observed adverse effect level
MEHP	mono-2-ethylhexyl phthalate
MMTV	mammary tumor virus
NOAEL	no observed adverse effect level
NOEL	no observed effect level
PCB	polychlorinated biphenyl
PCDD	polychlorinated dibenzo-p-dioxin
PCDF	polychlorinated dibenzofuran
PH	pseudo hermaphrodite
	pseudo hermaphroditism
PME	paper mill effluent
PND	postnatal day
PPS	preputial separation
RAR	retinoic acid receptor
RXR	retinoid receptor
SD	Sprague-Dawley
SG	shell gland
TCDD	2,3,7,8-tetrachlorodibenzo-p-dioxin
TEQ	toxic equivalent
TH	true hermaphrodite, true hermaphroditism

1
Introduction

Currently, the potential effects of "endocrine disrupting chemicals" (EDCs) on human health and the proven effects of EDCs on wildlife are a major focus among the scientific community. In 1996 the US Environmental Protection Agency (USEPA) was given a mandate under the Food Quality Protection Act (FQPA) and Safe Drinking Water Act to develop test protocols to screen for endocrine effects. The initial impetus for these actions arose from a Work Session in 1991 on "Chemically Induced Alterations in Sexual Development: The Wildlife/Human Connection" [1]. The consensus of the group was that "Many compounds introduced into the environment by human activity are capable of disrupting the endocrine system of animals, including fish, wildlife, and humans. Endocrine disruption can be profound because of the crucial role hormones play in controlling development" [2]. Among these chemicals are pesticides and industrial chemicals, pharmaceuticals, phytochemicals, and other anthropogenic products. These scientists also "estimated with confidence" that developmental impairments in humans have resulted from exposure to endocrine disruptors that are present in the environment from human activities. Laboratory studies corroborate the abnormalities of reproductive development observed in the field and, in some cases, provide mechanisms to explain the effects.

Recent findings have contributed to these concerns; for example, it has been suggested that in utero exposure to environmental estrogens, antiandrogens, or chemicals like 2,3,7,8-tetrachlorodibenzo-p-dioxin (TCDD) could be responsible for the reported 50% decline in sperm counts in some areas and the apparent increases in cryptorchid testes, testicular cancer, and hypospadias [3–5]. In females, exposure to EDCs during development could contribute to earlier age at puberty [6] and to increased incidences of endometriosis [7] and breast cancer [8]. Concerns about the effects of exposure to endocrine disruptors during development led the National Academy of Sciences to release a report on "Pesticides in the Diets of Infants and Children", which suggested that the young are a special concern with respect to pesticide exposures [9].

In the past, the discussion of "endocrine disruptors" was focused on toxicants reported to possess estrogenic activity, with little consideration given to other mechanisms of endocrine toxicity; mechanisms that, in fact, may be of equal or greater concern. In addition, there has been a great deal of misinformation communicated on issues concerning endocrine disruptors; for example, nonestrogenic chemicals (i.e., phthalates and p,p'-DDE) are repeatedly reported to be estrogenic. There is a lack of appreciation for the fact that many endocrine disruptors (i.e., TCDD, ethinyl estradiol) are very potent reproductive toxicants. In addition, there has been a tendency to dismiss the wildlife data as correlative, ignoring examples of clear cause and effect relationships between chemical exposure and reproductive alterations (e.g., DDT metabolite effects in birds, PCB effects in fish, and environmental estrogen effects in domestic animals). There is a lack of recognition that subtle, low dose reproductive effects seen in laboratory studies with endocrine disruptors will be difficult, if not impossible, to detect in typical epidemiological studies because of high variability normally seen in human reproductive function (e.g., time to fertility, fecundity, and sperm measures), the delayed appearance of the reproductive lesions, and a lack of exposure data. There is a lack of appreciation for the complexity of the multiple mechanisms by which a single chemical can alter the endocrine milieu, to say nothing of the complex endocrine alterations induced by mixtures of chemicals.

Although several reviews, including our own, have been published within the last decade [1, 3, 10–22], recent scientific developments within the last year or so are changing the "landscape" of this issue. Several new studies have refined our understanding of some issues. For example, the increased incidence of testicular cancer in many areas of the world appears real [3, 23]. While it appears more certain that hypospadias and cryptorchidism are increasing in the industrialized world as well, one cannot rule out some confounding variables [23, 24]. The differences in sperm counts between regions are so large that they cannot be explained by methodological biases and "environmental effects are entirely plausible" [23]. In addition, new trends in human health have been reported including altered sex ratios. Toxicant exposure has been associated with fewer human male births and skewed sex ratios in rodents in the wild. New classes of toxicants have been shown to alter reproductive function via mechanisms of action beyond those originally considered by the scientific community. It is now apparent that entire classes of chemicals, like the phthalates which display significant antiandrogenic activity, have not been adequately tested, such that neither claims of "safety" nor "eminent hazard to children" can be adequately supported by the available toxicological data. In other cases, as adequate dose-response data are developed, adverse effects are being noted in rodents at dosage levels below former lowest observed adverse effect levels (NOAELs) which, in some cases, appear to overlap with human exposure levels. Scientists from several laboratories are reporting the discovery of androgens in the environment. We are now not only concerned about pesticides and other toxic substances in the environment, but the issue has broadened considerably due to the growing awareness that the list of EDCs present in the environment from human activities includes potent pharmaceutical products and phytosterols. Among these drugs are estrogens, antibiotics, betablockers, antiepileptics, and lipid regulating agents. One study reported that the pharmaceuticals found as contaminants of aquatic systems included 36 of 55 pharmaceuticals and 5 of 9 metabolites measured including antiepileptic drugs [25]. Another reported that several lipid regulating agents were present in the effluents near mg/l concentrations [26], while a third reported concentrations of antibiotics in hog lagoons in North Carolina, USA as high as 700 µg/l [27].

In the area of wildlife toxicology and ecosystem health, one challenge in interspecies extrapolation is to link laboratory mechanistic studies with EDCs to both individual and population effects in the field. Furthermore, the value of wildlife as sentinel species, forewarning potential adverse effects in humans, will be enhanced by mechanistic linkages among the vertebrates. In this area, it is even more apparent that clear cut cause and effect relationships exist between exposure to EDCs and adverse effects in several vertebrate classes from fish to mammals. In Lake Apopka, for example, normal agricultural pesticide use from the 1940s to the 1970s, coupled with the ongoing Lake Apopka Restoration Project, which contaminated the lake with pesticides by flooding muck farms at the northern end, resulted in the deaths of thousands of fish-eating birds from contaminant exposures that were underestimated in the ecological risk assessment process. It also appears that the effect of DDT metabolites on vertebrate eggshell thinning (ET) is mediated via endocrine mechanisms that do not involve estrogenicity or the estrogen receptor (ER). Many of the effects seen in lower vertebrates have been linked to estrogens in the environment, while other effects arise from different mechanisms of action.

Finally, the legislation of 1996 mandated that the USEPA establish validated screening and testing procedures for estrogens and other substances as deemed appropriate. In addition, the FQPA of 1996 mandated that the risk assessment process considers combinations of chemicals rather than evaluates the potential risk on a chemical by chemical basis. The development of EDC screening and testing procedures is underway and remains contentious. Reports of U-shaped (nonmonotonic), ultra-low dose effects and non-threshold effects for EDCs are challenging some of the basic assumptions of risk assessment for noncancer endpoints. While the focus of this debate has centered on the low dose effects of bisphenol A [28, 29], U-shaped dose response curves are known from other in vitro and in vivo studies. For example, administration of testosterone produces a well-characterized and reproducible U-shaped dose response for spermatogenesis [30–32] in the intact adult male rat. In vitro studies show that such responses do not always involve multiple mechanisms of action. Several androgen receptor (AR) ligands, antagonists at low to moderate concentrations, became AR agonists at high concentrations [33]. Furthermore, the basic tenet of toxicology from Paracelcus (1564) (cited in [34]) that "dose alone determines the poison" is too limited for some EDCs because the timing of exposure dictates not only the effect but also whether the effects are adverse or beneficial. Even when administered during adult life, drugs with EDC-activity can simultaneously have a beneficial effect on one tissue and an adverse effect on another.

While progress has been made on several fronts, considerable research needs to be done. In several areas it is unclear if research funding is adequate or will be long-term enough to provide answers to some of the more currently intractable problems. In particular, linking human health effects to EDCs would require studies over several decades as the exposures of concern occur early in life but the effects may not appear until after puberty or old age. While global research plans for EDCs have been developed, it is not clear that current funding levels for EDC research on human and wildlife health will be sustained long enough or if the current approach to funding EDC research provides a coordinated program that addresses the highest priority issues.

In the following review, we will provide an update on many of the issues presented above, focusing on new data and concerns about EDCs. The review will focus primarily, but not exclusively, on effects of androgenic and antiandrogenic chemicals.

2
Effects of Endocrine Disrupting Chemicals (EDCs)

The list of chemicals that are known to affect humans, domestic animals, and/or wildlife via functional developmental toxicity or endocrine mechanisms includes TCDD, polychlorinated biphenyls (PCBs) and polychlorinated dibenzofurans (PCDFs), methylmercury, ethinyl estradiol (EE), alkylphenols, plant sterols, fungal estrogens, chlordecone, dibromochloropropane (DBCP), o,p'-DDD (Mitotane), o,p'-DDT, and p,p'-DDE. In addition to these xenobiotics, more than 30 different drugs taken during pregnancy have been found to alter human development as a consequence of endocrine disruption [35]. These drugs are not limited to estrogens like diethylstilbestrol (DES). EDCs are known to alter human development via several mechanisms besides the estrogen receptor (ER): this includes binding to the AR or retinoic acid (RAR, RXR) receptors, and by inhibition of steroidogenic enzymes or the synthesis of thyroid hormones [35]. Recent findings on the effects of background levels of PCBs on the neurobehavioral development of the child have contributed to the concerns about the effects of EDCs on human health via alteration of hormone function including the thyroid [36].

Wildlife populations from contaminated ecosystems display a variety of reproductive alterations including cryptorchidism in the Florida panther [37, 38], small baculum in young male mink and otter [39], small penises in alligators, sex reversal in fish, and malformations, failed reproduction, and altered social behavior in birds. The existence of cause-and-effect relationships for EDCs and reproductive alterations in wildlife continues to be a subject of debate because consistent criteria for the effects of contaminants of wildlife reproductive success have not been rigorously applied on a case-by-case basis. To this end, we developed a list of criteria that can be applied to different examples to determine the strength of the association between an EDC and observed effects in wildlife [22]. In several instances, most of the criteria are satisfied while in other cases additional research is needed to clarify the relationship between the contaminants and the observed reproductive alterations.

2.1
Cause-and-Effect Relationships for EDCs

Several examples of cause-and-effect relationships for EDCs in wildlife and humans are listed in Table 1. Details on these examples can be found in published reviews [10–17, 40].

Table 1. Selected examples of cause-and-effect relationships for EDCs and effects in wildlife and humans

---

A. Wildlife
Cause-and-effect established

Minks and PCBs
Fish-eating birds and PCBs and TCDD
Estrogens and vitellogenin production and intersex gonads
Kraft mill effluent and reduced gonad function and masculinization in fish
p,p'-DDE and eggshell thinning in birds and reptiles

Strong correlation with exposure and effects at the population level

Sex reversal of bloater chub (99% female) in Lake Michigan associated with o,p'- and p,p'-
    DDTs and their metabolites

B. Domestic Animals
Cause-and-effect established

Phytoestrogens
Fungal mycotoxins
PCBs and TCDD

C. Humans
Cause-and-effect established

DES
Androgenic and antiandrogenic drugs
Retinoids
PCBs and development
Phytoestrogens (miroestrol), menstrual cyclicity and fertility
Phytoantiandrogens (permixon) and prostatic function
Mitotane (o,p'-DDD) and adrenal function and fertility
DDTs and adrenal function
Ethinyl estradiol and reproductive function
Chlordecone (kepone), DBCP, and amsonic acid (DAS), and infertility and lower testosterone
    in workers

2.2
Known Effects of Drugs on Human Sexual Differentiation

In humans (and rodents), exposure to hormonally active chemicals during sex differentiation can produce pseudohermaphroditism [21, 35]. In addition, exposure to the androgenic substances, danazol or methyltestosterone, masculinizes human females (i.e., "female pseudohermaphroditism"). Progestins act both as androgen antagonists, demasculinizing males such that they display ambiguous genitalia with hypospadias [35], and as androgen agonists, masculinizing females. Laboratory studies demonstrate that these chemicals alter sex differentiation in rodents as well [21]. The drug aminoglutethimide, which alters steroid hormone synthesis in a manner identical to many fungicides, also masculinizes human females following in utero exposure.

Diethylstilbestrol (DES) provides an unfortunate example of how in utero exposure to a potent endocrine disruptor with estrogenic activity can alter reproductive development in humans. Although a few cases of masculinized females were noted in the late 1950s, most of the effects of DES were not apparent until after the children attained puberty. Transplacental exposure of the developing fetus to DES causes clear cell adenocarcinoma of the vagina, as well as gross structural abnormalities of the cervix, uterus, and fallopian tube. These women are more likely to have an adverse pregnancy outcome, including spontaneous abortions, ectopic pregnancies, and premature delivery [41]. Some of the pathological effects that develop in males following fetal DES exposure appear to result from an inhibition of androgen action or synthesis (hypospadias, underdevelopment or absence of the vas deferens, epididymis, and seminal vesicles) and anti-Mullerian duct factor (persistence of the Mullerian ducts) [35, 41, 42]. DES also causes epididymal cysts, hypotrophic testes, and infertility in males. Some males have reduced ejaculate volume with reduced numbers of motile sperm, and some also experience difficulty in urination [41].

It has also been reported that DES can alter sex differentiation of the human brain. Several behavioral alterations have been observed in some DES daughters [43, 44]. Meyer-Bahlburg et al. [43] reported that women exposed to DES in utero were found to have less well established sex-partner relationships, and to be lower in sexual desire and enjoyment, sexual excitability, and coital functioning. In addition, Hines and Shipely [45] found that DES-exposed women showed a more masculine pattern of cerebral lateralization on a verbal task than did their sisters.

2.3
Known Effects of Plant and Fungal Estrogens in Animals and Humans

The phytoestrogens and fungal mycotoxins provide some of the most conclusive data demonstrating that environmental estrogens are toxic to mammalian reproductive function under natural conditions. Naturally occurring compounds with estrogenic and other endocrine activities are widespread in nature. Farnsworth et al. [46] listed over 400 species of plants that contain potentially estrogenic isoflavonoids or coumestans or were suspected of being estrogenic based on biological grounds. Plants contain many other compounds in addition to estrogens that can affect reproductive performance [47], such as the antiandrogenic activity in the oil from saw palmetto.

Although most environmental estrogens are relatively inactive, when compared to steroidal estrogens or DES, the phytoestrogen miroestrol or ethinyl estradiol (EE) are as potent as estradiol in vitro and even more potent than estradiol when administered orally. In addition, many plant estrogens occur in such high concentrations that they induce reproductive alterations in domestic animals [48]. "Clover disease", which is characterized by dystocia, prolapse of the uterus and infertility, is observed in sheep grazed on highly estrogenic clover pastures. Permanent infertility (defeminization) can be produced in ewes by much lower amounts of estrogen over a longer time period than are needed to produce "clover disease". In domestic animals, feeds contaminated with the zearalenone-producing fungus (*Fusarium* sp.) induce adverse reproductive effects in a wide variety of domestic animals, including impaired fertility in cows and hyperestrogenism in swine and turkeys [48].

In World War II, people in the Netherlands consumed large quantities of tulip bulbs, containing high levels of estrogenic activity, and many women displayed signs of estrogenism, including uterine bleeding and other abnormalities of the menstrual cycle [48–50]. In addition, an extract from the roots of *Pueraria mirifica* in Thailand yielded miroestrol, an estrogenic substance that was as potent in increasing uterine weight in rats as DES when administered orally (both of which are orders of magnitude more potent than oral estradiol) and about 70% as active as estradiol when administered by subcutaneous injection [51]. A limited clinical trial with miroestrol was carried out to examine the utility of this plant estrogen for treatment of amenorrhea in women. When miroestrol was administered orally at doses of 1 or 5 mg per day for several days, marked estrogenic responses were seen, but the onset of effects was slow and in some cases persisted after the cessation of treatment. More recently, Cassidy et al. [52] reported that addition of 45 mg of dietary isoflavones given for one month to six regularly cycling women increased the duration of the follicular phase or delayed menstruation. Mid-cycle luteinizing hormone (LH) and follicle-stimulating hormone (FSH) surges also were suppressed, suggesting that dietary isoflavones display estrogenic activity on the hypothalamic-pituitary axis.

2.4
Known Effects of Plant Antiandrogens in Humans and Rodents

Permixon is an antiandrogenic extract of the saw palmetto that is used in some European countries to treat androgen-dependent prostatic diseases [53]. Clinical studies indicate that it is effective in this regard. Mechanistically, it has been reported to act as an AR antagonist [54] and to inhibit the activity of 5α-reductase activity, which metabolizes testosterone to dihydrotestosterone (DHT) [55]. Rhodes et al. [56] reported that permixon failed to inhibit testosterone- or DHT-stimulated prostate growth in castrated rats, suggesting that it was not antiandrogenic, while, in contrast, Paubert-Braquet et al. [57] reported that the lipidosterolic extract of *Serenoa repens* (permixon) inhibited the effects of estradiol in combination with testosterone on prostatic growth in castrated male rats. The developmental effects of this plant extract have not been determined.

2.5
Known Effects of Pesticides and Toxic Substances on Endocrine Function and/or Reproduction in Humans

2.5.1
Known Effects of PCBs, PCDDs, and PCDFs in Humans

In addition to drugs and plant substances, several pesticides and toxic substances have been shown to alter human reproductive function (listed above). An accidental high dose in utero exposure to PCBs and PCDFs has been associated with reproductive alterations in boys, increased stillbirths, low birth weight, malformations, and IQ and behavioral deficits [58]. In addition to the effects associated with this inadvertent exposure, relatively subtle adverse effects were seen in infants and children exposed to relatively low levels of PCBs, PCDDs, and PCDFs [36]. The authors reported that the LOAELs for developmental neurobehavioral and reproductive endpoints, based on body burdens of toxic equivalents (TEQs) in animals, are within the range of current human body burdens.

2.5.2
Known Effects of DDTs in Humans

One metabolite of DDT was found to alter adrenal function with sufficient potency to be used as a drug to reduce adrenal androgen production. o,p'-DDD (mitotane) is used to treat adrenal steroid hypersecretion associated with adrenal tumors. In addition to this usage, lower doses of mitotane restored menstruation in 13 of 15 women with spanomenorrhea associated with hypertrichosis. Pregnancies occurred in 5 of 15 women during the treatment period [34, 59]. In addition to direct effects on adrenal steroid producing cells, mitotane is also useful in controlling excessive androgen levels in the blood by increasing androgen metabolism.

2.5.3
Occupational Exposure to Chlordecone and Dibromochloropropane

Occupational exposure to pesticides and other toxic substances, i.e., chlordecone and dibromochloropropane (DBCP), in the work place have been associated with reduced fertility, lowered sperm counts, and/or endocrine alterations in male workers. Workers in Hopewell, Virginia, exposed to high levels of chlordecone, an estrogenic and neurotoxic organochlorine pesticide, displayed obvious signs of intoxication which included severe neurotoxicity and abnormal testicular function [60]. As this cohort was not followed, it is not known if the effects of chlordecone were completely reversible. DBCP is a pesticide that appears to alter the endocrine system indirectly. DBCP-exposed workers were infertile, displaying reduced sperm numbers and altered Sertoli cell function, as indicated by an elevation of serum FSH (presumably through a decline in Sertoli cell inhibin secretion) [61]. When administered to rats during perinatal life, the effects of DBCP are more complex. Prenatal DBCP-treatment alters sexual and testicular differentiation [62]. Male rat offspring, whose dams were treated with 25 mg DBCP/kg on days 14 to 19 of pregnancy, displayed a 90 % reduction in testis weight, altered hypothalamic morphometry, a lack of male mating behavior, and increased female-like sex behavior. Taken together, these results indicate that DBCP-treatment during sexual differentiation interferes with testicular development and testosterone production, resulting in antiandrogenic effects in the male offspring.

2.5.4
Occupational Exposure to Amsonic Acid

It is surprising to learn that occupational exposures to potential EDCs at effective concentrations apparently have not been eliminated from the workplace. A series of publications from about 1990 to 1996 presented documentation of sexual impotence in chemical factory workers exposed to a DES-like stilbene derivative. The National Institute of Occupation Safety and Health conducted two studies in response to complaints of impotence and decreased libido among male workers involved in the manufacture of 4,4′-diaminostilbene-2,2′-disulfonic acid (DAS), a key ingredient in the synthesis of dyes and fluorescent whitening agents. Both current and former workers had lower serum testosterone levels [63] and reduced libido [64] as compared to control workers. In addition, duration of employment was negatively correlated with testosterone levels. These studies replicated the observations reported by Quinn et al. [65] who reported low levels of serum testosterone and problems with impotence in male workers. In an uterotropic assay, while DAS was only weakly to negligibly estrogenic [66, 67], a single subcutaneous 30 mg/kg dose of 4-nitrotoluene, a precursor of DAS, increased uterine weights without producing overt toxicity. Samples of DAS from the workplace displayed estrogenic activity. In addition, 4-nitrotoluene in the feed impairs testicular function and alters estrous cycles in rats [68].

3
Environmental Androgens Revealed

Within the last decade, several pesticides and toxic substances have been identified as "environmental antiandrogens" which are of sufficient potency to induce hypospadias and ambiguous genitalia in rats. Although environmental chemicals with androgenic activity had not previously been detected, it is now evident that there are "environmental androgens". Three independent research groups have detected androgenic activity in paper mill effluent (PME) [69–71]. Two groups have found that kraft mill effluent (KME) from sites on the Fenholloway river in Florida, which contain masculinized female fish (*Gambusia holbrooki*), contains a chemical mixture that binds AR and induces androgen-dependent (but not glucocorticoid-dependent) gene expression in vitro. Water samples collected from three sites downstream from the discharge point of the PME all display androgenic activity while, in contrast, water samples upstream of the plant or from a nearby river do not display androgenicity. Female mosquito fish from the contaminated sites display an anal fin that is enlarged into a male-like gonopodium [72–74]. Masculinization can also be achieved in female mosquito fish and killifish in the laboratory with exposures to KME and from microbial (*Mycobacterium smegmata*) metabolites of phytosterols present in the KME [73]. Similar in vivo and in vitro responses have been reported from a paper mill on Jackfish Bay, Lake Ontario, by another group of investigators. Using cytosolic preparations of fish testes, gonad, and brain, this group found substances in PME that bound AR and they also identified this activity in PME constituents (β sitosterol, flavonol, and flavone) [70]. In addition, male and female white sucker fish (*Catostomus commersoni*) exposed to effluent in the field displayed a variety of reproductive effects, including an increase in size and number of reproductive tubercles, which are an androgen-dependent sex trait of males.

4
Potential Effects of Toxicants on Sex Ratio in Humans and Animals

The effects of exposure to endocrine disruptors during sex differentiation are of special concern for a number of reasons: this process is very sensitive to the effects of relatively low doses of endocrine disruptors, the effects are irreversible, i.e., the system is "imprinted" by the initial hormonal environment, functional alterations of the sex differentiation are often not apparent until after puberty or even later in life, and the abnormalities, which include malformations and infertility, cannot be predicted from the transient alterations in hormone levels produced by similar exposure in adult animals. It is important to understand the key role that animal models play in this research as developmental reproductive toxicity data are often critical in the assessment of noncancer health effects of EDCs. Furthermore, when similar congenital reproductive abnormalities have been detected in laboratory species, they have dramatically facilitated our understanding of genetic errors of steroid metabolism and receptor function in humans. For this reason, rodent and other models have great utility for evaluating the potential of xenobiotics to alter human reproductive development. We do not have to wait until adverse effects appear in the human population, as we did with DES and other drugs, to take precautionary action.

The role of hormones in sex differentiation in mammals is well understood. Even the most severe alterations of this process are not lethal. The normal development of the sexual phenotype from an indifferent state entails a complex series of events [75]. Genetic sex is determined at fertilization and this normally governs the expression of the "male factor" and the subsequent differentiation of gonadal sex. Prior to sex differentiation, the embryo has the potential to develop a male or female phenotype. Following gonadal sex differentiation, testicular secretions induce differentiation of the male duct system and external genitalia. The development of phenotypic sex includes persistence of either the Wolffian (male) or Mullerian (female) duct system, and differentiation of the external genitalia and the central nervous system (CNS). Other organ systems, like the liver, muscles, and brain, are "imprinted" as well. The male phenotype arises due to the action of testicular secretions, testosterone, and Mullerian inhibiting substance. Testosterone induces the differentiation of the Wolffian duct system into the epididymis, vas deferens, and seminal vesicles, while its metabolite DHT induces the development of the prostate and male external genitalia. It has been suggested that in the absence of these secretions, the female phenotype is expressed (whether or not an ovary is present). This hypothesis is supported by the observation that when estrogen receptor α and β genes are both knocked out ($\alpha\beta$ERKO mouse) the female mouse still develops a complete reproductive tract [76]. However, there is some indication of intersex in the ovaries as Sertoli-like cells, normally found in the testis of the male, are found within the ovary. In the CNS, testosterone is aromatized (via the steroidogenic enzyme aromatase) to estradiol and reduced via 5α-reductase to DHT in a species-specific manner. It has been suggested for some species that all three hormones (testosterone, DHT, and estradiol) play a role in the masculinization of the CNS.

Reports of EDC-induced alterations of sex ratios and possible production of intersex in wildlife presents a new scientific challenge. A decline in the number of males born has been reported in the zone of Seveso contaminated with highest levels of TCDD [77], and the general decline in the percentage of males born in industrialized nations world-wide also has been linked to the EDC issue [78–82]. However, Jacobsen et al. [83] and Biggar et al. [84] suggested that this trend was more likely related to family size, not EDCs, while Vartiainen et al. [85] reported that the changes in male proportions did not correlate with industrialization or the use of pesticides or hormonal drugs, rendering a causal association unlikely. In addition, such a trend was not observed in Ireland, which industrialized at about the same time at the rest of Europe [86].

Data from animal studies suggest that alterations of the hormonal environment can result in changes in sex ratios in the offspring [87, 88]. Vandenbergh and Huggett [89] reported that in mice the mother's intrauterine position during fetal development affects the sex ratio of her offspring. Females adjacent to two males, presumably exposed to higher androgen levels, produced first litters that were 58% males vs females that were adjacent to two females in utero, which had fewer (42%) males in the first litter. Likewise, the intrauterine position of gerbils appears to alter the sex ratio in a similar fashion [87]. Asynchronous mating, in which the time of mating varies from the time of ovulation, has been shown to alter sex ratios in several mammalian species [90, 91]. In the field, mammalian populations can show significant changes in sex ratio that are related to environmental factors. Kruuk et al. [92] reported that dominant female red deer (*Cervus elaphus*) produce more males at low population levels, an effect that disappeared as the population increased to carrying capacity, implying a social influence on sex determination which could be mediated via endocrine mechanisms. In a human study, Sas and Szollosi [93] reported that the sex ratio of children born to fathers treated with methyltestosterone was 67% male. While such observations appear in conflict with the concept of Mendelian genetics, mechanisms that could explain altered sex ratios have been identified. For example, hormones or other factors that regulate genes present on the Y chromosome that control sperm motility by modulating the activity of sperm motility kinases could alter the probability of Y sperm fertilizing the egg [94].

In most cases, mammalian hermaphrodites are pseudohermaphrodites (PHs) rather than true hermaphrodites (THs) [95–97]. In PHs, varying degrees of intersex of the hormone-dependent tissues is achieved but gonadal differentiation is consistent with the genetic sex of the individual. The degree of natural PHs among species can vary greatly with some species normally having masculinized females like the spotted hyaena (*Crocuta crocuta*) [98] or polled goats [95]. Genetic mapping of the autosomal region in XX sex-reversal and horn development revealed that in goats, abnormalities in sex determination are intimately linked to a dominant Mendelian gene coding for the "polled" (hornless) character [99]. This species provides an interesting animal model for cases of SRY-negative XX males. In the absence of information about what is "normal" it is difficult to interpret whether or not a female PH rate of 1.5% for the polar bear is abnormal [100].

In THs, the "intersex" individual has both ovarian and testicular tissue [96, 97]. Unilateral THs have an ovary on one side and a testis on the other or an ovotestis on one side and an ovary or a testis on the other. Bilateral THs have ovarian and testicular tissue on both sides, usually in the form of an ovotestis. The nature of the gonad influences the differentiation of the ipsilateral reproductive tract and only a very small percentage of TH humans are fertile. In addition, only a handful (reported as four) [101] of TH individuals, including all mammalian species, have been shown to have bilateral TH with separate ovary and testis on both sides and a complete male and female reproductive tract.

In humans, TH is generally related to one of several different errors of genetic sex determination: There are a number of genetic errors involving sex determining mechanisms which result in either TH or PH including complete and incomplete sex reversals (XX males and XY females); sex chromosome anomalies [97], single gene defects coding for a defective steroidogenic enzyme, which leads to reduced synthesis of sex steroids (20,22-desmolase; 17-ketosteroid reductase; and 5α-reductase deficiency), defective steroid receptor, resulting in abnormal handling of androgens in the target tissues (complete androgen insensitivity syndrome, Reifensten syndrome), and various other genetic defects (LH deficiency and lack of responsiveness to human chorionic gonadotropin) [97].

In wildlife there are some rather unusual examples of altered sex ratios in lemmings. Lyon [96] reviewed the sex ratios of different types of "female" wood lemmings (*Myopus schisticolor*), in which the percentage of females born ranges from 50% in one type (XX) to 100% in another type of female (XY). In the latter case, XY females develop carrying mutation on their single deviant X chromosome, which prevents development of the male phenotype. These XY females are fertile but they produce only daughters (50% XX and 50% XY). A similar condition exists in the varying lemming (*Dicrostonyx torquatus*).

TH is also common in some mammalian species. TH has been identified in four species of European moles (insectivora, mammalia). In *Talpa occidentalis*, *T. europaea*, *T. romana*, and *T. stankovici* all XX individuals are intersex, whereas XY individuals have a normal male phenotype. Intersex XX females had bilateral ovotestes with a small portion of histologically normal ovarian tissue and a variably sized large mass of dysgenetic testicular tissue, accompanied by a small epididymis. SRY gene was found to be present in males but not females [102].

TH has been identified in a St Lawrence beluga whale (*Delphinapterus leucas*) by De Guise et al. [101]. This animal had two testes, two separate ovaries, and the complete ducts of each sex, but the cervix, vagina, and vulva were absent. Unilateral TH has been noted in an FVB/N mouse [103] with male and female gonads and reproductive tissues on opposite sides.

Our interest in the ability of EDCs to induce TH was stimulated by reports of possible TH appearing simultaneously in 1998 in 29/87 small mammals at two sites in California, including a former wildlife refuge, which was closed due to extensive selenium contamination. In the process of collecting animals for assessment of contaminant levels by overnight snap-trapping, it was noted that four species (house mouse – *Mus musculus*, deer mouse – *Peromyscus maniculatus*, California vole – *Microtus californicus*, and the western harvest mouse – *Reithrodontomys megalotis*) appeared to be bilateral THs, displaying tissues which were tentatively identified as paired separate ovaries and scrotal testes and uterine tissue. Externally, the animals were phenotypic males. A collection of animals several years prior to the present sampling in 1995 revealed a lower incidence (3 of 105) of possible intersex small mammals [104]. In this regard, we examined reproductive tissues from six animals (four *Mus musculus* and two *Reithrodontomys megalotis*) from one site (not Kesterson) identified as possible intersex based upon the tentative identification of the presence of ovarian, testicular, and uterine tissues. The entire reproductive tract plus gonads were preserved as a unit in fixative in the laboratory after trapping [104]. These tissues were trimmed, embedded in paraffin, sectioned, stained with hematoxilin and eosin, and evaluated by a veterinary pathologist (Pathco, Inc, Research Triangle Park) for the microscopic appearance of the tissues and tissue identification. In general, the tissues submitted were in various stages of advanced autolysis, which accounts for the difficulty in identification of organs during gross necropsy. However, the microscopic appearance of the tissues was adequate for identification purposes. Each animal was confirmed as a male, and no female tissues were identified from any of the animals. Male tissues confirmed were testis, epididymis, seminal vesicles, and other sex accessory tissues. In some cases the testes were mature and sperm was present in the epididymis. Serial sections through the enlarged cephalic portions of the seminal vesicles (initially identified as possible ovaries) confirmed only seminal vesicle tissue.

The fact that TH was not identified in these animals, which had been initially identified as possible "intersex", is consistent with our initial expectations. Not only is the occurrence of TH animals with paired, separate ovaries and testis, along with both a male and female duct system extremely rare, but also the simultaneous appearance of this form of TH in four species of rodents in less than a decade would seem to be of very low probability. However, as indicated above, marked deviations from the standard mammalian plan for sex determination have been noted in rodents (voles and moles). For this reason, it was critical that these gross observations be confirmed or refuted with a histological examination. Initially, the preliminary diagnosis of TH was linked by the press to contaminants at the former wildlife refuge, and had such a profound effect been confirmed, there was concern for the induction of similar alterations in other species, including humans.

5
Vinclozolin

5.1
In Vitro and In Vivo Effects

Of the antiandrogenic EDCs, the cellular and molecular mechanisms of action of the fungicide vinclozolin is one of the most thoroughly characterized (see elsewhere in this volume). Vinclozolin metabolites, M1, and M2 but not vinclozolin itself, competitively inhibit the binding of androgens to the mammalian AR. M1 and M2 also inhibit DHT-induced transcriptional activity in cells transfected with the human AR. More recently, Kelce et al. [105] demonstrated that vinclozolin treatment altered gene expression in vivo in an antiandrogenic manner. In contrast to their ability to bind to the AR, neither vinclozolin nor its antiandrogenic metabolites display affinity for the ER, although they do have weak affinity for the progesterone receptor. Vinclozolin, M1, and M2 do not inhibit 5α-reductase activity in vitro, the enzyme required for the conversion of testosterone to the more active androgen DHT. Androgen-induced gene expression is a multistep process. Agonist-bound AR undergoes conformational changes, loses heat-shock proteins, forms homodimers, is imported from a perinuclear region into the nucleus, and binds androgen response elements on regulatory sequences on the DNA "upstream" from androgen-responsive genes, activating mRNA synthesis. Binding of antagonists to AR results in conformations that differ from that obtained with the natural ligands such that AR-DNA binding and gene expression are blocked. In addition, vinclozolin inhibits growth of androgen-dependent tissues in the castrate-immature-testosterone treated and pubertal male rat. In the intact pubertal and adult male rat, vinclozolin treatment also alters hypothalamic-pituitary-gonadal function. Oral treatment with vinclozolin causes elevations of serum LH and testosterone. In contrast to vinclozolin, treatment with some other antiandrogens like p,p'-DDE [106] and methoxychlor [107, 108] fail to induce any significant change in serum LH or testosterone levels.

5.2
Dose-Response of Developmental Effects

Oral administration of vinclozolin at 100 or 200 mg/kg/day to pregnant rats during sexual differentiation (gestational day 14 to postnatal day 3) demasculinizes and feminizes the male offspring. Vinclozolin-treated male offspring display female-like anogenital distance (AGD) at birth, retained nipples, cleft phallus with hypospadias, suprainguinal ectopic testes, a blind vaginal pouch, epididymal granulomas, and small to absent sex accessory glands. In contrast, the testis is a relatively insensitive target for antiandrogens as compared to the external genitalia and sex accessory glands, and female offspring do not display malformations or permanent functional alterations. A comparison of the in vitro dosimetry data with the biological effects of vinclozolin suggest that when M1 and M2 concentrations in maternal serum approach their respective K_i values for AR, male offspring would display hypospadias [109].

When we administered vinclozolin by gavage to the dam at 0, 3.125, 6.25, 12.5, 25, 50, or 100 mg/kg/day from gestational day (GD) 14 to postnatal day 3 [110], doses of 3.125 mg/kg/day and above reduced AGD in newborn male offspring and increased the incidence of nipples/areolas in infant male rats. These effects were associated with permanent alterations in other androgen-dependent tissues. Ventral prostate weight in one year old male offspring was reduced in all treatment groups (significant at 6.25, 25, 50, and 100 mg/kg/day) and permanent nipples were detected in males at 3.125 (1.4%), 6.25 (3.6%), 12.5 (3.9%), 25 (8.5%), 50 (91%), and 100 (100%) mg/kg/day. To date, permanent nipples in adult male offspring (not to be confused with areolas or what some authors incorrectly described as "retained nipples" in infant male rats) have never been observed in a control male from any study in our laboratory. Vinclozolin-treatment at 50 and 100 mg/kg/day induced reproductive tract malformations, reduced ejaculated sperm numbers, and fertility. Even though all of the effects of vinclozolin likely result from the same initial event (AR binding), the different endpoints displayed a wide variety of dose-response curves and ED50s and some of these dose response curves failed to display an obvious threshold.

These data demonstrate that vinclozolin produces subtle alterations in sex differentiation of the external genitalia, ventral prostate, and nipple tissue in male rat offspring at dosage levels below the previously described no-observed-effect-level (NOEL). Some of the functional and morphological alterations were evident at dosage levels one order of magnitude below that required to induce malformations and reduce fertility. Hence, multigenerational reproduction studies of antiandrogenic chemicals conducted under the "old" multigenerational test guidelines that did not include endpoints sensitive to antiandrogens at low dosage level could yield a NOEL that is at least an order of magnitude too high.

Another study was designed to determine the most sensitive period of fetal development to the antiandrogenic effects of vinclozolin [111]. Pregnant rats were dosed orally with vinclozolin at 400 mg /kg/day on either GD 12–13, GD 14–15, GD 16–17, GD 18–19, or GD 20–21, with corn oil vehicle (2.5 ml/kg). Malformations and other effects were seen in male rat offspring dosed with vinclozolin on GD 14–15, GD 16–17 and GD 18–19, with the most pronounced effects resulting from exposure on GD 16–17. These effects include reduced AGD, increased number areolas and nipples, malformations of the phallus, and reduced levator ani/bulbocavernosus weight. The fetal male rat is most sensitive to antiandrogenic effects of vinclozolin on GD 16 and 17, although effects are more severe and 100% of male offspring are affected with administration of vinclozolin from GD 14 through GD 19.

5.3
Effects During Puberty

Peripubertal administration of EDCs can alter the onset of pubertal landmarks in male [107, 112] and female rats [20, 107]. Androgens play a key role in pubertal maturation in young males [113] and antiandrogens like vinclozolin produce predictable alterations in this process. The ease with which a delay in preputial separation (PPS), a landmark of puberty in the rat, can be measured enables us to use this endpoint to evaluate chemicals for this form of endocrine activity. A "pubertal male assay" including an assessment of PPS is being considered by the USEPA and others [114] for screening chemicals for endocrine activity, as mandated by 1996 US legislation (the Food Quality Protection Act and Safe Drinking Water Act).

Peripubertal treatment with the antiandrogen p,p'-DDE [105] or methoxychlor [107], which is not only antiandrogenic but also displays estrogenic activity, delays the onset of androgen-dependent PPS. Pubertal delays have also been detected following exposure to weakly antiandrogenic toxicants like linuron and di-n-butyl phthalate [115]. In contrast, reproductive toxicants like carbendazim that indirectly alter FSH levels without affecting serum testosterone fail to delay PPS even at dosage levels that cause profound alterations of testicular and hypothalamic-pituitary (FSH secretion) function [116].

We conducted a study to examine the effects of peripubertal oral administration of vinclozolin (0, 10, 30, or 100 mg/kg/day) on morphological landmarks of puberty, hormone levels, and sex accessory gland development in male rats [112]. Since binding of the M1 and M2 to AR alters the subcellular distribution of AR by inhibiting AR-DNA binding, we examined the effects of vinclozolin on AR distribution in the target cells after in vivo treatment, and measured serum levels of vinclozolin, M1, and M2 in the treated males so that these could be related to the effects on the reproductive tract and AR distribution. Vinclozolin treatment delayed pubertal maturation, and retarded sex accessory gland and epididymal growth (at 30 and 100 mg/kg/day). Serum LH (significant at all dosage levels), testosterone and 5α-androstane-3α,17β-diol (at 100 mg/kg/d) levels were increased. Testis size and sperm production, however, were unaffected. It was apparent that these effects were concurrent with subtle alterations in the subcellular distribution of AR. In control animals, most AR was in the high salt cell fraction, apparently bound to the natural ligand and DNA. Vinclozolin treatment reduced the amount of AR in the high salt (bound to DNA) fraction and increased AR levels in the low salt (inactive, not bound to DNA) fraction. M1 and M2 were found in the serum of animals from the two highest dosage groups at levels well below their Ki values. These results suggest that when the vinclozolin metabolites occupy a small percentage of AR, this prevents maximal AR-DNA binding, alters in vivo androgen-dependent gene expression and protein synthesis, which in turn results in obvious alterations of morphological development and serum hormone levels.

5.4
Effects on the Liver and Adrenal

Although vinclozolin and its metabolites do not bind the glucocorticoid receptor [109], treatment has been shown to alter the pituitary-adrenal axis in several mammalian species. Also noteworthy are the effects of vinclozolin and flutamide on liver function. Although short-term (7 day) and chronic treatments with these two chemicals increase liver size, the mechanism of action for the hepatic effects of these antiandrogens has not been elucidated, and the role of AR in this process, if any, is unknown.

5.5
Effects in Lower Vertebrates

It is generally held that mammals possess a single AR [117] as evidenced by the complete phenotypic sex reversal displayed in humans with complete androgen insensitivity syndrome as a consequence of a single base substitution in the AR [117]. In contrast, two marine fish species, the kelp bass (*Paralabrax clathratus*) and Atlantic croaker (*Micropogonas undulatus*), have at least two ARs, termed AR1 and AR2 [118]. AR1 in the brain displays binding affinities for ligands quite distinct from AR2. AR2 from gonadal tissues of the Atlantic croaker and kelp bass has similar ligand affinities to mammalian AR. AR2 has been shown to bind *p,p*-DDE and vinclozolin metabolites, M1 and M2, demonstrating the homology of AR function in vitro among diverse classes of vertebrates. In vivo, vinclozolin treatment induces intersex in the Medaka (*Oryzias latipes*), a fish species with a "mammalian type" sex differentiation process (male heterogametic – androgen mediated). In contrast, the fact that Makeynen et al. [119] did not obtain sex reversal in the fathead minnow with vinclozolin treatment may be related to several factors including a lack of metabolic activation of vinclozolin and undefined role for androgens in the sex differentiation process of this species. However, they also reported that M1 and M2 did not bind AR in the fathead minnow. The effects of vinclozolin also were studied in ovo in both birds and reptiles. Crain et al. [120] reported that in ovo treatment with vinclozolin failed to induce sex reversal in either male or female alligators, a lack of effect that is not too surprising given the key role that estrogen plays in sexual differentiation in this species. In contrast, 17β-estradiol and tamoxifen caused sex reversal from male to female, with a corresponding increase in aromatase activity. In the Japanese quail, however, male behavior was partially demasculinized by in ovo vinclozolin-treatment [121]. It is noteworthy, that in these species, as in many other vertebrates, the role of androgens in sexual differentiation is not well defined and differs from species to species.

6
Procymidone

6.1
In Vitro Effects

Procymidone is a dicarboximide fungicide structurally related to vinclozolin. In vitro, procymidone inhibits DHT-induced transcriptional activation at 0.2 µmol/l in CV-1 cells cotransfected with the human AR and a mouse mammary tumor virus (MMTV)-luciferase reporter gene; at 10 µmol/l, DHT-induced transcriptional activity is completely inhibited [122]. Using a Chinese hamster ovary (CHO) cell promoter interference assay, we demonstrated that

1 μmol/l procymidone also blocked DHT-induced AR-DNA binding. The fact that these effects were seen at a concentration more than a 1000-fold below the Ki of the parent material for AR suggests that a procymidone metabolite(s) is the active antiandrogenic compound(s).

6.2
Dose-Response of Developmental Effects

When administered by gavage at 100 mg/kg/day on GD 14 to day 3 after birth, procymidone reduces AGD in male pups and induces retained nipples, hypospadias, cleft phallus, a vaginal pouch, and reduced sex accessory gland size in male rat offspring [123]. When administered at lower dosage levels using the same protocol (25, 50, 100, 200 mg/kg/day), effects were noted at all dosage levels [122]. Procymidone exposure reduced AGD (at 25 mg/kg/day and above), induced nipples (25 and above), permanently reduced the size of several androgen-dependent tissues (levator ani and bulbocavernosus muscles (25 and above), prostate (50 and above), seminal vesicles (100 and above), Cowper's gland (100 and above), and glans penis (100 and above)) and induced malformations (hypospadias (50 and above), cleft phallus (50 and above), exposed os penis, vaginal pouch (50 and above), and ectopic, undescended testes (200)). Procymidone had a marked effect on the histology of the dorsolateral and ventral prostatic and seminal vesicular tissues (at 50 mg/kg/day and above). The effects consisted of fibrosis, cellular infiltration, and epithelial hyperplasia [122]. In contrast to the developmental effects, procymidone has less effect on the reproductive tract of the adult male rat (two weeks at dosage levels as high as 2000 ppm in the diet) [124].

Since the role of androgens in mammalian sexual differentiation is highly conserved, it is likely that humans would be adversely affected by vinclozolin or procymidone in a predictable manner if the human fetus were exposed to sufficient levels of the active metabolite(s) during critical stages of intrauterine life.

7
Linuron

7.1
In Vitro and Short-Term In Vivo Effects

Linuron is a urea-based herbicide with an acute oral LD50 for rats of 4000 mg/kg. Existing in vitro data demonstrate that linuron is a weak AR ligand [125, 126] with an EC50 of 64 – 100 μmol/l. However, it has not been determined if linuron is an AR agonist or antagonist, and the antiandrogenic potential of linuron has not been studied using sensitive in vivo assays at nontoxic dosage levels. Lambright et al. [127] utilized a battery of in vivo and in vitro assays to assess whether linuron altered AR function. Linuron competed in vitro with an androgen for rat prostatic AR (EC50 = 100 – 300 μmol/l) and human AR (hAR) in a COS cell whole cell binding assay (EC50 = 20 μmol/l), and linuron inhibited DHT-hAR induced gene expression in CV-1 and MDA-

MB-453 cells (EC_{50} = 10 μmol/l). Linuron treatment (100 mg/kg/day oral for 7 days) reduced testosterone- and DHT-dependent tissue weights in the Hershberger assay [128] (using castrate-immature-testosterone propionate-treated male rats), and linuron treatment (100 mg/kg/day oral for 4 days) altered the expression of androgen-regulated genes in ventral prostate in situ.

7.2
Developmental Effects

The antiandrogenic effects of linuron are difficult, if not impossible, to detect in adult animals, but are quite apparent in the offspring when administered during gestation. In a modified multigenerational study, the only effects seen in P0 male rats when linuron was administered from weaning through puberty, breeding, and lactation at 0, 20, or 40 mg/kg/day by gavage in oil was a 2.5 day delay in PPS and a small reduction in seminal vesicle and cauda epididymal weights. Fertility and serum hormone levels were unaffected in the P0 generation at dosage levels up to 40 mg/kg/day [115]. In contrast, dramatic effects were seen in the F1 generation, including malformations and subfertility. The F1 pairs sired fewer pups under continuous breeding conditions (63 pups vs 104, mated continuously over 12 breeding cycles) and the F1 males had reduced testes and epididymal weights, and lower testes spermatid numbers. These developmental effects were surprising because it has been reported that when linuron was administered in the diet at concentrations up to 125 ppm over three generations, reproductive malformations were not reported [129] and Khera et al. [130] reported that linuron was not teratogenic in the rat at dosage levels up to 100 mg/kg/day.

To resolve this apparent discrepancy, we administered linuron by gavage in oil at 100 mg/kg/day from days 14–18 of gestation [115]. AGD in male offspring, adjusted by analysis of covariance for body weight, was reduced by about 30% (pup weight was down 20%), and the incidence of areolas (with and without nipples) seen in the male offspring as infants was increased from 0% in controls to more than 44% in the linuron-treated males. Linuron treatment also induced epispadias in 1/13 males (partial hypospadias with the urethral opening half way down the phallus) and several androgen-dependent tissues were reduced in size in linuron-treated male rats, including the seminal vesicles, ventral prostate, levator ani/bulbocavernosus muscles, and epididymides. While the above effects are consistent with the action of a relatively weak AR antagonist, the high incidences of epididymal and testicular malformations (>50% of the linuron-treated males displaying agenesis or atrophy of one or both organs) were surprising. The epididymal malformations seen in treated male offspring included agenesis of the caput and/or corpus epididymides, while some testes were atrophic, fluid filled, and flaccid [115, 127]. These malformations are also produced at lower dosage levels. McIntyre et al. [131] detected malformations in male rat offspring at dosage levels of linuron as low as 12.5 mg/kg/day (GD 10–22), the lowest dose examined. These data demonstrate that linuron is an AR antagonist both in vivo and in vitro. However, linuron produces a profile of malformations that differs from the standard AR antagonist, one that curiously resembles the effects seen with di-*n*-butyl phthalate or di-(ethylhexyl) phthalate treatment (see below). It remains to be determined if linuron alters sexual differentiation by additional mechanisms of action in addition to AR antagonism or if tissue specific metabolites are formed.

8
DDT and DDE

8.1
Background

Although use of DDT has been banned in some countries, it is still in use in many parts of the world and all wildlife and humans are exposed, with some exposures in the high ppm range. A world-wide ban of this pesticide is currently being considered, but this has become very controversial because DDT is used to control vectors of malaria, a disease which accounts for many deaths. Although agricultural usages of DDT is declining and will eventually end, human exposure from DDT use continues. Hence, it is now more important than ever to determine the potential effects of continued usage of this pesticide on humans. In addition, high concentrations of DDT and its metabolites, especially *p,p'*-DDE, persist in North American fields, farms, orchards, and Superfund sites. As a result, some wildlife populations still display incredibly high total DDT residues [i.e., 16, 132, 133]. Over the decades of use, some orchards may have had more than 1000 kg DDT applied per hectare [132], while around Lake Apopka it was reported that DDT was sometimes applied on a daily basis [133]. In the orchards and fields sampled by Elliott et al. [132], both migratory and nonmigratory birds had high tissue levels of *p,p'*-DDE (up to 103 mg/kg), while at Lake Apopka fat samples in birds were orders of magnitude higher. Elliott et al. [132] concluded that levels of 100 ppm in passerines were sufficiently high to present a considerable toxic hazard to birds of prey.

8.2
In Vitro and Developmental/Reproductive Effects of *p,p'*-DDE

In 1995 Kelce et al. [106] reported that *p,p'*-DDE displayed antiandrogenic activity both in vivo and in vitro that was similar to vinclozolin, both being AR antagonists. In vitro, *p,p'*-DDE binds to the AR and prevents DHT-induced transcriptional activation in cells transfected with the human AR and inhibits androgen-dependent gene expression in vivo [105]. Interestingly, Wakeling and Visek [134] reported that several chlorinated pesticides including dieldrin and *o,p'*-DDT inhibited binding of DHT to proteins in the rat prostate cytosol (they did not examine *p,p'*-DDT or DDE).

When *p,p'*-DDE is administered by gavage in oil during gestation, treatment at 100 mg/kg/day on GD 14–18 reduces AGD and induces hypospadias, retained nipples, and smaller androgen-dependent tissues in treated Long Evans Hooded (LE) and Sprague-Dawley (SD) male rat offspring [115]. While the alterations were evident in both rat strains, the SD strain appeared to be more affected: only the SD strain displayed hypospadias, and other effects were of a greater magnitude in the treated SD than in the LE rats. It is uncertain if this reflects a true strain difference in sensitivity or if it merely results from experiment to experiment variation. You et al. [135] studied the effects of *p,p'*-DDE on the male offspring using the same protocol, and they also found that *p,p'*-DDE induced antiandrogenic effects on AGD and areola development in both LE and SD rat strains. Following oral treatment with *p,p'*-DDE at 100 mg/kg/day as above, fetal rat tissue *p,p'*-DDE levels ranged from 1 to 2 μg/g during sexual differentiation in this dosage group [136], a concentration well below that seen in human fetal tissues from the late 1960s in areas of normal DDT use in the USA and Israel [137, 138].

When *p,p'*-DDT, which also is antiandrogenic [106], was administered to Dutch Belted rabbits during gestation (does treated) and lactation (pups treated) at a weekly dose of 25 mg/kg (dams) and 10 mg/kg (pups), reproductive abnormalities were displayed by male offspring [139]. Infantile exposure alone resulted in delays in testicular descent, while combined gestational plus lactational exposure induced uni/bilateral cryptorchidism. Serum levels of *p,p'*-DDT and DDE in offspring were 208 ppb *p,p'*-DDT and 38 ppb *p,p'*-DDE. These levels are well below the concentrations of *p,p'*-DDT and DDE seen in human fetal tissues in the USA, in areas of normal pesticide use during the late 1960s [138]. Taken together, these data indicate that adverse developmental reproductive effects are seen in rats and rabbits at levels (based on tissue residues) that are within the range reported for the human fetus in the late 1960s, exposed to DDT at this time through legal applications [137, 138].

This antiandrogen produces subtle changes in pubertal development in the male rat. When *p,p'*-DDE is administered at 0, 30, or 100 mg/kg/day from weaning until about 50 days of age, PPS was delayed about five days in male rats treated with the high dose, but sex accessory weights and serum hormone levels were not significantly altered [106, 112].

8.3
Potential Antiandrogenic Effects of *p,p'*-DDE in Wildlife

The antiandrogenic activity of *p,p'*-DDE may be one of the mechanisms by which it produces adverse effects in wildlife. The high incidence of undescended testes in the Florida panther could be the result of the high levels of *p,p'*-DDE in the food chain [37, 38]. In the alligator, *p,p'*-DDE exposure has been linked to developmental reproductive abnormalities (small penis, abnormal hormone levels, skewed sex ratio with an increase in the percentage of intersex and decrease in the percent male offspring) in Lake Apopka [15, 120, 140–144]. Although alligator eggs from Lake Apopka contained levels of *p,p'*-DDE (5.8 ppm) and several other contaminants [15], *p,p'*-DDE concentrations alone were above the concentration that block AR function in vitro, and some of the effects are likely the consequence of this chemical acting via multiple endocrine mechanisms.

8.4
Endocrine Mechanisms of Eggshell Thinning

The most widely know adverse endocrine effect of *p,p*-DDE is its ability to induce eggshell thinning (ET) in avian and reptilian oviparous vertebrates. Reviews indicate that DDE induces ET by an endocrine mechanism via an inhibition of prostaglandin synthesis in the shell membrane [40, 145]. Many avian species are susceptible to p,p'-DDE-induced ET. Toxicant-induced ET, especially in predatory birds, can result in cracked or broken eggs and other adverse reproductive effects [10, 146–151]. Although p,p'-DDE exposures were accompanied by exposures to other pollutants, careful field studies indicated that p,p'-DDE rather than PCBs or dieldrin was responsible for most, if not all, of the ET. These observations were verified in several laboratory studies using a variety of susceptible avian species. Although several toxicants (e.g., PCBs, lindane, dieldrin, dicofol with DDT as a byproduct) [145, 152] can also induce ET via a variety of mechanisms, these effects typically differ from that induced by p,p-DDE, being of lesser magnitude (up to 10%) and persisting, in some cases, for only a few days after cessation of treatment. Some of these effects of toxicants (lindane) or dietary restriction of calcium on ET can be reversed by estradiol administration [153]. In addition, laboratory studies also demonstrate transgenerational effects of estrogenic chemicals [154]. In ovo treatment with DES, estradiol, or o,p'-DDT (the estrogenic isomer) results in female progeny that lay eggs without shells on a regular basis and some females have extra oviductal tissue.

The site of action of p,p'-DDE on ET appears to be located in the mucosal cell of the shell gland (SG) of the oviduct and the probable endocrine mode of action involves altered calcium metabolism in the shell gland (reviewed by [145]) within the SG mucosa. In some of the early work it was observed that dietary p,p'-DDT inhibited carbonic anhydrase and reduced serum estradiol levels [148]. However, since p,p'-DDE injections also caused ET and reduced carbonic anhydrase without lowering serum estradiol, it was proposed that reduced serum estradiol was not responsible for p,p'-DDE-induced ET. More recently, it was proposed that inhibition of Ca^{2+}-Mg^{2+}-ATPase activity in the SG by p,p'-DDE was responsible for ET; however, Lundholm [145] reported that this effect, although prominent in vitro, was only occasionally affected in vivo. Rather, Lundholm proposed [145, 155, 156] that p,p'-DDE induced ET by inhibition of prostaglandin synthesis in the SG mucosa. He demonstrated that decreased PGE_2 blocks HCO_3^- transport, which in turn reduces Ca^{2+} transport. However, he also observed that the reduction in HCO_3^- transport did not result from an inhibition of HCO_3^- stimulated ATPase activity, as proposed earlier by other investigators. In support of his hypothesis, Lundholm and Bartonek [157] found that indomethacin treatment, which inhibits prostaglandin (PGF_2 alpha and PGE_2) synthesis in the SG at the onset of shell calcification, reduced eggshell thickness by 21% 14 h after treatment while calcium level in the SG mucosa was increased by 153%.

8.5
p,p'-DDE and Lake Apopka Alligators and Birds

Although we generally consider that the problems of some chemicals like DDT are behind us, the situation in Lake Apopka demonstrates that this is not the case. Furthermore, ecological risk assessments sometimes fail to consider adequately how contaminants from normal agricultural uses ("brown-farms") that ended nearly 30 years ago can still elicit adverse effects on wildlife. In Lake Apopka, Florida, reproductive problems in alligators (*Alligator mississippiensis*) and other species have been well documented. This 12,500-hectare lake is polluted with contaminants from agricultural activities around the lake, sewage treatment from a nearby town, and a "spill" in 1980 at Tower Chemical Company. Although this spill reportedly resulted in contamination of the lake with dicofol, DDT, and sulfuric acid, some confusion exists about the actual contaminants released into the lake [158]. Shortly after this spill, there was a 90 % decline in the alligator population that persisted from 1980–1984 and the population remains depressed to date. Alligator egg clutch viability was depressed due to an embryonic mortality rate of 80% in the first month after fertilization [140]. In addition, juvenile alligators continue to display endocrine and morphological abnormalities [15, 140]. Lake Apopka female alligators displayed polyovular follicles in their ovaries with multinucleated oocytes, males exhibited plasma testosterone levels one-third those of Lake Woodruff males (the reference site), phalli were 24% smaller than Lake Woodruff male alligators, and abnormal structures were observed within the seminiferous tubules of the testes [143]. In addition, testicular estradiol synthesis was elevated threefold in Lake Apopka males over males from Lake Woodruff. Freshwater turtles from this lake show developmental abnormalities of the gonad and plasma sex steroids similar to those seen in alligators [141].

An initial analysis of the contaminants in alligator eggs from Lake Apopka indicate that these reptiles are exposed to several pesticides, with p,p'-DDE being the most abundant, occurring at levels (5.8 ppm in alligator eggs) [159] above those known to reduce embryo viability in avian eggs [140]. In addition to p,p'-DDE, these eggs contained not detectable (ND) to 1.8 ppm p,p'-DDD, 0.02–1.0 ppm dieldrin, and ND to 0.25 ppm chlordane. More recently, Guillette et al. [16] detected 16 (of 18 measured) organochlorine pesticides or metabolites and 23 (of 28 measured) congener-specific PCBs in juvenile alligator serum from Lake Apopka, Orange Lake, and Lake Woodruff National Wildlife Refuge (the two reference sites). Lake Apopka alligators had higher levels of p,p'-DDE, dieldrin, mirex, endrin, chlordane, total DDTs, and total PCBs vs alligators from the other lakes, and the Lake Apopka alligators had lower serum testosterone and smaller phalli (adjusted for body size) than animals from Lake Woodruff, replicating effects seen in earlier studies in a different cohort of animals [143].

The fact that p,p'-DDE displays antiandrogenic activity in mammalian in vitro and in vivo assays led Guillette et al. [141] and Kelce et al. [106] to hypothesize that the small phallus size could result from an antiandrogenic effect of p,p'-DDE in juvenile male alligators. On the other hand, Guillette et al. [141] proposed that the increase in testicular abnormalities could result from an "estrogenic environment" created by the mixture of pesticides in the tissues through inhibition of testosterone action via the AR, by lowering plasma levels of testosterone, and increasing synthesis of testicular estradiol. They also noted that polyovular follicles with multinucleated oocytes is pathognomonic of in utero DES exposure in female mice, suggesting that some estrogenic effector may be present in the Lake Apopka ecosystem.

Beginning in 1998, wildlife populations at Lake Apopka suffered a new, major and unanticipated insult from pesticide contaminants that had persisted in the soil of muck farms since their use was banned [133, 160]. As part of the Lake

Apopka Restoration Project, muck farms on the northern end of the lake were flooded in 1998, which inadvertently mobilized pesticide residues that had accumulated for decades. Since the flooding, thousands of piscivorous birds (pelicans, wood stork, herons, etc.) have been found dead or dying around the lake. Tissue residue analyses from the moribund animals reveals lethal levels of pesticides including, p,p'-DDE, toxaphene, dieldrin, and chlordane. One great blue heron fat sample contained 2.3% (23,000 ppm) p,p'-DDE alone. The effects on other fish and wildlife species or humans who fish in the lake have not been rigorously evaluated at this time. It seems likely that other species will be adversely affected as well. If some way is found to reduce the release of pesticides into the lake and partially remediate this problem, one would anticipate that reproductive problems in fish and wildlife will persist well after the residues decline below lethal levels. If the restoration project continues to mobilize these pesticides into the biota of the lake, it is unclear how these contaminants can be controlled. Clearly, the lack of concern about the EDC issue [160], which was raised by some scientists prior to the flooding, and lack of consideration of the lethal and reproductive effects of pesticides on fish and wildlife populations in the risk assessment process, coupled with the failure to adequately sample these farms for residues, led to this disaster, one that should not be repeated, especially as consideration is being given to restoration of the nearby Everglades ecosystem.

9
Antiandrogenic Effects of Phthalate Esters During Development

Recently, concerns about exposure of children to phthalates in toys and other products have resulted in a ban of phthalates in certain toys by the European Union. Although industry has repeatedly assured the safety of these chemicals [161], most of them, including di-2-ethylhexyl phthalate (DEHP), have never been rigorously examined by the manufacturers for multigenerational effects.

The phthalates represent a class of toxicants which alter reproductive development via a mechanism of action that does not appear to involve AR or ER binding. Although many of the same effects are seen in animals exposed in utero to AR antagonists, like vinclozolin, in vitro studies, conducted to determine the biochemical mechanism responsible for the adverse developmental effects of DEHP, found that neither DEHP nor the primary metabolite mono-2-ethylhexyl phthalate (MEHP) compete with androgens for binding to the androgen receptor [162]. While some have suggested that di-n-butyl phthalate (DBP) was estrogenic [163], this activity is only displayed in vitro. For example, we have found that DBP did not produce any signs of estrogenicity in the ovariectomized female rat [115]. DBP (by subcutaneous injection at 200 or 400 mg/kg/day or by gavage at 1000 mg/kg/day, administered for two days, followed on the third day by 0.5 mg progesterone subcutaneously) did not induce a uterotropic response or estrogen-dependent sex behavior (lordosis). In addition, phthalate-treatment did not increase uterine weight [164] in juvenile female rats, and oral DBP-treatment (250, 500, or 1000 mg/kg/day from weaning through adulthood) failed to accelerate vaginal opening, or to induce constant estrus in intact female rats [115].

Recent publications demonstrate that perinatal exposure to a number of phthalate esters alters development of the male reproductive tract in an antiandrogenic manner, causing underdevelopment and agenesis of the epididymis at relative low dosage levels. Arcardi et al. [165] reported that administration of DEHP in the drinking water to the dam during pregnancy and lactation (estimated LOAEL of 3 mg/kg/day) produced testicular histopathological alterations in male rat offspring. Although DEHP is not an AR antagonist in vitro at concentrations up to 10 µmol/l, it inhibits fetal Leydig cell testosterone synthesis in vivo when orally administered to the dam at 0.75 g/kg/day starting at day 14 of pregnancy. As a consequence, fetal testosterone concentrations are reduced in males to female levels from day 17 of gestation to 2 days after birth. This reduction in testosterone levels results in a wide range of malformations of the androgen-dependent tissues in male rats including reduced AGD, retained nipples, hypospadias, cleft phallus, vaginal pouch, agenesis of the gubernacula cords and sex accessory tissues, underdevelopment of levator ani muscles, undescended testis, hemorrhagic testes, epididymal agenesis, and testicular atrophy. Mylchreest et al. [166] observed similar malformations in male rat progeny after prenatal oral exposure (GD 10–22) to DBP with effects occurring at dosage levels as low as 100 mg/kg/day. In our multigenerational assessment of the reproductive effects of DBP on the male and female parents and their progeny, daily oral administration of 500 mg/kg/day by gavage delayed puberty in male rats and reduced fertility in both male and female rats [115] while 250 mg/kg/day induced reproductive tract malformations and reduced fecundity in the offspring. In addition, when dams were dosed by gavage with 500 mg DBP/kg in oil or an equimolar dose of DEHP (750 mg/kg/day) during sexual differentiation (GD 14 – PND 4) the male offspring were profoundly malformed. More limited dosing in "pulses" during 4-day periods of gestation demonstrated that DBP at this dose was most effective on days 16–19 [115].

In a recent investigation [167] we examined several phthalate esters to determine if they also altered sexual differentiation in an antiandrogenic manner. We hypothesized that the phthalate esters that altered testes function in the pubertal male rat [168] would also alter testis function in the fetal male and produce malformations of androgen-dependent tissues. In this regard, we expected that benzyl butyl phthalate (BBP) and DEHP would alter sexual differentiation, while dioctyl terephthalate (DOTP), diethyl phthalate (DEP), and dimethyl (DMP) phthalate would not. We also expected that the phthalate mixture, diisononyl phthalate (DINP), would be weakly active due to the presence of some phthalates with a 4–7 carbon side chain. DEHP, BBP, DINP, DEP, DMP, or DOTP were administered orally to the dam at 0.75 g/kg from gestational day 14 to postnatal day 3. Male, but not female, pups from the DEHP and BBP groups displayed shortened AGDs (25%) and reduced testis weights (30%). As infants, males in the DEHP, BBP, and DINP groups displayed female-like areolas/nipples (86%, 70%, and 22%, respectively) vs 0% in other groups and they displayed reproductive malformations. The percentage of males with malformations was 91% for DEHP, 84% for BBP, and 7.7% ($p < 0.04$) in the DINP group. These phthalate esters produced a wide range of malformations of the external genitalia, sex accessory glands, epididymides, and testes. In the DINP group, 2/52 males (from 2/12 litters) displayed nipples, another male from one of the above litters displayed bilateral testicular atrophy, and a fourth male in the DINP group, from a third litter, displayed unilateral epididymal agenesis with hypospermatogenesis and scrotal fluid-filled testis, devoid of spermatids. In summary, DEHP, BBP, DBP, and DINP all induce antiandrogenic effects on sexual differentiation. DEHP and BBP were of equivalent potency while DINP was about an order of magnitude less active.

It is evident that the phthalate esters are high production volume toxic substances that have not been adequately tested for transgenerational effects. It is also clear that standard developmental toxicity studies are inadequate in their assessment of the effects of antiandrogens like the phthalates on the reproductive system [169, 170]. Developmental toxicity studies, conducted in several reputable laboratories using dosage levels well above those used herein, have failed to detect these malformations. There are many reasons for the insensitivity of the developmental toxicity test to detect reproductive tract malformations in the offspring. In studies conducted under regulatory test guidelines prior to 1998, exposure to the dams was on GD 6 to 15 in the rat, which is prior to the development of the fetal reproductive tract. In 1998, the period of exposure was extended to GD 19. However, this change did not resolve the major limitation of this protocol. Even when dosing is continued to GD 19, very few of the reproductive malformations cited above will be detected in a routine teratologic evaluation. Not only are some of these tissues very small and only detectable histologically, but also the process of sexual differentiation continues through perinatal life in the rat and some reproductive tissues are not fully differentiated until puberty is complete. For these reasons, the reproductive system cannot be fully evaluated during perinatal life.

10
Polychlorinated Aromatic Compounds

10.1
TCDD: A Putative Antiandrogen?

Exposure to aromatic hydrocarbon (Ah) receptor agonists such as TCDD, PCBs, and PCDFs is causally linked to developmental/reproductive toxicity in humans, nonhuman primates, rodents, mink, fish, and other wildlife species [171]. Effects are expressed in vertebrates at ppt concentrations. TCDD and other agonists bind a cellular steroid hormone-like receptor, termed the Ah receptor, forming a complex that acts as a transcriptional factor by binding to specific dioxin response elements (DREs) on specific genes. Some, if not all, of the toxicity of TCDD appears to result from activation of the Ah receptor. TCDD is an "endocrine disruptor" that acts on multiple components of the endocrine axis. TCDD exposure alters the levels of many hormones, growth factors, and their receptors.

In addition to the effects of these chemicals seen in humans (discussed earlier), exposure to very low doses of TCDD produces infertility in rodent progeny [172–180]. Exposure to a single low dose of TCDD ranging from 50 ng to

2 µg/kg during sex differentiation of the rat or Syrian hamster results in a number of unusual reproductive alterations in male and female progeny [181–184]. In the female rat, oral treatment with 0.2–1 µg TCDD/kg on GD 15 induced clefting of the phallus with a mild degree of hypospadias in females and a permanent "thread" of tissue across the opening of the vagina of the progeny [184]. Female progeny treated earlier in gestation with 1 µg TCDD/kg displayed reduced fecundity, a high incidence of constant estrus, and cystic endometrial hyperplasia at middle age. Female hamsters, treated by gavage on GD 11 with 2 µg TCDD/kg, also display clitoral clefting, reduced fertility as a result of several functional reproductive problems, but they did not display the vaginal thread. In TCDD-treated male rat and hamster offspring (dosing as per female offspring above), puberty was delayed, ejaculated and epididymal sperm numbers are reduced, while the reductions in ventral prostate, seminal vesicle, and testis size, displayed during peripubertal life (49–63 days of age), are attenuated with age. Mating behavior was normal in male hamsters, while male rats had some difficulty achieving intromissions. No malformations were noted in male rat or hamster offspring at these dosage levels, although epididymal agenesis has been reported in males exposed orally to doses above 1 µg TCDD/kg on GD 15 [185].

Studies indicate that development of the reproductive system is seriously altered when fetal TCDD concentrations reach 50 ppt, produced by dosing the dam with 1 µg TCDD/kg on GD 15 [186]. Our two lowest dosage levels (0.2 and 0.05 µg/kg on GD 15), which lower sperm counts and induce anomalies in female offspring, occur at fetal TCDD levels of about 5–10 ppt [187]. Similar or slightly higher levels are toxic to some embryonic fish [188] and avian [189] species in the field and in the laboratory.

10.2
PCB 169: A Putative Antiandrogen?

The PCB congener 169 is an Ah receptor agonist with a toxic equivalency factor of about 0.001, as compared to the potency of TCDD. PCB 169 treatment during pregnancy (administered as a single dose of 1.8 mg/kg to the dam on GD 8) alters reproductive development of LE hooded male and female rats in a manner almost identical to TCDD [115]. However, the sensitive androgen-dependent measures (AGD, areolas, and nipples) were not altered in treated males. Similar to TCDD treatment, PCB 169 treatment accelerated the age at eye opening and induced vaginal threads and mild hypospadias (urethral opening separate from the vaginal canal with cleft phallus) in female offspring, without reducing AGD or inducing areolas or nipples in males. In this regard, these organs were more affected in PCB 169-treated males at 65 days of age than in middle-aged males. PCB 169 treatment in utero increased the incidence of prostatitis in the dorsolateral lobe of the prostate of old male rats from 2/15 in controls to 7/10 in the treated males ($p < 0.01$) [115].

It is evident that the overall profile of effects seen in TCDD- and PCB 169-treated male and female offspring bears only limited resemblance to that seen in animals exposed to known antiandrogens. The fact that PCB 169 treatment fails to reduce AGD and induce areolas, retained nipples, and male hypospadias, hallmarks of antiandrogenic action, at a dosage level that produces significant reproductive toxicity suggests that the Ah receptor agonists may be affecting these tissues via alternative pathways, ones that do not involve the androgens or their receptor. Many other growth factors (i.e., epidermal growth factor) and hormones (i.e., prolactin, thyroid, and growth hormones) also are required for maximal development of these tissues. In addition, PCB 169 and TCDD induce many effects that are clearly unrelated to androgen action (eye opening, and female effects like cleft phallus and vaginal threads).

11
Inhibitors of Steroid Hormone Synthesis

11.1
Ketoconazole

Several fungicides inhibit fungal membrane synthesis and growth by inhibiting specific cytochrome P450 enzymes, especially 14α-demethylation of lanosterol in the sterol pathways. The process of steroidogenesis is sufficiently conserved that these chemicals also inhibit mammalian steroidogenesis. In general, however, at relatively high concentrations these fungicides are nonspecific inhibitors of CYP450 enzymes. Hence, effects in vertebrates may not be limited to the reproductive system and include adrenal and liver steroid metabolism and ecdysteroid synthesis in invertebrates.

Goldman et al. [190] produced male PH in rats with inhibitors of steroid 17α-hydroxylase and C17–20 lyase. Developmental alterations can also be obtained from in utero treatment with drugs that inhibit 5α-reductase, which is not a P450 enzyme, blocking the conversion of testosterone to DHT [191, 192]. The antifungal imidazole derivative, ketoconazole, inhibits various enzymes which belong to the cytochrome-P450-dependent mono-oxygenases in rodents and humans such as side chain cleavage of cholesterol, 11β-hydroxylase in the adrenal, and 17α-hydroxylase and C17–20 lyase in rat and human testes. For example, human testicular mono-oxygenase activities in vitro are reduced by 50% by 3.1 µmol/l ketoconazole. Schurmeyer and Nieschlag [193] demonstrated that ketoconazole and other imidazole fungicides inhibited testosterone production in males, while Pepper et al. [194] reported that ketoconazole was useful in the treatment of ovarian hyperandrogenism in women. Ketoconazole also has been shown to alter hepatic testosterone metabolism [195]. Four hours after male CD-1 mice were orally treated with ketoconazole from 0 to 160 mg/kg, serum testosterone levels, gonadal testosterone production, and hepatic testosterone hydroxylase activities were decreased in a dose-related manner.

We found that administration of ketoconazole by gavage from GD 14 in the rat at 100 mg/kg/day leads to reduced maternal weight gain and whole litter loss within a few days of the initiation of treatment [115]. These effects are consistent with the ability of this fungicide to inhibit progesterone synthesis. When the dosage level of ketoconazole was lowered to 0, 12.5, 25, or 50 mg/kg/day from GD 14 to PND 3, treatment delayed the onset of parturition by as much as three days and reduced the numbers of live pups at all but the lowest dosage level. Surviving male pups in the 12.5 and 25 mg/kg/day dose groups (there was only one pup at 50 mg/kg/day) did not display any indication of being demasculinized or feminized [115].

11.2
Fenarimol

Aromatase inhibitors prevent the conversion of androgens to estrogens. This P450 enzyme is highly conserved in a wide variety of tissues and many species, but the overall homology of the gene with other cytochrome P450s is only about 30%. Hence, this enzyme is considered to be in a separate gene family within the overall superfamily. As a consequence of the lack of sequence homology with other P450 enzymes in the steroid pathway, inhibitors of aromatase can display more specific activities than drugs like ketoconazole. In this regard, aromatase inhibitors have significant clinical uses and often present fewer untoward effects on other endocrine organs. Several fungicides like fenarimol inhibit aromatase activity [196] in mammals, resulting in infertility in both sexes. In our study, fenarimol was administered daily by gavage from weaning, through puberty, mating, gestation and lactation to the P0 generation (at 0, 17.5, 35, or 70 mg/kg/day) while the F1 offspring were only exposed indirectly through the placenta and milk. Fenarimol treatment inhibits male rat mating behavior (some males would not mount a sexually receptive female rat) in the P0 generation, presumably by inhibiting the conversion of androgens to estrogens in the brain [20]; it also inhibits parturition in P0 female rats because of the critical role of estrogens near term in the induction of labor in this species. Although fertility in the P0 generation was reduced by fenarimol treatment, the F1 generation, exposed indirectly only during pregnancy and during lactation, was unaffected. This pesticide also has effects in invertebrates, inhibiting ecdysteroid synthesis [197], while in reptiles, other aromatase inhibitors inhibit female gonadal sex determination [120].

12
Conclusions

It is well documented that substances present in the environment from human activities (pesticides, other toxic substances, phytochemicals, pharmaceuticals) with EDC activity have altered sexual differentiation and/or reproductive function in fish, wildlife, and domestic animals in the field [1, 2, 10–17, 48, 70–74, 140–151, 188, 189]. In many cases, such observations are supported by laboratory investigations and the mechanisms of action are known. As discussed herein, EDCs act via several distinct mechanisms. In the laboratory, we can alter sexual differentiation in rodents with some of these EDCs and, in some cases, the dosage levels that produce effects in laboratory animals overlap with background levels seen in the human population [36, 136–139, 183, 184, 186, 187]. In other cases, our low dose animal studies are relevant to the adverse effects of synthetic EDCs seen only in more highly exposed workers [60, 61, 63–66], accidental exposures [58, 77], or medical usages. Fortunately, in other cases, adverse effects from an EDC in an animal study occur only at dosage levels that exceed any real-world exposure. More often than not, however, we do not have sufficient field exposure or laboratory dose response or exposure data to determine in which of the above categories the EDC resides. One question that needs to be considered is: if humans are being affected by EDCs, how are these effects likely to be expressed? We can make some predictions based on what we know from (i) the malformations produced in humans by drugs with EDC activity [35] and (ii) effects in animals in the field and the laboratory. We know that exposure to individual EDCs and mixtures of EDCs varies over several orders of magnitude among animal and human populations. For this reason, rather than expecting global changes or "trends" in reproductive health, where much of the current attention is focused, we would expect that the prevalence and severity of the reproductive and developmental effects displayed would vary considerably from group to group based on historical and current EDC exposures. Except for high dose accidental or occupational exposures [58, 171], the effects seen in humans exposed to EDCs would often be latent, subtle (both hard to detect, but still potentially serious), functional reproductive alterations in highly exposed individuals or their progeny, with rare malformations occurring in only the most susceptible individuals. Compared to fish and wildlife, the effects of EDCs on human health are more the subject of debate than investigation. In regards to the current knowledge base about the effects of EDCs on human health, we must consider that "The absence of evidence is not evidence of absence" [198]. Some additional uncertainties in the risk assessment for EDCs are (i) what are the characteristics of the dose-response curves in the low dose range (threshold or not, monotonic or U-shaped), (ii) how will real-world mixtures of EDCs interact, and (iii) what other important mechanisms of endocrine action will be displayed by synthetic chemicals besides those (estrogen, antiandrogen, and antithyroid) that we are currently focused on?

13
References

1. Colborn T, Clement C (1992) Chemically-induced alterations in sexual and functional development: the wildlife/human connection. Princeton Scientific Publishing, Princeton
2. Alleva E, Brock J, Brouwer A, Colborn T, Fossi MC, Gray E, Guillette L, Hauser P, Leatherland J, MacLusky N, Mutti A, Palanza P, Parmigiani S, Porterfield S, Santti R, Stein SA, vom Saal F, Weiss B (1998) Toxicol Ind Health 14: 1
3. Toppari J, Larsen JC, Christiansen P, Giwercman A, Grandjean P, Guilette LJ Jr, Jegou B, Jensen TK, Jouannet P, Keiding N, Leffers H, McLachlan JA, Meyer O, Muller J, Rajpert-De Meyts E, Scheike T, Sharpe R, Sumpter J, Skakkebaek NE (1996) Environ Health Perspect 104(Suppl 4): 741
4. Carlsen E, Giwercman A, Keiding N, Skakkebaek NE (1992) Brit Med J 305: 609
5. Giwercman A, Skakkebaek NE (1992) Int J Androl 15: 373
6. Hannon W, Hill F, Bernert J, Haddock L, Lebron G, Cordero J (1987) Arch Environ Contam Toxicol 16: 255
7. Koninckx PR (1999) Gynecol Obstet Invest 47(1): 47
8. Davis DL, Bradlow HL, Wolff M, Woodruff T, Hoel DG, Anton-Culver H (1993) Environ Health Perspect 101: 372

9. National Research Council (1993) Pesticides in the diets of infants and children
10. Giesy JP, Bowerman WW, Mora MA, Verbrugge DA, Othoudt RA, Newsted JL, Summer CL, Aulerich RJ, Bursian SJ, Ludwig JP, Dawson GA, Kubiak TJ, Best DA, Tillitt DE (1995) Arch Environ Contam Toxicol 29: 309
11. Giesy JP, Snyder EM (1998) Xenobiotic modulation of endocrine function in fishes. In: Kendall R, Dickerson R, Giesy JP, Suk W (eds) Principles and processes for evaluating endocrine disruption in wildlife. SETAC Press, Pensacola, p 155
12. Ankley GT, Giesy JP (1998) Endocrine disruptors in wildlife: a weight of evidence perspective. In: Kendall R, Dickerson R, Giesy JP, Suk W (eds) Principles and processes for evaluating endocrine disruption in wildlife. SETAC Press, Pensacola, p 349
13. Van der Kraak G, Munkittrick KR, McMaster ME, MacLatchy D (1998) A comparison of bleached kraft mill effluent, 17β-estradiol, and β-sitosterol effects on reproductive function in fish. In: Kendall R, Dickerson R, Giesy JP, Suk W (eds) Principles and processes for evaluating endocrine disruption in wildlife. SETAC Press, Pensacola, p 294
14. Monosson E (1997) Reproductive and developmental effects of contaminants in fish populations: establishing cause and effect. In: Rolland R, Gilbertson M, Peterson RE (eds) Chemically induced alterations in functional development and reproduction in fishes. SETAC Press, Pensacola, p 177
15. Guillette LJ Jr, Woodward AR, Crain DA, Pickford DB, Rooney AA, Percival HF (1999) Gen Comp Endocrinol 116: 356
16. Guillette LJ Jr, Brock JW, Rooney AA, Woodward AR (1999) Arch Environ Contam Toxicol 36: 447
17. Guillette LJ Jr, Guillette EA (1996) Toxicol Ind Health 12: 537
18. Gray LE Jr (1991) Delayed reproductive effects following exposure to toxic chemicals during critical developmental periods. In: Cooper RL, Goldman JM, Harbin TJ (eds) Aging and toxicology: biological and behavioral perspectives. Johns Hopkins University Press, Baltimore, p 101
19. Gray LE Jr (1998) Toxicol Lett 102/103: 331
20. Gray LE Jr, Ostby JS (1998) Toxicol Ind Health 14: 159
21. Gray LE Jr (1992) Chemical-induced alterations of sexual differentiation: a review of effects in humans and rodents. In: Colborn T, Clement C (eds) Chemically-induced alterations in sexual and functional development: the wildlife/human connection. Princeton Scientific Publishing, Princeton, p 203
22. Gray LE Jr, Ostby JS, Wolf CJ, Lambright CR, Kelce WR (1998) Environ Toxicol Chem 17: 109
23. Jegou B, Auger J, Multigner L, Pineau C, Thonneau P, Spira A, Jouannet P (2000) The saga of sperm count decrease in humans and wild and farm animals. In: Gagnon C (ed) The male gamete. From basic science to clinical applications. Cache River Press, Montreal, p 446
24. Paulozzi L (1999) Environ Health Perspect 107: 297
25. Ternes T (1999) SETAC 20th Annual Meeting, Philadelphia, Abstract book, p 110
26. Metcalfe C, Keonig B, Ternes T, Hirsch R (1999) SETAC 20th Annual Meeting, Philadelphia, Abstract book, p 111
27. Meyer M, Bumgarner J, Thurman E, Hostetler K, Daughtridge J (1999) SETAC 20th Annual Meeting, Philadelphia, Abstract book, p 111
28. Nagel SC, vom Saal FS, Thayer KA, Dhar MG, Boechler M, Welshons WV (1997) Environ Health Perspect 105: 70
29. vom Saal FS, Timms BG, Montano MM, Palanza P, Thayer KA, Nagel SC, Ddhar MD, Ganjam VK, Parmigiani P, Welshons WV (1997) Proc Natl Acad Sci USA 94: 2056
30. Ewing LL, Desjardins C, Irby DC, Robaire B (1977) Nature 269: 409
31. Robaire B, Ewing LL, Irby DC, Desjardins C (1979) Biol Reprod 21: 455
32. Robaire B, Duron J, Hales BF (1987) Biol Reprod 37: 327
33. Wong C, Kelce WR, Sar M, Wilson EM (1995) J Biol Chem 270: 19,998
34. Hayes WJ, Laws ER (eds) (1991) Handbook of pesticide toxicology. Academic Press, San Diego

35. Schardein JL (1993) Hormones and hormonal antagonists. In: Schardein JL (ed) Chemically induced birth defects. Marcel Dekker, New York, p 271
36. Brouwer A, Ahlborg UG, Van den Berg M, Birnbaum LS, Boersma ER, Bosveld B, Denison MS, Gray LE Jr, Hagmar L, Holene E (1995) Eur J Pharmacol 293: 1
37. Facemire C, Gross T, Guillette LJ Jr (1997) SETAC 17th Annual Meeting, Washington DC, Abstract book, p 19
38. Facemire C, Gross T, Guillette LJ Jr (1995) Environ Health Perspect 103: 79
39. Harding LE, Harris ML, Stephen CR, Elliot JE (1999) Environ Health Perspect 107: 141
40. Fairbrother A, Gray LE Jr, Van der Kraak G, Francis B, Ankley G (1999) Reproductive and developmental toxicity of contaminants in oviparous animals. In: Di Guilio RT, Tillitt D (eds) Reproductive and developmental effects of contaminants in oviparous vertebrates. SETAC Press, Boca Raton, p 283
41. Steinberger E, Lloyd JA (1985) Chemicals affecting the development of reproductive capacity. In: Dixon RL (ed) Reproductive toxicology. Raven Press, New York, p 1
42. McLachlan JA (1981) Rodent models for perinatal exposure to diethylstilbestrol and their relation to human disease in the male. In: Herbst AL, Bern HA (eds) Developmental effects of diethylstilbestrol in pregnancy. Thieme-Stratton, New York, p 48
43. Meyer-Bahlburg HF, Ehrhardt AA, Feldman JF, Rosen LR, Veridiano NP, Zimmerman I (1985) Psychosom Med 47: 497
44. Hines M, Green R (1991) Rev Psychiat 10: 536
45. Hines M, Shipley C (1984) Develop Psychol 20: 81
46. Farnsworth NR, Bingel AS, Cordell GA, Crane FA, Fong HS (1975) J Pharmaceut Sci 64: 717
47. Salunkhe DK, Adsule RN, Bhonsle KI (1989) Antifertility agents of plant origin. In: Cheeke PR (ed) Toxicants of plant origin. Vol IV: Phenolics. CRC Press, Boca Raton, p 53
48. Adams RR (1989) Phytoestrogens. In: Cheeke PR (ed) Toxicants of plant origin. Vol IV: Phenolics. CRC Press, Boca Raton, p 23
49. Labov JB (1977) Comp Biochem Physiol 57: 3
50. Coussens R, Sierens G (1949) Arch Int Pharmacodyn Ther 78: 309
51. Cain JC (1960) Nature 188: 774
52. Cassidy A, Bingham S, Setchell KDR (1994) Am J Clin Nutr 60: 333
53. Wilt TJ, Ishani A, Stark G, MacDonald R, Lau J, Mulrow C (1998) J Am Med Assoc 280: 1604
54. Carilla E, Briley M, Fauran F, Sultan C, Duvilliers C (1984) J Steroid Biochem 20: 521
55. Bayne CW, Donnelly F, Ross M, Habib FK (1999) Prostate 40: 232
56. Rhodes L, Primka RL, Berman C, Vergult G, Gabriel M, Pierre-Malice M, Gibelin B (1993) Prostate 22: 43
57. Paubert-Braquet M, Richardson FO, Servent-Saez N, Gordon WC, Monge MC, Bazan NG, Authie D, Braquet P (1996) Pharmacol Res 34: 171
58. Guo YL, Lai TJ, Ju SH, Chen YC, Hsu CC (1993) Organohalogen Compds 14: 235
59. Klotz HP, Thibaut E, Russo F (1971) Ann Endocrinol 32: 763
60. Guzelian PS (1982) Ann Rev Pharmacol Toxicol 22: 89
61. Whorton D, Krauss R, Marshall S, Milby T (1977) Lancet 1259
62. Warren DW, Ahmad N, Rudeen PK (1988) Biol Reprod 39: 707
63. Grajewski B, Whelan EA, Schnorr TM, Mouradian R, Alderfer R, Wild DK (1996) Am J Ind Med 29: 49
64. Whelan EA, Grajewski B, Wild DK, Schnorr TM, Alderfer R (1996) Am J Ind Med 29: 59
65. Quinn M, Wegman D, Greaves I, Hammond S, Ellenbecker M, Spark R, Smith E (1990) Am J Med 18: 55
66. Smith E, Quinn M (1992) J Toxicol Environ Health 36: 13
67. Hostetler KA, Leach MW, Hyde TE, Wei LL (1996) J Toxicol Environ Health 48: 141
68. Dunnick JK, Elwell MR, Bucher JR (1994) Fund Applied Toxicol 22: 411
69. Parks LG, Lambright CS, Orlando EF, Guillette LJ Jr, Ankley GT, Gray LE Jr (2001) Toxicol Sci 62: 257
70. Wells K, Biddscombe S, Van der Kraak G (1999) SETAC 20th Annual Meeting, Philadelphia, Abstract book, p 189

71. Howell WM, Angus R (1999) Estrogens in the Environment Meeting, Tulane University, New Orleans (abstract available on ECME web site,
    http: //www.mcl.tulane.edu/ cbr/ecme/eehome/
72. Orlando EF, Davis W, Guillette LJ (1999) Estrogens in the Environment Meeting. Tulane University, New Orlean (abstract available on ECME web site,
    http: //www.mcl.tulane.edu/cbr/ecme/eehome/
73. Davis WP, Bortone SA (1992) Effects of mill effluent on fish sexuality. In: Colborn T, Clement C (eds) Chemically-induced alterations in sexual and functional development: the wildlife/human connection. Princeton Scientific Publishing, Princeton, p 113
74. Howell WM, Black DA, Bortone SA (1980) Copeia 4: 676
75. Wilson JD (1978) Ann Rev Physiol 40: 279
76. Couse JF, Hewitt SC, Bunch DO, Sar M, Walker VR, Davis BJ, Korach KS (1999) Science 286: 2328
77. Mocarelli P, Brambilla P, Gerthoux PM, Patterson DG Jr, Needham LL (1996) Lancet 10: 409
78. Davis DL, Gottlieb MB, Stampnitzky JR (1998) J Am Med Assoc 279: 1018
79. Astolfi P, Zonta LA (1999) Hum Reprod 14: 3116
80. James WH (1999) Reprod Toxicol 13: 235
81. James WH (1999) Fertil Steril 71: 775
82. James WH (1999) Am J Ind Med 35: 664
83. Jacobsen R, Moller H, Mouritsen A (1999) Hum Reprod 14: 3120
84. Biggar RJ, Wohlfahrt J, Westergaard T, Melbye MA (1999) J Epidemiol 150: 957
85. Vartiainen T, Kartovaara L, Tuomisto J (1999) Environ Health Perspect 107: 813
86. Moynihan JB, Breathnach CS (1999) Acta Pathol Microbiol Immunol Scand 107: 365
87. Clark MM, Karpiuk P, Galef BG Jr (1993) Nature 364: 712
88. Clark MM, Galef BG Jr (1995) Physiol Behav 57: 297
89. Vandenbergh JG, Huggett CL (1994) Proc Natl Acad Sci USA 9: 11055
90. Hornig LE, McClintock MK (1994) Anim Behav 47: 1224
91. Krackow S (1995) Biol Rev 70: 225
92. Kruuk LE, Clutton-Brock TH, Albon SD, Pemberton JM, Guinness FE (1999) Nature 399: 459
93. Sas M, Szollosi J (1980) Orvosi Hetilap 121: 2807
94. Herrmann BG, Koschorz B, Wertz K, McLaughlin KJ, Kispert A (1999) Nature 402: 141
95. Benirschke K (1981) Hermaphrodites, freemartins, mosaics and chimaeras in animals. In: Austin R, Edwards RG (eds) Mechanisms of sex differentiation in animals and man. Academic Press, New York, p 421
96. Lyon M, Cattanach B, Charlton H (1981) Genes affecting sex differentiation in mammals. In: Austin R, Edwards RG (eds) Mechanisms of sex differentiation in animals and man. Academic Press, New York, p 327
97. Polani P (1981) Abnormal sex development in man. 1. Anomalies of sex determining mechanisms, In: Austin R, Edwards RG (eds) Mechanisms of sex differentiation in animals and man. Academic Press, New York, p 465
98. Licht P, Frank L, Pavgi S, Yalcinkaya T, Siiteri P, Glickman S (1992) J Reprod Fertil 95: 463
99. Vaiman D, Koutita O, Oustry A, Elsen JM, Manfredi E, Fellous M, Cribiu EP (1996) Mamm Genome 7: 133
100. Wiig O, Derocher AE, Cronin MM, Skaare JU (1998) J Wildl Dis 34: 792
101. De Guise S, Lagace A, Beland P (1994) J Wildl Dis 30: 287
102. Sanchez A, Bullejos M, Burgos M, Hera C, Stanntopoulos C, Dias D, Jimenez R (1996) Mol Reprod Dev 44, 289
103. Sztein J, Raber J (1994) Lab Animal 19
104. CH2MHILL (1999) Kesterson Program. Kesterson Reservoir: 1998 Biological Monitoring Report: Prepared for the US Bureau of Reclamation, Mid-Pacific Region (SAC\147841\.DOC) by CH2MHILL, Sacramento
105. Kelce WR, Lambright CR, Gray LE Jr, Roberts K (1997) Toxicol Appl Pharmacol 142: 192
106. Kelce WR, Stone CR, Laws SC, Gray LE Jr, Kemppainen JA, Wilson EM (1995) Nature 375: 581

107. Gray LE Jr, Ostby JS, Ferrell J, Rehnberg G, Linder R, Cooper RL, Goldman JM, Slott V, Laskey J (1989) Fund Applied Toxicol 12: 92
108. Gray LE Jr, Ostby JS, Cooper RL, Kelce WR (1999) Toxicol Ind Health 15: 37
109. Kelce WR, Monosson E, Gamcsik MP, Laws S, Gray LE Jr (1994) Toxicol Appl Pharmacol 126: 275
110. Gray LE Jr, Ostby JS, Monosson E, Kelce WR (1999) Toxicol Ind Health 15: 48
111. Wolf CJ, LeBlanc GA, Ostby JS, Gray LE Jr (2000) Toxicol Sci 55: 152
112. Monosson E, Kelce WR, Lambright CR, Ostby JS, Gray LE Jr (1999) Toxicol Ind Health 15: 65
113. Korenbrot CC, Huhtaniemi I, Weiner R (1977) Biol Reprod 17: 298
114. Ashby J, Lefevre PA (1997) Regul Toxicol Pharmacol 26: 330
115. Gray LE Jr, Wolf CJ, Lambright CR, Mann PC, Price M, Cooper RL, Ostby JS (1999) Toxicol Ind Health 15: 94
116. Gray LE Jr, Ostby JS, Linder R, Goldman JM, Rehnberg G, Cooper RL (1990) Fund Applied Toxicol 15: 281
117. Quigley CA, De Bellis A, Marschke KB, El-Awady MK, Wilson EM, French FS (1995) Endocrine Rev 16: 271
118. Sperry TS, Thomas P (1999) Biol Reprod 61: 1152
119. Makeynen E, Kahl M, Jensen K, Tietge J, Wells K, Van Der Kraak G, Ankley GT (1999) SETAC 20th Annual Meeting, Philadelphia, Abstract book, p 193
120. Crain DA, Guillette LJ Jr, Rooney AA, Pickford D (1997) Environ Health Perspect 105: 528
121. McGary S, Ottinger M, Henry P (1999) SETAC 20th Annual Meeting, Philadelphia, Abstract book, p 174
122. Ostby JS, Kelce WR, Lambright CR, Wolf CJ, Mann PC, Gray LE Jr (1999) Toxicol Ind Health 15: 80
123. Gray LE Jr, Kelce WR (1996) Toxicol Ind Health 12: 515
124. Hosokawa S, Murakami M, Ineyama M, Yamada T, Yoshitake A, Yamada H, Miyamoto Y (1993) J Toxicol Sci 18: 83
125. Cook JC, Mullin LS, Frame SR, Biegel LB (1993) Toxicol Appl Pharmacol 119: 195
126. Waller C, Juma B, Gray LE Jr, Kelce WR (1996) Toxicol Appl Pharmacol 137: 219
127. Lambright C, Ostby J, Bobseine K, Wilson V, Hotchkiss AK, Mann PC, Gray LE Jr (2000) Toxicol Sci 56: 389
128. Hershberger L, Shipley E, Meyer R (1953) Proc Soc Exp Biol Med 83: 175
129. Hodge HC, Downs WL, Smith DW, Maynard EA (1968) Food Cosmet Toxicol 6: 171
130. Khera KS, Whalen C, Trivett G (1978) Toxicol Appl Pharmacol 45: 435
131. McIntyre BS, Barlow NJ, Wallace DG, Foster PM (1999) European Teratology Society Meeting, Oxford, Abstract book, p 14
132. Elliott JE, Martin PA, Arnold TW, Sinclair PH (1994) Arch Environ Contam Toxicol 26: 435
133. Williams T (1999) Audubon 10: 64
134. Wakeling A, Visek W (1973) Science 181: 659
135. You L, Casanova M, Archibeque-Engle S, Sar M, Fan LQ, Heck HA (1998) Toxicol Sci 45: 162
136. You L, Gazi E, Archibeque-Engle S, Casanova M, Conolly RB, Heck HA (1999) Toxicol Appl Pharmacol 157: 134
137. Wassermann M, Wassermann D, Zellermayer L, Gon M (1967) Israel Pesticide Monit 1: 15
138. Curley A, Copeland F, Kimbrough R (1969) Arch Environ Health 19: 628
139. Veeramachanei D, Tesari J, Nett T, Plamer J, Miller C, Kinonen C, Arguello J, Cosma G (1995) Toxicologist 30: 143
140. Guillette LJ Jr, Gross TS, Masson GR, Matter JM, Percival HF, Woodward AR (1994) Environ Health Perspect 102: 680
141. Guillette LJ Jr, Gross TS, Gross DA, Rooney AA, Percival HF (1995) Environ Health Perspect 103(4): 31
142. Guillette LJ Jr (1995) Human Ecol Risk Assess 1: 25
143. Guillette LJ Jr, Pickford DB, Crain DA. Rooney AA, Percival HF (1996) Gen Comp Endocrinol 101: 32

144. Crain DA, Guillette LJ Jr, Pickford DB, Percival HF, Woodward AR (1998) Environ Toxicol Chem 17: 446
145. Lundholm CE (1987) Comp Biochem Physiol C 88: 1
146. Ratcliffe DA (1967) Nature 215: 208
147. Ratcliffe DA (1970) J Applied Ecol 7: 67
148. Peakall DB (1970) Science 168: 592
149. Anderson DW, Jehl JR, Risebrough LA, Deweese LR, Edgecomb WG (1975) Science 190: 806
150. Grier JW (1982) Science 218: 1232
151. Wiemeyer SN, Lamont TG, Bunck CM, Sindelar CR, Gramlich FJ, Fraser JD, Byrd MA (1984) Arch Environ Contam Toxicol 13: 529
152. Schwarzbach SE, Shull L, Grau CR (1988) Arch Environ Contam Toxicol 17: 219
153. Chakravarty S, Lahiri P (1986) Toxicology 42: 245
154. Bryan EE, Gildersleeve RP, Wiard RP (1989) Teratology 39: 525
155. Lundholm CE, Bartonek M (1992) Arch Toxicol 66: 387
156. Lundholm CE (1994) Comp Biochem Physiol C 109: 57
157. Lundholm CE, Bartonek M (1992) Comp Biochem Physiol C 102: 379
158. Tilman (1995) Rohm and Haas letter from A Tilman to J. Loranger of the USEPA, Aug 5, 1995. In reference to Pesticide spill at Tower Chemical Company
159. Heinz GH, Percival HF, Jennings ML (1991) Environ Monit Assess 16: 277
160. FOLA (Friends of Lake Apopka) (1999) www.fola.org/e.conc/e.prob/
161. Koop CE, Juberg DR (1999) Medscape Gen Med J 22: 1
162. Parks L, Ostby JS, Lambright CR, Abbott B, Klinefelter GR, Gray LE Jr (2000) Toxicol Sci 58: 339
163. Jobling S, Reynolds T, White R, Parker MG, Sumpter JP (1995) Environ Health Perspect 103: 582
164. Meek M, Clemons J, Wu Z, Fielden M, Zacharewski T (1997) Toxicologist 36: 295
165. Arcardi FA, Costa C, Impertore C, Marchese A, Rapisarda A, Salemi M, Trimarchi GR, Costa G (1998) Food Chem Toxicol 36: 963
166. Mylchreest E, Sar M, Cattley RC, Foster PM (1999) Toxicol Appl Pharmacol 156: 81
167. Gray LE Jr, Ostby JS, Furr J, Price MG, Parks LG (2000) Toxicol Sci 58: 350
168. Foster PM, Thomnas L, Cook M, Gangolli S (1980) Toxicol Appl Pharmacol 54: 392
169. Ema M, Amano H, Itami T, Kawasaki H (1993) Toxicol Lett 69: 197
170. Ema M, Amano H, Ogawa Y (1994) Toxicology 86: 163
171. Golub M, Donald J, Reyes J (1991) Environ Health Perspect 94: 245
172. Khera KS, Ruddick JA (1973). Polychlorodibenzo-*p*-dioxins: perinatal effects and the dominant lethal test in Wistar rats. In: Blair EH (ed) Chlorodioxins – origin and fate. American Chemical Society, Washington DC, p 70
173. Murray FJ, Smith FA, Nitschke KD, Humiston CG, Kociba RJ, Schwetz BA (1979) Toxicol Appl Pharmacol 50: 241
174. Gray LE Jr, Kelce WR, Monosson E, Ostby JS, Birnbaum LS (1995) Toxicol Appl Pharmacol 131: 108
175. Mably T, Moore RW, Goy RW, Peterson RE (1992) Toxicol Appl Pharmacol 114: 108
176. Mably T, Bjerke DL, Moore RW, Gendron-Fitzpatrick A, Peterson RE (1992) Toxicol Appl Pharmacol 114: 118
177. Mably T, Moore RW, Peterson, RE (1992) Toxicol Appl Pharmacol 114: 97
178. Bjerke DL, Peterson RE (1994) Toxicol Appl Pharmacol 127: 241
179. Bjerke DL, Brown TJ, MacLusky NJ, Hochberg RB, Peterson RE (1994) Toxicol Appl Pharmacol 127: 258
180. Bjerke DL, Sommer RJ, Moore RW, Peterson RE (1994) Toxicol Appl Pharmacol 127: 250
181. Wolf CJ, Ostby JS, Gray LE Jr (1999) Toxicol Sci 51: 259
182. Gray LE Jr, Ostby JS (1995) Toxicol Appl Pharmacol 133: 285
183. Gray LE Jr, Ostby JS, Kelce WR (1997) Toxicol Appl Pharmacol 146: 11
184. Gray LE Jr, Wolf CJ, Mann PC, Ostby JS (1997) Toxicol Appl Pharmacol 146: 237
185. Wilker C, Johnson L, Safe S (1996) Toxicol Appl Pharmacol 141: 68

186. Hurst C, DeVito M, Abbott B, Birnbaum LS (1996) Toxicologist 30: 198
187. Hurst C, DeVito M, Abbott B, Birnbaum LS (1997) Toxicologist 36: 257
188. Zabel E, Walker M, Hornung M, Clayton M, Peterson RE (1995) Toxicol Appl Pharmacol 134: 204
189. Bellward G, Janz D, Sanderson J (1995) Organohalogen Compds 25: 99
190. Goldman AS, Eavey RD, Baker MK (1976) J Endocrinol 71: 289
191. Imperato-McGinley J, Sanchez RS, Spencer JR, Yee B, Vaughn ED (1992) Endocrinology 131: 1149
192. Weir PJ, Conner M, Johnson CM (1990) Teratology 41: A599
193. Schurmeyer T, Nieschlag E (1984) Acta Endocrinol 105: 275
194. Pepper G, Brenner S, Gabrilove J (1990) Fertil Steril 54: 38
195. Wilson V, LeBlanc G (2000) Toxicol Sci 54: 125
196. Hirsch KS, Adams ER, Hoffman DG, Markham JK, Owen NV (1986) Toxicol Appl Pharmacol 86: 391
197. Warren JT, Rybczynski R, Gilbert LI (1995) Insect Biochem Mol Biol 25: 679
198. Foster PM, Mylchreest E, Wallace D, Sar M (1999) Symposia 5. Endocrine Disruptors. 27th Annual European Teratology Society Meeting, Oxford, Abstract S5.3

CHAPTER 9

Developmental and Reproductive Abnormalities Associated with Endocrine Disruptors in Wildlife

Edward F. Orlando[1], Louis J. Guillette Jr[2]

[1] St. Mary's College of Maryland, Department of Biology, 18952 E. Fisher Road, St. Mary's City, MD 20686, USA
E-mail: *eforlando@osprey.smcm.edu*
[2] University of Florida, Department of Zoology, PO Box 118525, Gainesville, FL 32611–8525, USA
E-mail: *ljg@zoo.ufl.edu*

Since the middle part of this century, humans have become increasingly aware of their impact on the environment. Some examples of human impact on the environment include the developmental deformities and population declines of several species of birds around the Great Lakes, acid rain-associated loss of fish in northeast USA lakes, and human reproductive health problems attributed to industrial dumping at Love Canal, New York, USA. These and other examples have changed the way humans view their impact on the environment. Concern over the health and welfare of humans, as well as other organisms, has driven mankind to try to understand and manage this impact. One of the areas that scientists are focusing on is the alteration of normal endocrine system function by chemicals in the environment. In this chapter we present a brief review of the developmental and reproductive abnormalities associated with exposure to endocrine disrupting chemicals (EDCs). Although this chapter focuses on data from wildlife research, where necessary, we also provide examples from research conducted on domesticated animals from aquaculture, agriculture, and laboratory studies. We begin with a definition and a historical perspective of the field of toxicology. Following a definition of the endocrine disruption hypothesis, we present a synthesis of some of the data. This presentation is organized around four hierarchical levels of endocrine system function that include the control, production, availability, and action of hormones. Next, we discuss areas of cutting edge research, suggest areas for future study, and underscore the need to explore possible population-level effects of exposure to EDCs.

Keywords. Signal modification, Reproduction, Development, Population effects

1	**Introduction**	251
1.1	Toxicology and Past Perspectives	251
1.2	Brief History of the Field of Endocrine Disruption Research	251
1.3	Definition of the Endocrine Disruption Hypothesis	252
1.4	Traditional Methods of Investigation	252
2	**Normal and Altered Endocrine System Function**	253
2.1	Normal Endocrine System Function	253
2.2	Altered Endocrine System Function	256
2.2.1	Brain – Hormone Synthesis Control	258
2.2.2	Gonad – Hormone Production	258
2.2.3	Liver – Hormone Availability	259
2.2.4	Receptors in Brain, Gonad, and Liver – Hormone Action	261

The Handbook of Environmental Chemistry Vol. 3, Part M
Endocrine Disruptors, Part II
(ed. by M. Metzler)
© Springer-Verlag Berlin Heidelberg 2002

3   **Possible Future Areas of Research** 261
3.1   Other Hormone Axes and Possible Levels of Disruption 261
3.2   Methods of Investigation 263
3.3   Methods of Analysis 263
3.4   Going Beyond the Organism 265

4   **Conclusion: The Field of Signal Modification?** 266

5   **References** 266

List of Abbreviations

A	androgen
ACTH	adrenocorticotropic hormone
B	corticosterone
BKME	bleach kraft mill effluent
CBG	corticosteroid-binding globulin
CRH	corticotropin releasing hormone
DDE	dichlorodiphenyldichloroethene
DDT	dichlorodiphenyltrichloroethane
DES	diethylstilbestrol
E	estrogen
E2	estradiol-17β
EDC	endocrine disrupting chemical
EROD	ethoxyresorufin-O-deethylase
F	cortisol
FSH	follicle-stimulating hormone
GH	growth hormone
GHRH	growth hormone releasing hormone
GnRH	gonadotropin releasing hormone
GtH	gonadotropin
IGF	insulin-like growth factor
IQ	intelligence quotient
LC	lethal concentration
LD	lethal dose
LH	luteinizing hormone
NIEHS	National Institute of Environmental Health Sciences
NOAEL	no observable adverse effect level
P	progestin
PAH	polycyclic aromatic hydrocarbon
PCB	polychlorinated biphenyl
PCDF	polychlorinated dibenzofuran
SBP	serum binding protein
sGnRH	synthetic gonadotropin releasing hormone
SHBG	sex hormone binding globulin

SXR   steroid-xenobiotic receptor
T   testosterone
T_3 triiodothyronine
T_4 thyroxine
TBG   thyroxine-binding globulin
TRH   thyrotropin releasing hormone
TSH   thyroid-stimulating hormone
Vtg   vitellogenin

1
Introduction

1.1
Toxicology and Past Perspectives

Toxicology has been described as the study of the adverse effects of xenobiotics. Xenobiotics are chemicals that an organism is usually not exposed to, and can be of natural or anthropogenic origin. The discipline dates back to the earliest humans who made use of natural venom and poisons taken from animals and plants to assist them in hunting or in battle (for a historical perspective of toxicology, see [1]). The science of toxicology developed from the desire to safeguard human and environmental health. Modern toxicology was born in the early 1960s in the USA with the publication of Rachel Carson's *Silent Spring* and the development of sensitive analytical tools. Following establishment of the US Environmental Protection Agency and related agencies worldwide, the field of toxicology grew to include risk assessment [2].

Traditionally, research by human toxicologists focused on high concentration exposures and examined endpoints such as death (LD-50 or LC-50 – the dose or concentration of a chemical necessary to cause the death of 50% of the experimental organisms), birth defects (teratology), and cancer (uncontrolled cell division) [3]. The main concern was to prevent the death of the individual human. In contrast, wildlife toxicologists are concerned with a lower level of certainty, and their concern is to safeguard the population or keep a population from going extinct [4, 5].

1.2
Brief History of the Field of Endocrine Disruption Research

The understanding that certain chemicals can interact with the nervous and endocrine systems of an organism has been known since at least the early 1950s. Pesticides like dichlorodiphenyltrichloroethane (DDT) were and are currently used because of their ability to act as neurotoxins and interfere with membrane ion transport (e.g., calcium), inhibit selective enzyme activities, or alter the release or removal of neurotransmitters at neuronal synapses. In the early 1960s, Rachel Carson wrote about the relationship between DDT, eggshell thinning, and the demise of the bald eagle and other bird populations around the Great Lakes.

During that same period, doctors were prescribing a synthetic estrogen, diethyl-stilbestrol (DES), to pregnant women desiring to reduce spontaneous abortions and other complications associated with high risk pregnancies [6]. Simultaneously, the growth of the petrochemical industry was occurring, promising their customers extended lives full of convenience, health, and plentiful food.

In the 1970s, concern mounted over the widespread use of DES as an anabolic agent in animal husbandry, and its identification as the causative agent in the transgenerational effects observed in the reproductive health of daughters and sons of DES mothers [7]. Due in large part to this concern, a meeting was convened in 1979 by the US National Institute of Environmental Health Sciences (NIEHS), called "Estrogens in the Environment" [8]. Discussions from this first meeting, as well as a subsequent NIEHS sponsored meeting in 1985 [9] and a related 1991 meeting organized by the World Wildlife Fund, called "Chemically Induced Alterations in Sexual Development: The Wildlife/Human Connection" [10] were the beginnings of the field of endocrine disruption research (for an in-depth historical discussion, see [11]). It was at these first meetings that researchers from various disciplines including wildlife biology, reproductive physiology, endocrinology, neurology, immunology, molecular biology, and toxicology came together and realized there was a common thread running through all of their research. That common element can be summarized in the consensus statement of the 1991 meeting: "We are certain of the following: A large number of man-made chemicals that have been released into the environment, as well as a few natural ones, have the potential to disrupt the endocrine systems of animals, including human" [10].

1.3
Definition of the Endocrine Disruption Hypothesis

The endocrine disruption hypothesis suggests that certain chemicals in the environment, some naturally occurring and others produced by humans, have the ability to alter the development and reproduction of organisms by disrupting the normal functioning of the endocrine system [5, 10]. This could be caused by xenobiotics changing the normal internal hormonal environment of an organism.

Anthropogenic chemicals, such as pesticides, surfactants, industrial compounds, and heavy metals are widely distributed throughout the planet [12–16]. Given the interaction between the nervous, endocrine, and immune systems [17], and the ability of many of these compounds effectively to mimic, inhibit, or block the actions of native hormones and other signaling systems, there is concern for wildlife, domestic-animal, and human health [3, 10].

1.4
Traditional Methods of Investigation

By definition, an endocrine disrupting chemical (EDC) is a compound that is foreign to the organism, yet can bind to the hormone receptors within the organism or alter other aspects of normal endocrine system function. EDCs are termed agonists if they bind the receptor and induce hormone action (for a comparative discussion of short-term assays for certain EDCs see [18]). EDCs are antagonists if they bind the native receptor, yet block hormone action. Traditionally, investigators have used receptor-binding assays to measure relative binding affinity of a suspected EDC [19]. Here, a graded concentration series of the suspected EDC are combined with radiolabeled forms of the native hormone. Competition for the receptor will yield information about the relative affinity of the EDC compared to the native hormone. Yet, information from binding affinity assays is incomplete, because these assays do not demonstrate if the EDC bound to the hormone receptor will function as an agonist, resulting in hormone action or as an antagonist, actually inhibiting that action. To address these functional questions, endocrinologists have traditionally used in vivo tissue response assays (e.g., prostate growth [20]), or in vitro cell culture assays (e.g., E-screen, MCF-7 human breast cancer cells [21]). More recently, transgenic organisms (e.g., yeast with human estrogen or androgen receptor [22, 23]) and knock-out organisms (e.g., mice without functional estrogen receptors [24]) have become available to elucidate hormone or EDC action. These assays have produced some interesting results. Xenobiotics, such as chlordecone, phenol, and o,p'-DDT, were considered estrogen agonists because they bound the estrogen receptor. These results were obtained from experiments where each compound was examined singly. In mixtures, some xenoestrogens functioned as antagonists or anti-estrogens in a gene expression assay when combined with the native hormone, estradiol [25].

2
Normal and Altered Endocrine System Function

2.1
Normal Endocrine System Function

The timing of reproduction is thought to be under the direct control of the hypothalamus of the brain (for an in-depth discussion see [26, 27]) (Fig. 1). Pulsatile secretions from the hypothalamus of gonadotropin releasing hormone (GnRH) are known to induce the synthesis and release of gonadotropins (GtHs) from the pituitary. In most vertebrates, these gonadotropins are called follicle-stimulating hormone and luteinizing hormone (FSH and LH, respectively) because of their original association with ovarian follicle development and the beginning of the luteal or postovulatory phase in mammals. GtHs are carried in the blood to the gonads, where they induce production of steroid hormones, such as progestins, androgens, and estrogens, as well as gamete growth and differentiation. Androgens and estrogens play a major role during sexual differentiation of the embryo [28]. In the adult, these steroid hormones are responsible for the formation of gametes and development of secondary sex characteristics including reproductive behavior [26, 27]. Steroid hormones are lipophilic and do not dissolve easily in water. A very small percentage of the hormones are free in the blood, and most are carried bound to serum binding proteins (SBPs), such as sex hormone binding globulin. Hormones bound to SBP are protected

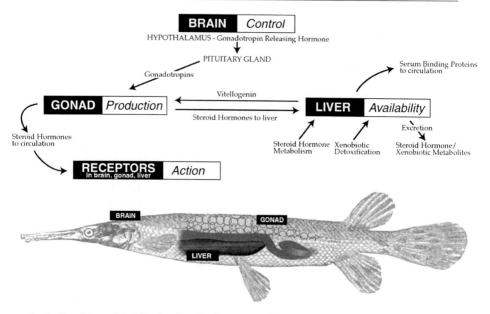

Fig. 1. Four hierarchical levels of endocrine system function include *control* (hypothalamus and pituitary of the brain), *production* (gonad), *availability* (steroid hormone metabolism/ xenobiotic detoxification and serum binding protein synthesis in the liver), *action* (receptors in the brain, gonad, and liver). The potential for alteration of steroid hormone concentration exists at each of these levels

from degradation by the liver and are also thought to function as a reserve of available hormone. The concentration of SBPs and the rate of degradation and excretion from the organism regulate the availability of steroid hormones to their target tissue [22, 29, 30]. Hormone action, thus a biological response, results from the binding of a hormone with its receptor (Fig. 2). Most commonly, steroid hormones move through the cell membrane and bind intracellular receptors. This hormone-receptor complex attaches to a particular site on the DNA (hormone response elements) and induces transcription [26]. Recently, membrane-bound steroid hormone receptors have been discovered [31–35]. Various nongenomic functions, such as ion channel regulation and enzyme activation, have been proposed as additional functions for steroid hormones. In summary, the brain controls steroid synthesis, steroid production is in the gonads, the liver regulates steroid availability, and steroid action is a function of receptor location.

There are several axes that are under the control of the hypothalamus – pituitary regions of the brain. These include the reproductive, growth, thyroid, and stress axes (Fig. 3). Any chemical that can mimic an endogenous compound has the potential to interact with and potentially disrupt the normal functioning of any of these axes. In addition, there are two nonsteroid hormone members of the nuclear receptor superfamily: retinoic acid and vitamin D_3. Both systems have been shown to have the potential for disruption by xenobiotics [36].

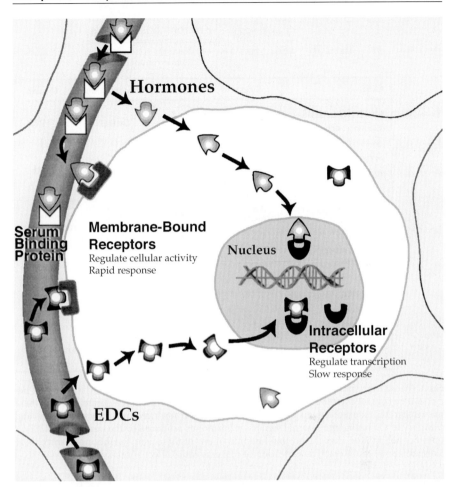

Fig. 2. Steroid hormone action depicted at the cellular level. Most of the steroid hormones are transported in the blood bound to serum binding proteins, whereas most endocrine disrupting chemicals (EDCs) that have been examined circulate freely. Both steroid hormones and EDCs can bind to membrane-bound or intracellular receptors. Membrane-bound receptors help regulate cellular activity and provide a rapid response to changes in the extracellular environment. Intracellular receptors regulate transcription and the response period is relatively slower compared to the action of the membrane-bound receptors

These hormone systems do not work in isolation, but cross talk or interaction between these axes is known to occur, and together they control development and reproduction in an organism [37, 38].

The potential for endocrine system disruption is complex and can occur in different organs (e.g., the hypothalamus, pituitary, and gonad and liver), and at different biochemical levels (e.g., receptors, serum binding proteins, enzymes for synthesis or degradation, hormone response elements) [35, 36]. Given the fact that endocrine axes are related by their interactive effects on both develop-

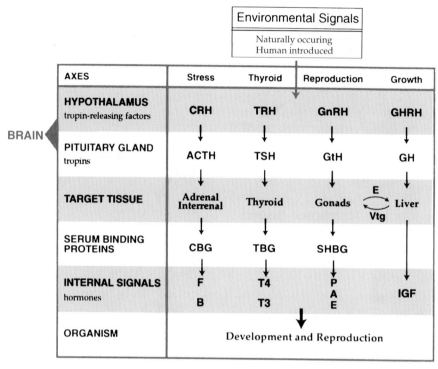

Fig. 3. These four endocrine axes (stress, thyroid, reproductive, and growth) are related because they control the development and reproduction in a typical vertebrate. Each of these axes integrates external signals from the environment, transduced by the hypothalamus and pituitary of the brain, into internal signals or hormones produced by the target tissues. EDCs have the potential to alter the normal function of any of these axes. Furthermore, since these axes are known to interact, modification of one axis could affect another axis

ment and reproduction, the disruption of one axis could impact another axis (e.g., the stress axis and its effect on the reproductive axis [39]).

2.2
Altered Endocrine System Function

Chemicals that disrupt the endocrine system can have profound effects on the development and reproduction of organisms (for examples see Table 1). A classic example is the in utero exposure of young boys and girls to the pharmaceutical estrogen, diethylstilbestrol (DES). The majority of the children born of mothers given DES during pregnancy had measurable alterations of their reproductive organs [40]. Many of the DES daughters exhibit lower fertility or infertility, and sons have elevated occurrence of cryptochidism and testicular cancer. EDCs, such as DES, have the potential to disrupt the endocrine system by modifying hormone synthesis, control, production, availability, and action. We will briefly document how EDCs can act at each of these levels in the following sections.

Table 1. Examples of developmental and reproductive abnormalities associated with endocrine disrupting chemicals

Taxa	Animal	Putative EDC	Noted effect	Citation
Invertebrates				
	Mud snail	Tributyl tin	Masculinization of females	[73]
	Mussel	Tributyl tin	↓ Aromatase activity	[74]
	Whelk	Tributyl tin	Masculinization of females	[75]
Vertebrates Fish	Atlantic croaker	PCBs/ Cadmium	Arochlor 1254: ↓ GtH, ovary growth and E2. Cd: ↑ ovary growth and E2	[76]
	Common carp	Sewage	↓ Plasma T, Vtg induction	[42]
	Common carp	4-*tert*-pentyl-phenol	↓ PGCs, Vtg induction, ↓ spermatogenesis, developed oviducts	[77, 78]
	Largemouth bass	Industrial effluent	↓ Plasma T, ↑ liver size, Vtg induction	[43]
	Mosquito-fish	Paper mill effluent	Masculinization of females	[79]
	Rainbow trout	Sewage effluent	Vtg induction in males ↓ testes size	[80]
	Rainbow trout	Surfactants	↓ Growth/ovosomatic index	[81]
	Roach	Sewage effluent	Vtg induction in males, intersexuality in gonads	[82]
	White sucker	Paper mill effluent	↓ Gonad size, age to maturation, sex steroids and GtH-II, ↑ liver size, ↑ EROD	[41, 83]
	Flounder	Sewage effluent	Vtg induction in males, malformed testes	[84]
Amphibian	Cricket frog	PCBs and PCDFs	Sex ratio reversal	[85]
	Frogs and toad	Agricultural runoff	Limb deformities	[86]
	Leopard frog	Acetochlor	↑ T_3-induced metamorphosis	[87]
	Senegal walking frog	DDT	Corticosterone-like developmental abnormalities	[88]
Reptiles	Alligator	Dicofol, *p,p'*-DDE, atrazine	↓ T, and phallus size in males; ↑ aromatase activity in males, ↑ E2 and ovarian abnormalities in females	[49, 44]
	Red-eared slider	PCBs, nona-chlor, *p,p'*-DDE, chlordane	Sex reversal	[89, 90]
	Snapping turtle	Organo-chlorines	Alteration of secondary sex characteristics	[91]

Table 1 (continued)

Taxa	Animal	Putative EDC	Noted effect	Citation
Birds	Bald eagles	DDT	Eggshell thinning, embryo mortality	[92, 93]
	Cormorants	Organo-chlorines	Crossbills, altered reproductive behavior	[94, 95]
	Gulls	DDT	Altered reproductive behavior	[96]
	Gulls	Organo-chlorines	↓ Metabolic enzyme activity, ↓ plasma B	[97]
Mammals	Florida panther	Organo-chlorines	Cryptorchidism	[98]
	Seal	PCBs	↓ In reproductive success	[99]
	Rat	Vinclozolin, p,p'-DDE	↓ T, developmental/behavioral alterations	[72, 100]
	Rat	PCB mixture	↓ Thyroid hormones	[101]
	Seal	PCBs	↓ In reproductive success	[99]

2.2.1
Brain – Hormone Synthesis Control

There are several endocrine axes that are controlled by the hypothalamus-pituitary regions of the brain, such as the reproductive, growth, thyroid, and stress axes. Within the reproductive axis, gonadotropins from the pituitary, such as gonadotropin II (GtH-II) found in fish, alter steroid hormone production in the gonad [26]. Experimental exposure of feral white sucker fish (*Catastomus commersoni*) to bleach kraft mill effluent (BKME) stimulated a 30- to 50-fold decrease in GtH-II concentrations and fish given a synthetic gonadotropin releasing hormone (sGnRH-A) released GtH-II at a significantly lower rate when compared to controls [41]. Further, ovulation in BKME-exposed females did not occur after injection of the GnRH analog as occurred in all control fish. BKME-exposed fish showed a transitory increase in testosterone in response to a sGnRH-A injection, whereas controls had higher concentrations of plasma testosterone and did not respond to treatment with a further elevation [41]. These data suggest that endocrine disruption can occur at the hypothalamic-pituitary axis. Further, alterations can occur in gonadotropin release or response. To date, few studies have examined endocrine disruption at this level of organization; that is, the hypothalamus-pituitary control of steroid hormone production.

2.2.2
Gonad – Hormone Production

A common observation in wildlife exposed to various forms of environmental contamination is an alteration in gonadal hormone production. These alterations can occur in adults or be due to developmental alterations in the gonad.

Male carp (*Cyprinus carpio*) and largemouth bass (*Micropterus salmoides*) exposed to sewage effluent and effluents from various industrial sources, respectively, exhibit a significant depression in circulating testosterone concentrations [42, 43]. This is assumed to be due to altered gonadal steroidogenesis, but the mechanism has not been clarified.

Male American alligators (*Alligator mississippiensis*) exposed to endocrine disrupting contaminants in ovo exhibit altered plasma concentrations of testosterone at hatching [44], and juvenile alligators from a contaminated lake had reduced penis size compared to alligators from a reference lake [45]. These abnormalities apparently persist, as depressed plasma testosterone concentrations and reduced phallus size has been observed in animals as old as 8–9 years [46–48]. The depressed androgen levels are due, in part, to an altered gonadotropin response [49, 50]. As described above for the white suckers, gonadal responsiveness to gonadotropin stimulation can be altered by exposure to environmental chemicals. BKME-exposed fish showed a transitory increase in testosterone in response to an sGnRH-A injection [41]. Testosterone synthesis is not the only gonadal hormone affected by contaminants. Plasma estradiol-17β synthesis was also observed to be altered in alligators exposed to pesticides. Hatchling females have elevated plasma estradiol concentrations whereas juvenile females have significantly lower plasma concentrations [47–49].

As observed in alligators, white suckers exposed to BKME in Jackfish Bay, Lake Superior had decreased plasma sex steroid concentrations [51, 52]. In additional studies, white suckers collected from 12 areas receiving effluents from paper mills using chlorine-based or sulfite-based processes were also examined. Fish exposed to sulfite-based mill effluents exhibited decreased steroid levels [53].

Other steroid synthetic pathways in other organs can also be affected, as plasma cortisol concentrations were suppressed in BKME-exposed fish from Jackfish Bay subjected to stress during sampling [54]. Fish exposed to environments contaminated with polycyclic aromatic hydrocarbons (PAHs), polychlorinated biphenyls (PCBs), or heavy metals also show a suppressed stress response as measured by plasma cortisol concentrations [54, 55]. In contrast, fish exposed to pesticides can exhibit an augmented stress response and elevated plasma cortisol concentrations compared to reference fish (Orlando, Binczik, Kroll, and Guillette, unpublished data). Likewise, alligators living in a pesticide-contaminated lake exhibit similar basal concentrations of corticosterone [47] but show a more rapid response to stressful stimuli when compared to reference animals [56].

2.2.3
Liver – Hormone Availability

Research in rats indicates that embryonic steroid exposure establishes sexual dimorphism of steroid-metabolizing enzymes in the liver. Male and female rodents exhibit different patterns of hepatic steroid metabolism, and these patterns are organized during development by exposure to androgens [57–59]. For instance in rats, gonadectomy of males results in female-pattern steroid degradation, whereas administration of testosterone to a gonadectomized male results in male-pattern steroid degradation [60]. It is clear that sex steroids affect liver metabolism, such that exogenous androgens can masculinize a female liver and exogenous estrogens can feminize a male liver [59].

Recent evidence indicates that exposure to contaminants can alter hormone metabolism in laboratory mammals [61, 62] and various wildlife species such as the fathead minnow (*Pimephales promelas*) [63, 64] and American alligator (Gunderson, LeBlanc, and Guillette, unpublished data). Although many previous researchers have observed that toxins induce hepatic enzyme induction, these recent studies demonstrate that specific enzymes associated with hormone degradation and clearance are altered in specific ways. For example, in the fathead minnow, exposure to industrial effluent altered testosterone biotransformation processes [64]. Exposed minnows eliminated a variety of testosterone metabolites at a faster rate compared to nonexposed fish. Interestingly, many of the hepatic enzymes used for testosterone biotransformation and elimination are used in the gonad for 'normal' steroidogenesis presenting an interesting evolutionary dichotomy (for discussion, see [65, 66]). Differential induction of these enzymes in the liver and gonad occurs but little is known about the factors controlling the differential up-regulation of these enzymes in wildlife.

In addition to steroid hormone metabolism, the liver is responsible for another factor that affects the availability of steroid hormones to the cells. Serum binding proteins (SBPs), produced in the liver, transport the lipophilic steroid hormones in the plasma. Examples of SBPs include sex hormone binding globulin (SHBG), corticosteroid and thyroid-retinoic acid hormone binding globulins [27]. Most (>95%) of the steroid hormone titer is bound and not free in the plasma, yet only the free fraction of the steroid can produce a biological effect. The bound fraction is thought to function as a storage mechanism for steroid hormones. As long as the hormone is bound to the SBP it is protected from metabolism and excretion [67]. To date, little research has been conducted on the interaction of EDCs with SBPs. Certain contaminants seem to have very low affinity for SBPs. For example, the estrogenic activity of DES, octylphenol, and *o,p'*-DDT were poorly inhibited by the presence of SHBG or albumin (nonspecific SBP) in an in-vitro assay when compared to the total inhibition of estradiol activity [22]. This suggests that certain EDCs have their effective plasma concentration increased because they do not bind to the SBPs. This enables most of the compound to remain free in the plasma and therefore more available to the cells. Some EDCs could alter availability of native hormones by actually binding to the SBP, causing displacement, metabolism, and excretion of the hormone. Hydroxylated non-*ortho* PCB metabolites were shown to bind thyroid-retinoic acid hormone binding globulin, displacing the native thyroid hormone [68]. This increase in the free fraction of the hormone resulted in a decrease in plasma concentration (thus bioavailability) due to an increase in metabolism and excretion. Another way that EDCs could alter the availability of hormones in an organism is to induce the production of SBPs, thus binding more free hormone and reducing their bioavailability to the cells. Some plant compounds, such as lignans and isoflavonoids, seem to stimulate SHBG synthesis in the liver in vivo and in vitro [69].

To summarize, many hormones are transported in the blood by SBPs. Because of a low binding affinity for the SBP, many EDCs are effectively more available to enter the cells. Also, the relationship of EDCs with these SBPs could affect the availability of hormones to cells by increasing their storage or their metabolism and excretion.

2.2.4
Receptors in Brain, Gonad, and Liver – Hormone Action

Action or biological effect of a hormone as well as an EDC are generally thought to depend on that compound binding to its receptor (for a more detailed discussion of contaminant interactions with steroid receptors see [70]). Larger, more hydrophilic hormones, such as the gonadotropins or adrenocorticotropic hormone, bind to membrane-bound receptors. Steroid hormone receptors belong to the nuclear receptor superfamily, which also includes thyroid hormones, retinoic acid, and vitamin D_3 [26, 27]. These hormones bind to their receptors in the cytoplasm or nucleus of the target cell, where they cause gene transcription (Fig. 2).

Most research in the field of endocrine disruption has focused on xenobiotics that mimic estrogens, although more recently some research has been carried out on environmental androgens, antiandrogens, and antiestrogens. Environmental xenobiotic compounds such as pesticides, surfactants from detergents, industrial waste, sewage treatment plant effluent, paper mill effluent, heavy metals, and phytosterols have been shown to be estrogenic [71]. Certain fungicides, such as the metabolites of vinclozolin, have been identified as antiandrogens in rats. Interestingly, metabolites of other compounds can have very different receptor-binding affinity from the parent compound. For example, o,p'-DDT is a known estrogen agonist, and one of its metabolites p,p'-DDE is antiandrogenic in rodents [72].

3
Possible Future Areas of Research

In this chapter we have presented research strongly suggesting an association between xenobiotics and altered development and reproduction in wildlife. Most of that research to date has focused on steroid hormone mimics and antagonists, especially estrogenic and androgenic EDCs. To begin to understand fully the possible impact of xenobiotics on wildlife and humans, we must broaden our research plans to include (1) other hormone axes and possible levels of disruption, (2) varied methods of investigation and analysis, (3) an emphasis on embryonic/pre-adult life stages, and (4) levels of biological organization beyond the organism.

3.1
Other Hormone Axes and Possible Levels of Disruption

There are other hormone axes similar in structure to the reproductive axis. These include the growth, thyroid, and stress axes (Fig. 3). The proper biological function of each axis depends on the concentration of hormones, whose control, production, availability, and action are similar to the reproductive axis. Because of these structural parallels, each of these endocrine axes has the potential for modulation by xenobiotics [102, 103]. Few studies have been designed to examine the possible effects of xenobiotics on these other axes. It is important that we examine these possible direct effects as well as the indirect effects that could occur through cross talk or interaction between axes.

As an example, the stress axis is known to affect the reproductive and growth axes in fish. Increases in stress hormones are associated with impaired reproductive function and decreases in growth rate [39]. There is still a great deal of research needed to understand the normal endocrine mechanisms involved in these relationships. Recently, glucocorticoid receptors were immunolocalized in the neurons in the part of the brain associated with the control of reproduction and the pituitary of rainbow trout (*Oncorhynchus mykiss*) [104]. This research suggests a mechanistic connection between stress hormones and the brain-level control of reproductive steroid hormones. A few studies have examined the modulation of the stress axis in fish by xenobiotics such as paper mill effluent and PCBs and heavy metals [54, 55, 105]. Even fewer studies have been implemented to look at the possible effects of the stress response on reproductive steroid hormones of contaminant-exposed fish [54]. Given the presence of the glucocorticoid receptor in the brain and pituitary of certain fish, it is important that we examine the role of xenobiotics as stressors in the modulation of reproduction.

As members of the nuclear superfamily, reproductive and stress steroid hormone receptors belong to a group of hormones that include thyroid hormones, retinoic acid, and vitamin D_3 [106]. In the past, these hormones were believed to act strictly as intranuclear transcription regulators. Recent studies have found the existence of membrane bound receptors for estrogens in fish and mammals [32, 33, 107]. Membrane bound receptors would add another function to the steroid hormone and possibly other members of the nuclear superfamily. In addition to their role as transcription regulators, estrogens acting through membrane receptors and secondary messenger systems could mediate cell activity (enzyme activation/inhibition and intracellular ion concentration) (Fig. 2). Future research should examine steroid membrane receptor nongenomic function alone or perhaps in conjunction with nuclear receptors of native hormones and xenobiotics.

Among the nuclear receptor superfamily there are receptors with unidentified ligands (hormones) called orphan receptors [108]. Some of these orphan receptors are now receiving names as their ligands are discovered. For example, the steroid-xenobiotic receptor (SXR) is a nuclear receptor that has been shown to activate transcription when bound by certain native steroid hormones or xenobiotics. SXR appears to induce transcription of xenobiotic and steroid hormone detoxification and metabolizing enzymes [37]. Steroidogenic factor-1 (SF-1) is still an orphan receptor because its ligand remains unidentified. However, it has been associated with regulation of some P450 steroidogenic enzymes, and more recently was shown to play an important role in the development of the gonad and adrenal glands in mice [109] and gonadal sex-differentiation in the chicken [110]. The obvious influence of former and present orphan receptors on development and reproduction in lab animals strongly suggests that more research is required into the relationship of xenobiotics with these receptors. One has to be curious about what future research will uncover concerning the roles of the remaining orphan receptors.

3.2
Methods of Investigation

Traditionally we have examined the effects of single compounds on in vitro cancer cell lines or transfected yeast cells. These assays are effective as a first screen for possible endocrine disruption as measures of receptor binding and gene transcription. Data from these assays typically are further tested using in vivo standardized assays on laboratory strains of rodents, with domestically bred fish (e.g., fathead minnow and medaka) or birds (e.g., chicken and quail). Each of these assays is appropriate for addressing certain questions.

For a thorough understanding of the effects of xenobiotics on the development and reproduction of *wildlife*, there is a need to complement the aforementioned assays with field and laboratory experiments done with wild animals that are native to the area of land or water being investigated. Studies carried out with wild animals, if designed correctly and backed up with controlled laboratory experiments, have the potential to give us the most accurate information about the health of the environment [111, 112]. Field studies can provide information about (1) the effects of greater genetic diversity, (2) the effects of contaminant mixtures and exposure duration in long-lived predators, (3) food chain effects such as biomagnification, and (4) the ability to measure effects on the population, community, and ecosystem levels. Laboratory studies on wildlife could be implemented to test the data collected from the field studies. This approach had been effectively implemented to study EDCs, but more work on a greater range of species is needed (for examples see [83, 113–115]). Treatment studies should be done during early developmental stages when the actual development of reproductive systems is in progress. We know that the most sensitive period to EDCs is during embryogenesis and early post-hatching life stages [7]. It is during these periods that ephemeral developmental windows occur. At these times, exposure to endogenous hormones regulates tissue and organ formation. For example, estrogen receptors and mERR-2 (another orphan receptor) gene transcription are known to be restricted to discrete developmental stages [116, 117]. Additionally, the concentration and type of xenobiotic(s) should mirror what was measured in the animals from the field study, i.e., appropriate xenobiotic mixtures and their concentrations must be used in treatment experiments so that laboratory conditions can mimic the environmental conditions the animal experiences.

3.3
Methods of Analysis

The needs of traditional toxicology were to safeguard humans against genotoxic effects of xenobiotic exposure leading to birth defects, cancer, and death. One of the classic tools of toxicology is to examine the effect of treatment of a laboratory rat using a range of concentrations of the xenobiotic under investigation [118, 119]. These concentrations are typically quite high, often 4–5 orders of magnitude above what would normally be experienced by a human. Using mathematical models, the data are extrapolated along a linear dose response curve to doses estimated to cause risks in 1 out of 1,000,000 people [119].

As a criterion of risk assessment for nongenotoxic xenobiotics, toxicologists use the estimate of no observable adverse effect level (NOAEL) [2, 120]. NOAEL is that level – or threshold – of exposure to a xenobiotic below which no adverse effect is observed. It has been argued that the threshold hypothesis may not be valid for EDCs, chemicals that operate through the same receptor-based pathways as the native hormone [121, 122]. With many of these compounds, the dose response curve is not linear, but actually approximates an inverted U shape, so extrapolations could yield inaccurate estimates of risk [20, 122]. In developing embryos steroid hormones at particular concentrations enable sexual differentiation of the organism [123]. If an embryo is already experiencing a threshold for the native hormone, then any amount of the EDC would be above the threshold. There may be no threshold concentration of an EDC in a developing embryo [120, 122]. Future studies need to examine the possible nonlinear dose response curves and test the concept of the no threshold concentration for EDCs.

Another consideration is the way data are analyzed. The differences between experimental and reference groups could very well be obscured by the natural and induced variation in natural populations exposed to xenobiotics. If we are only examining endpoint means from each group and the variance about the mean is great enough to eliminate a statistical difference, we may be missing a subtle yet important difference between the groups [124]. Important differences may lie at the extremes of the sample distributions. Differences in performances on neurophysiological tests tend to be modest, where nonsignificant differences based on p values proves to be an inappropriate method of analyzing the data and in this case could lead to an inaccurate conclusion of the risks of exposure. Instead of looking for measures of central tendency, in cases where subtle differences can lead to unacceptable adverse effects, perhaps we should be examining distributions [124]. For example, only a shift of five IQ (intelligence quotient) points, associated with lead exposure in humans, can be amplified greatly at the extremes of the distribution. In a population of 100 million a shift in mean IQ from 100 to 95 *decreases* the number of people with scores above 130 from 2.3 million to 990,000. The number of people with scores below 70 (which mandates special classrooms and other services) *increases* from 2.3 to 3.6 million [124].

There may be another important reason to reevaluate how we analyze data when looking at the health of our ecosystems. Contaminate-induced increases in variation of a measured endpoint may have important consequences for the population [125]. That is, evolution of a population works not on the mean but on the extremes. The increased variation that has been observed in some studies could be a mechanism for evolution. Organisms continue to live in polluted environments. The xenobiotic may be a selection force effectively removing those organisms from the population that are unable to survive or breed. If we examine populations by using only broad-based measures such as number of organisms or sex ratio, we may be missing a more subtle change that could be occurring in the genetic makeup of the population.

3.4
Going Beyond the Organism

All of the past, present, and future research in the field of endocrine disruption will forever beg the question of "so what?" Do the alterations in control, production, availability, and action of hormones have an effect at the population, community and ecosystem levels? Future research must rapidly move to address this question whatever model or system is being investigated.

There is a growing need for a blending of the strengths of the fields of ecology and toxicology. There is a dearth of studies that have examined the effects of endocrine disruption at the population level. A new way to ask questions about the effects of xenobiotics on the environment will need to include an integration of the best of both disciplines and recognition of each field's weaknesses [4, 126]. Toxicologists bring an understanding of controlled laboratory experiments and the focus on the organismal, cellular, and molecular levels of biological hierarchy. In comparison, ecologists bring an aptitude for examining the larger picture and how each individual organism relates to the population, community, and ecosystem levels. These disparate pursuits must integrate to enable us to understand truly the full impact of EDCs on the long-term welfare of wildlife and humans.

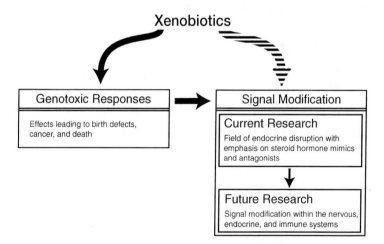

Fig. 4. A model of xenobiotic action in an organism. Traditionally, toxicology was concerned with safeguarding human health and measured genotoxic effects of xenobiotics, including birth defects, cancer, and death. More recently, certain xenobiotics have been identified as mimics and antagonists of steroid hormones, especially estrogens. These xenobiotics are known to alter the normal development and reproduction of wildlife by modifying endocrine system function. Crosstalk between the nervous, endocrine, and immune systems integrate these systems and enable maintenance of homeostasis in a dynamic environment. Future research should expand to include examination of possible xenobiotic modification of these signaling systems

4
Conclusion: The Field of Signal Modification?

The importance of endocrine hormones to the integration of a multicellular organism is widely accepted. Organisms live in a dynamic external environment yet must maintain a relatively constant internal environment to enable proper enzyme action, which underpins metabolism. We know that a range of animal and plants use signaling systems to control other organisms or synchronize reproduction. Certain legumes use flavonoids to activate nitrogen fixing bacteria, sheep are rendered infertile by the phytochemicals produced in the clover plants they graze upon, and the timing of gamete release in certain corals has been associated with the release of estrogen (for a discussion of estrogens as environmental signals, see [102]). These studies suggest that we have just begun to uncover the intricate signaling systems used by organisms. We know that the nervous, endocrine, and immune systems are integrated to support homeostasis [17]. All of these systems deliver a set of signals, whether they are called neurotransmitters, hormones, or cytokines, whose biological action is realized when that signal is received by a receptor of some kind. Traditional toxicology began the investigation of the dramatic genotoxic effects of high concentration exposure to xenobiotics. The field of endocrine disruption has emphasized the more subtle adverse effects of steroid hormone mimicry by xenobiotics (Fig. 4). We suggest future research should expand its questions to include not just the endocrine but also the nervous and immune systems. Perhaps the current exploration of the phenomenon of *endocrine disruption* should likewise be expanded to *signal modification*?

5
References

1. Gallo MA (1996) History and scope of toxicology. In: Klaassen CD, Amdur MO, Doull J (eds) Casarett and Doull's toxicology: the basic science of poisons. McGraw-Hill, New York, p 3
2. Faustman EM, Omenn GS (1996) Risk assessment. In: Klaassen CD, Amdur MO, Doull J (eds) Casarett and Doull's toxicology: the basic science of poisons. McGraw-Hill, New York, p 75
3. Colborn T (1994) Environ Health Perspect 102: 55
4. Banks JE, Stark JD (1998) Integr Biol 1: 195
5. Ankley GT, Giesy JP (1998) Endocrine disruptors in wildlife: a weight-of-evidence perspective. In: Kendall RJ, Dickerson RL, Giesy JP, Suk WA (eds) Principles and processes for evaluating endocrine disruption in wildlife. Society of Environmental Toxicology and Chemistry, Pensacola, p 349
6. Newbold RR (1999) Diethylstilbestrol (DES) and environmental estrogens influence the developing female reproductive system. In: Naz RK (ed) Endocrine disruptors. CRC Press, Boca Raton, p 39
7. Bern H (1992) The fragile fetus. In: Colborn T, Clement C (eds) Chemically-induced alterations in sexual and functional development: the wildlife/human connection. Princeton Science Publishing Company, Princeton, p 9
8. McLachlan JA (1980) Estrogens in the environment. Elsevier, New York
9. McLachlan JA (1985) Estrogens in the environment II: influences on development. Elsevier, New York

10. Colborn T, Clement C (1992) Chemically-induced alterations in sexual and functional development: the wildlife/human connection. Princeton Science Publishing Company, Princeton

11. Krimsky S (2000) Hormonal chaos: the scientific and social origins of the environmental endocrine hypothesis. Johns Hopkins University Press, Baltimore

12. Bason CW, Colborn T (1992) US application and distribution of pesticides and industrial chemicals capable of disrupting endocrine and immune systems. In: Colborn T, Clement C (eds) Chemically-induced alterations in sexual and functional development: the wildlife/human connection. Princeton Science Publishing Company, Princeton, p 335

13. Cohen M, Commoner B, Eisl H, Bartlett P, Cooney P (1997) Report to the W. Alton Jones Foundation, Charlotteville

14. Iwata H, Tanabe S, Sakal N, Tatsukawa R (1993) Environ Sci Technol 27: 1080

15. Pereira WE, Domagalski JL, Hostettler FD, Brown LR, Rapp JB (1996) Environ Toxicol Chem 15: 172

16. Smith C, Root D (1999) Int J Occup Environ Health 5: 141

17. Besedovsky HO, Rey AD (1996) Endocr Rev 17: 64

18. Andersen HR, Andersson AM, Arnold SF, Autrup H, Barfoed M, Beresford NA, Bjerregaard P, Christiansen LB, Gissel B, Hummel R, Jorgensen EB, Korsgaard B, Le Guevel RL, Leffers H, McLachlan JA, Moller A, Nielson JB, Olea N, Oles-Karasko A, Pakdel F, Pedersen KL, Perez P, Skakkebaek NE, Sonnenschein C, Soto AM, Sumpter JP, Thorpe SM, Grandjean P (1999) Environ Health Perspect 107: 89

19. Lamb JC, Balcomb R, Bens CM, Cooper RL, Gorsuch JW, Matthiessen P, Peden-Adams MM, Voit EO (1998) Hazard identification and epidemiology. In: Kendall RJ, Dickerson RL, Giesy JP, Suk WA (eds) Principles and processes for evaluating endocrine disruption in wildlife. Society of Environmental Toxicology and Chemistry, Pensacola, p 17

20. vom Saal FS, Timms BG, Montano MM, Palanza P, Thayer KA, Nagel SC, Dhar MD, Ganjam VK, Parmigiani S, Welshons WV (1997) Proc Natl Acad Sci USA 94: 2056

21. Soto AM, Sonnenschein C, Chung KL, Fernandez MF, Olea N, Olea Serrano F (1995) Environ Health Perspect 103(7): 113

22. Arnold SF, Robinson MK, Notides AC, Guillette LJ Jr, McLachlan JA (1996) Environ Health Perspect 104: 544

23. Kelce WR, Wilson EM (1997) J Mol Med 75: 198

24. Ryffel B (1997) CRC Crit Rev Toxicol 27: 135

25. Laws SC, Carey SA, Kelce WR (1995) Toxicologist 15: 294

26. Norris DO (1997) Vertebrate endocrinology, 3rd edn. Academic Press, San Diego

27. Bentley PJ (1998) Comparative vertebrate endocrinology, 3rd edn. Cambridge University Press, Cambridge

28. Gilbert SF (1997) Developmental biology, 5th edn. Sinauer Associates, Sunderland

29. Guillette LJ Jr, Arnold SF, McLachlan JA (1996) Anim Reprod Sci 42: 13

30. Nagel S, vom Saal FS, Thayer KA, Dhar MG, Boechler M, Welshons W (1997) Environ Health Perspect 105: 70

31. Orchinik M, Murray TF, Moore FL (1991) Science 252: 1848

32. Watson CS, Pappas TC, Gametchu B (1995) Environ Health Perspect 103: 41

33. Fortunati N, Fissore F, Fazzari A, Becchis M, Combi A, Catalano MG, Berta L, Frairia R (1996) Endocrinology 137: 686

34. Nakhla AM, Rosner W (1996) Endocrinology 137: 4126

35. Thomas P (1999) Nontraditional sites of endocrine disruption by chemicals on the hypothalamus-pituitary-gonadal axis: interactions with steroid membrane receptors, monoaminergic pathways, signal transduction systems. In: Naz RK (ed) Endocrine disruptors. CRC Press, Boca Raton, p 3

36. Van Der Kraak GJ, Zacharewski T, Janz DM, Sanders BM, Gooch JW (1998) Comparative endocrinology and mechanisms of endocrine modulation in fish and wildlife. In: Kendall R, Dickerson R, Giesy J, Suk W (eds) Principles and processes for evaluating endocrine disruption in wildlife. Society of Environmental Toxicology and Chemistry, Pensacola, p 97

37. Blumberg B, Sabbagh JW, Juguilon H, Bolado JJ, van Meter CM, Ong ES, Evans RM (1998) Genes Dev 12: 3195
38. Ignar-Trowbridge DM, Pimentel M, Teng CT, Korach KS, McLachlan JA (1995) Environ Health Perspect 103(7): 35
39. Pankhurst NW, Kraak GVD (1997) Effects of stress on reproduction and growth of fish. In: Iwama GK, Pickering AD, Sumpter JP, Schreck CB (eds) Fish stress and health in aquaculture. Cambridge University Press, Cambridge, vol 62, p 278
40. National Research Council (1999) Hormonally active compounds in the environment. National Academic Press, Washington DC, p 560
41. Van Der Kraak GJ, Munkittrick KR, McMaster ME, Portt CB, Chang JP (1992) Toxicol Appl Pharmacol 115: 224
42. Folmar LC, Denslow ND, Rao V, Chow M, Crain DA, Enblom J, Marcino J, Guillette LJ Jr (1996) Environ Health Perspect 104: 1096
43. Orlando EF, Denslow ND, Folmar LC, Guillette LJ Jr (1999) Environ Health Perspect 107: 199
44. Crain DA, Guillette LJ Jr, Rooney AA, Pickford DB (1997) Environ Health Perspect 105: 528
45. Guillette LJ Jr, Pickford DB, Crain DA, Rooney AA, Percival HF (1996) Gen Comp Endocrinol 101: 32
46. Guillette LJ Jr, Crain DA (1996) Comments Toxicol 5: 381
47. Guillette LJ Jr, Crain DA, Rooney AA, Woodward AR (1997) J Herp 31: 347
48. Guillette LJ Jr, Brock JW, Rooney AA, Woodward AR (1999) Arch Environ Contam Toxicol 36: 447
49. Guillette LJ Jr, Gross TS, Masson GR, Matter JM, Percival HF, Woodward AR (1994) Environ Health Perspect 102: 680
50. Guillette LJ Jr, Gross TS, Gross DA, Rooney AA, Percival HF (1995) Environ Health Perspect 103(4): 31
51. McMaster ME, Van Der Kraak GJ, Portt CB, Munkittrick KR, Sibley PK, Smith IR, Dixon DG (1991) Aquat Toxicol 21: 199
52. Munkittrick KR, Portt CB, Van der Kraak GJ, Smith IR, Rokosh DA (1991) Can J Fish Aquat Sci 48: 1
53. Munkittrick KR, Van der Kraak GJ, McMaster ME, Portt CB, van den Heuvel MR, Servos MR (1994) Environ Toxicol Chem 13: 1089
54. McMaster ME, Munkittrick KR, Luxon PL, Van der Kraak GJ (1994) Ecotoxicol Environ Saf 27: 251
55. Norris DO, Donahue S, Dores RM, Lee JK, Maldonado TA, Ruth T, Woodling JD (1999) Gen Comp Endocrinol 113: 1
56. Rooney AA (1998) Variation in the endocrine and immune systems of juvenile alligators: environmental influence on physiology. PhD Thesis, University of Florida, Gainesville
57. Lucier GW, Lui E, Powell-Jones W (1982) Regulation of hepatic enzyme development by steroids. In: Hunt VR, Smith MK, Worth D (eds) Environmental factors in human growth and development. Cold Spring Harbor Laboratory, Cold Spring Harbor, vol 11, p 137
58. Lucier GW, Sloop TC, Thompson CL (1985) Imprinting of hepatic estrogen action. In: McLachlan JA (ed) Estrogens in the environment II: influences on development. Elsevier, New York, p 190
59. Gustafsson JA (1994) Regulation of sexual dimorphism in the rat liver. In: Short RV, Balaban E (eds) The differences between the sexes. Cambridge University Press, Cambridge, p 231
60. Jansson JO, Ekberg S, Isaksson O, Mode A, Gustafsson JA (1985) Endocrinology 117: 1881
61. Wilson VS, McLachlan JB, Falls JG, LeBlanc GA (1999) Environ Health Perspect 107: 377
62. Wilson VS, LeBlanc GA (1998) Toxicol Appl Pharmacol 148: 158
63. Parks LG, LeBlanc GA (1998) Gen Comp Endocrinol 112: 69
64. Parks LG, LeBlanc GA (1998) In: SETAC 19th Annual Meeting, Society of Environmental Toxicology and Chemistry, Charlotte, p 146

65. Waxman DJ (1996) Steroid hormones and other physiologic regulators of liver cytochromes P450: metabolic reactions and regulatory pathways. In: Advances in molecular and cell biology. JAI Press, vol 14, p 341
66. Crain DA, Guillette LJ Jr (1997) Rev Toxicol 1: 47
67. Milgrom E (1990) Steroid hormones. In: Baulieu EE, Kelly PA (eds) Hormones. Hermann, Paris, p 387
68. Brouwer A, Van den Berg KJ (1986) Toxicol Appl Pharmacol 85: 301
69. Adlercreutz H (1995) Environ Health Perspect 103(Suppl 7): 103
70. Rooney AA, Guillette LJ Jr (1999) Contaminant interactions with steroid receptors: evidence for receptor binding. In: Crain DA, Guillette LJ Jr (eds) Environmental endocrine disruptors: an evolutionary perspective. Taylor and Francis, London, p 82
71. Colborn T, vom Saal FS, Soto AM (1993) Environ Health Perspect 101: 378
72. Gray LE, Monosson E, Kelce WR (1996) Emerging issues: the effects of endocrine disrupters on reproductive development. In: DiFiulio RT, Monosson E (eds) Interconnection between human and ecosystem health. Chapman and Hall, London, p 45
73. Matthiessen P, Gibbs PE (1998) Environ Toxicol Chem 17: 37
74. Morcillo Y, Albalat A, Porte C (1999) Environ Toxicol Chem 18: 1203
75. Oehlmann J, Fiorini P, Stroben E, Markert B (1996) Sci Tot Environ 188: 205
76. Thomas P (1989) Mar Environ Res 28: 499
77. Gimeno S, Komen H, Gerritsen A, Bowmer T (1998) Aquat Toxicol 43: 77
78. Gimeno S, Gerritsen A, Bowmer T (1996) Nature 384: 221
79. Davis WP, Bortone SA (1992) Effects of kraft mill effluent on the sexuality of fishes: an environmental early warning? In: Colborn T, Clement C (eds) Chemically-induced alterations in sexual and functional development: the wildlife/human connection. Princeton Science Publishing Company, Princeton, p 113
80. Purdom CE, Hardiman PA, Bye VJ, Eno NC, Tyler CR, Sumpter JP (1994) Chem Ecol 8: 275
81. Ashfield LA, Pottinger TG, Sumpter JP (1998) Environ Toxicol Chem 17: 679
82. Jobling S, Nolan M, Tyler CR, Brighty G, Sumpter JP (1998) Environ Sci Technol 32: 2498
83. Munkittrick KR, McMaster ME, McCarthy LH, Servos MR, Van der Kraak GJ (1998) J Toxicol Environ Health B Crit Rev 1: 347
84. Lye CM, Frid CLJ, Gill ME, McCormick D (1997) Mar Pollut Bull 34: 34
85. Reeder AL, Foley GL, Nichols DK, Hansen LG, Wikoff B, Faeh S, Eisold J, Wheeler MB, Warner R, Murphy JE, Beasley VR (1998) Environ Health Perspect 106: 261
86. Ouellet M, Bonin J, Rodrigue J, DesGranges JL, Lair S (1997) J Wildlife Diseases 33: 95
87. Cheek AO, Ide CF, Bollinger JE, Rider CV, McLachlan JA (1999) Arch Environ Contam Toxicol 37: 70
88. Hayes TB, Wu TH, Gill TN (1997) Environ Toxicol Chem 16: 1948
89. Bergeron JM, Crews D, McLachlan JA (1994) Environ Health Perspect 102: 780
90. Willingham E, Crews D (1999) Gen Comp Endocrinol 113: 429
91. deSolla SR, Bishop CA, Van der Kraak GJ, Brooks RJ (1998) Environ Health Perspect 106: 253
92. Bowerman WB, Best D, Kubiak T, Postupalsky S, Tillett DE (1991) Bald eagle reproduction impairment around the Great Lakes: association with organochlorine contamination. In: Schneider S, Campbell R (eds) Cause-effect linkages II: symposium abstracts. Michigan Audubon Society, Ann Arbor, p 31
93. Geisy JP, Ludwig JP, Tillitt DE (1994) Environ Sci Technol 28: 128
94. Fox GA (1992) Epidemiological and pathobiological evidence of contaminant-induced alterations in sexual development in free-living wildlife. In: Colborn T, Clement C (eds) Chemically-induced alterations in sexual and functional development: the wildlife/human connection. Princeton Science Publishing Company, Princeton, p 147
95. Fox GA, Gilman AP, Peakall DB, Anderka FW (1978) J Wildlife Management 42: 477
96. Fry DM (1995) Environ Health Perspect 103(Suppl 7): 165
97. Lorenzen A, Moon TW, Kennedy SW, Glen GA (1999) Environ Health Perspect 107: 179
98. Facemire CF, Gross TS, Guillette LJ Jr (1995) Environ Health Perspect 103(Suppl 4): 79

99. Reijnders PJH (1986) Nature 324: 456
100. Kelce WR, Stone CR, Laws SC, Gray LE, Kemppainen JA, Wilson EM (1995) Nature 375: 581
101. Gould JC, Cooper KR, Scanes CG (1997) Gen Comp Endocrinol 106: 221
102. Cheek AO, Vonier PM, Oberdörster E, Burrow BC, McLachlan JA (1998) Environ Health Perspect 106: 5
103. Crisp TM, Clegg ED, Cooper RL, Wood WP, Anderson DG, Baetcke KP, Hoffmann JL, Morrow MS, Rodier DJ, Schaeffer JE, Touart LW, Zeeman MG, Patel YM (1998) Environ Health Perspect 106: 11
104. Teitsma CA, Anglade I, Lethimonier C, Dréan GL, Saligaut D, Ducouret B, Kah O (1999) Biol Reprod 60: 642
105. Hontela A (1997) Reprod Toxicol 1: 1
106. Price PW (1996) The evolution of a gene superfamily. In: Biological evolution. Saunders College Publishing, Fort Worth, p 328
107. Loomis AK, Thomas P (1999) Biol Reprod 61: 51
108. McLachlan JA, Newbold RR, Teng CT, Korach KS (1992) Environmental estrogens: orphan receptors and genetic imprinting. In: Colborn T, Clement C (eds) Chemically-induced alterations in sexual and functional development: the wildlife/human connection. Princeton Science Publishing Company, Princeton, p 107
109. Luo X, Ikeda Y, Parker K (1994) Cell 77: 481
110. Smith CA, Smith MJ, Sinclair AH (1999) Gen Comp Endocrinol 113: 187
111. Tyler CR, Jobling S, Sumpter JP (1998) CRC Crit Rev Toxicol 28: 319
112. Guillette LJ Jr, Rooney AA, Crain DA, Orlando EF (1999) Steroid hormones as biomarkers of endocrine disruption in wildlife. In: Hensel DS, Black MC, Harrass MC (eds) Environmental toxicology and risk assessment: standardization of biomarkers for endocrine disruption and environmental assessment. American Society for Testing and Materials, West Conshohocken, vol 9, ASTM Standards Technical Publication No 1381
113. Guillette LJ Jr, Crain DA, Gunderson MP, Kools SAE, Milnes MR, Orlando EF, Rooney AA, Woodward AR (1999) Am Zool 40: 438
114. Peterson RE, Theobald HM, Kimmel GL (1993) CRC Crit Rev Toxicol 23: 283
115. Sumpter JP (1998) Toxicol Lett 102/103: 337
116. Gorski J, Hou Q (1995) Environ Health Perspect 103: 69
117. Pettersson K, Svensson K, Mattsson R, Carlsson B, Ohlsson R, Berkenstam A (1996) Mech Dev 54: 211
118. Eaton DL, Klaassen CD (1996) Principles of toxicology. In: Klaassen CD, Amdur MO, Doull J (eds) Casarett and Doull's toxicology: the basic science of poisons. McGraw-Hill, New York, p 13
119. Calabrese EJ, Baldwin LA (1999) Bioscience 49: 725
120. Sheehan DM, Willingham E, Gaylor D, Bergeron JM, Crews D (1999) Environ Health Perspect 107: 3
121. Gaylor DW, Sheehan DM, Young JF, Mattison DR (1988) Teratology 38: 389
122. Guillette LJ Jr, Crain DA, Rooney AA, Orlando EF (1997) Conference Proceedings: Advances in Comparative Endocrinology, Yokohama, Japan, p 1717
123. vom Saal FS, Montano MM, Wang MH (1992) Sexual differentiation in mammals. In: Colborn T, Clement C(eds) Chemically-induced alterations in sexual and functional development: the wildlife/human connection. Princeton Science Publishing Company, Princeton, p 17
124. Weiss B (1997) Neurotoxicology 18: 581
125. Bignert A, Olsson M, deWit C, Litzé K (1994) Fresenius J Anal Chem 348: 76
126. Munkittrick KR, McCarty LS (1995) J Aquat Ecosys Health 4: 77

CHAPTER 10

Endocrine Disruption in the Aquatic Environment

John P. Sumpter

Department of Biological Sciences, Brunel University, Uxbridge, Middlesex UB8 3PH, UK
E-mail: John.Sumpter@brunel.ac.uk

Although the words 'endocrine disruption' were introduced only relatively recently, some of what are now considered the clearest and best documented examples of endocrine disruption in aquatic organisms were first described 20 or more years ago. These include imposex in molluscs induced by tributyl tin (TBT) and masculinisation of some species of fish living downstream of where pulp mill effluent is discharged. There are now reasonably well-documented examples of what is probably endocrine disruption in a wide range of aquatic organisms from some invertebrates through to reptiles. However, despite this, much still has to be learnt about what chemicals cause these disturbances to the endocrine systems of exposed aquatic organisms, how they do so, and what the consequences of the effects are. The last issue is of particular importance: only in the case of TBT-induced imposex in molluscs has it been shown that exposure to an endocrine disrupting chemical (or mixture of chemicals) can lead to population declines. More research into the consequences of endocrine disruption to wildlife is required before the gravity of this type of toxicity can be determined. Only then can endocrine disruption be assessed in comparison to other factors, such as habitat loss and over-exploitation, which also undoubtedly adversely affect wildlife populations.

Keywords. Endocrine disruption, Aquatic environment, Fish, Oestrogen

1   Introduction 272

2   Endocrine Disruption in Aquatic Organisms 273

2.1   Invertebrates 273
2.2   Vertebrates 274
2.2.1   Fish 274
2.2.2   Amphibians 277
2.2.3   Reptiles 278

3   The Causative Chemicals 279

4   Laboratory Studies 282

5   Mechanisms of Action 286

6   Conclusions 287

7   References 288

The Handbook of Environmental Chemistry Vol. 3, Part M
Endocrine Disruptors, Part II
(ed. by M. Metzler)
© Springer-Verlag Berlin Heidelberg 2002

List of Abbreviations

p,p'-DDE	2,2-bis(*p*-chlorophenyl)-1,1-dichloroethylene
DDT	2,2-bis(*p*-chlorophenyl)-1,1,1-trichloroethane
EDC	endocrine disrupting chemical
OECD	Organisation for Economic Cooperation and Development
PCB	polychlorinated biphenyl
STW	sewage treatment work
TBT	tributyl tin
TIE	toxicity identification and evaluation

1
Introduction

Endocrine disruption has become one of the 'hottest' topics in environmental science today, and in the last few years major, multi-disciplinary research programmes on the issue have been initiated in many countries. However, despite the current intense activity, and the fact that the term 'endocrine disruption' was coined only very recently, for over two decades it has been apparent that exposure to anthropogenic chemicals present in the environment can cause sublethal, but still deleterious, effects to wildlife. In many cases reproduction has been adversely affected (such an effect is often quite noticeable, of course, which may well account for why reproduction has received much of the attention), and in at least some of these cases this disruption has been shown to be, or presumed to be, a consequence of disruption of the endocrine control of reproduction.

Most reported examples of what would now be considered cases of endocrine disruption of reproduction in wildlife have occurred in aquatic organisms, or species that feed on aquatic organisms. Thus, for example, some of the most widely reported incidences have been observed in fish (e.g. [1, 2]), alligators [3] and fish-eating birds such as gulls and terns [4]. This is probably not coincidental; it seems likely that, on average, aquatic species receive a higher exposure to most pollutants than do terrestrial organisms, for the simple reason that the aquatic environment is the ultimate sink for most wastes. Thus, for example, very high volumes of effluents of domestic and industrial origins are intentionally added to many rivers and inshore marine waters (Tokyo Bay receives about 4.7 km^3 of liquid wastes per year, for example), and unintentionally many chemicals (of both natural and man-made origin) are washed from the surrounding land into the aquatic environment.

Exactly which organisms could, or should, be considered 'aquatic' is a moot point. I have 'stretched' the interpretation of an aquatic species to include animals that either spend only part of their lives in water (e.g. many amphibians), or those that live both in water and on land (e.g. many reptiles), but have chosen to exclude other groups of organisms, such as aquatic birds, which other authors may have included. I have also excluded mentioning plants (except here!), even though many are truly aquatic; however, I know of no example of confirmed endocrine disruption in plants (which does not mean that none will ultimately be discovered: plants have hormones, and hence these could, at least theoretically, be disrupted by exposure to exogenous hormone-mimicking chemicals). Due to my particular interest in fish, I have undoubtedly given research into this group prominence. However, I have tried to balance this bias by including discussion of endocrine disruption in other groups of aquatic organisms, covering both vertebrates and invertebrates. My admittedly selective coverage has, nonetheless, allowed me to discuss most, if not all, of the major issues in this area of research.

I have chosen to consider reproductive effects only, despite the fact that it is inevitable that examples of endocrine disruption will include many, and varied, non-reproductive effects; in fact, some have already been reported, e. g. the effects of alkyl phenolic chemicals on growth of fish [5].

I have tried to provide a fairly balanced account. However, in a still very controversial area, with many different, and often powerful, stakeholders, what is balanced to one interested group of people will undoubtedly be biased to another. Such a situation is always going to occur when an issue, which could be very important, is in its infancy; only further research will allow a more informed and reasoned opinion to be reached.

2
Endocrine Disruption in Aquatic Organisms

2.1
Invertebrates

In 1970, Blaber [6] reported that many female dogwhelks (*Nucella lapillus*) in Plymouth Sound (UK) had a penis-like structure behind the right tentacle. This was the first report of what is now probably the clearest, and best documented, case of endocrine disruption in any organism. It is so because there is unequivocal evidence, from both field and laboratory studies, on the cause of the masculinisation, the effects are well described, and the consequences of these effects, at both the individual and population level, are known. It is also the only documented example of endocrine disruption where remedial action has been shown to lead to recovery of populations of affected organisms [7].

Following Blaber's report [6], Smith [8] reported similar structural masculinisation of the American mud snail (*Ilyanassa obsoleta*) along the Connecticut coast. He coined the term imposex to describe these female snails which had a penis and a vas deferens (the sperm duct), the latter often disrupting the oviducts and hence egg laying. Nevertheless, it was a decade later that it was first suggested [9–11] that this superimposition of male characteristics in females was caused by exposure to antifouling paints, in particular tributyltin (TBT) and related compounds used as biocides.

A large amount of subsequent research (reviewed in [12]) has supported the earlier reports. For example, there is strong evidence from studies in coastal European [13] and American [14] waters of imposex in gastropod molluscs caused by TBT. Imposex has been observed in areas where the concentration of TBT is only 1 ng/l; where concentrations have reached 6–8 ng/l, complete reproductive failure and local extinctions due to sterility of females has occurred. Even in the middle of the North Sea (between the UK and the Netherlands), tens of miles from land, the incidence of imposex in a whelk (*Buccinum undatum*) has been strongly correlated with the intensity of shipping traffic.

Laboratory studies have been used to investigate the species specificity of the phenomenon of intersexuality caused by TBT, the dose-response relationship, the critical period when exposure causes imposex, the permanence of the effect, and the mechanism of action of TBT (specific references can be found in [12]). The latter issue is not yet completely resolved, although it seems likely that TBT acts as a competitive inhibitor of aromatase (the enzyme that converts androgen to oestrogen). TBT also appears to inhibit the conjugation of androgens (which would normally de-activate these steroid hormones). Both mechanisms of action would lead to elevated androgen concentrations, and hence masculinisation.

Because the 'TBT story' is so well documented (and accepted), the pressure to ban the use of TBT-based antifouling paints has grown relentlessly, and it looks presently as though a world-wide ban can be achieved in the next few years. Hopefully, a ban would lead to the recovery of mollusc populations.

Ironically, other than this extremely well documented example of endocrine disruption in some species of molluscs, there is very little evidence available presently to suggest that any other invertebrates are adversely affected by endocrine disrupting chemicals (EDCs). Of course this absence of evidence does not imply that other groups of invertebrates are not experiencing endocrine disruption, but rather that the necessary studies do not appear to have been undertaken yet (though some are in progress). Some recent laboratory-based studies (e.g. [15]) have shown that reproduction of water fleas (daphnids) is affected by exposure to nonylphenol, an oestrogenic xenobiotic, but such studies are at a very early stage, and the significance of the results is presently unclear. Aquatic invertebrates are a very large, very heterogeneous group of organisms, often with unique hormones (e.g. the ecdysteroids, which are not found in vertebrates), that deserve serious study. I would not be at all surprised if other examples (besides TBT-induced imposex in some molluscs) of endocrine disruption in aquatic invertebrates came to light.

2.2
Vertebrates

2.2.1
Fish

More than twenty years ago, Howell et al. [16] reported the presence of masculinised female mosquito fish (*Gambusia affinis*) in a small creek downstream of a paper mill discharging bleached kraft mill effluent. Subsequently, masculinised females of other (but not all) species were found, and laboratory studies showed that exposure to bleached kraft mill effluent induced male secondary sexual characteristics [17]; thus, it appeared that the effluent contained an androgenic chemical, or mixture of chemicals. A comprehensive review of all the work on endocrine disruption in mosquito fish can be found in [1]. Very recently, spawning female channel catfish (*Tetalurus punctatus*) exhibiting male secondary sexual characteristics have been reported to be present in the Red River of the North in North Dakota, USA. [18]. There are no paper mills discharging effluent into this river, although it does receive waste water from a sewage treatment plant and a sugar beet processing plant; however, further investigation is required to identify the cause of masculinisation.

More recently, research has focused on the mechanisms whereby bleached kraft mill effluent has been shown to affect reproduction adversely by reducing plasma sex steroid hormone concentrations, delaying sexual maturity, reducing vitellogenin concentrations, and reducing gonad and egg size (e.g. [19, 20]). These multiple effects, which are undoubtedly effects of bleached kraft mill effluent on the endocrine system, are difficult, if not impossible, to classify as primarily androgenic or oestrogenic. This is unsurprising when one considers that the effluent is a highly complex mixture of natural and anthropogenic chemicals, many of which have been shown to alter the endocrine control of reproduction in various species.

It is also 20 years ago that grossly hermaphrodite (intersex) fish were found in the settlement lagoons of two sewage treatment works (STWs) in England. In the UK, and many other densely-populated, developed countries, effluents from STWs often contribute 50 % of the flow of rivers, a figure that can rise as high as 90% (or more!) in periods of low rainfall (when demand for water is highest). Thus, the fish in such rivers live in diluted, treated effluent, not clean water.

Research on the fisheries implications of effluents affecting sex determination began in the late 1980s. It was soon discovered that STW effluent was oestrogenic to fish [21]. Specifically, when caged trout were maintained in effluent channels, they responded by synthesising vitellogenin, which serves as a very sensitive and specific biomarker for exposure to 'oestrogens' [22]. Follow-up research in rivers receiving varying amounts of STW effluents showed that significant stretches of river downstream of major STWs were oestrogenic to caged fish [23, 24]. In the worse case, an entire 5-km stretch of river downstream of a large STW was extremely oestrogenic; maximum vitellogenesis occurred in the caged fish [24]. Further, not only were plasma vitellogenin concentrations extremely high, but the testes of the caged male trout were much smaller than those of control fish.

Extensive studies, such as those done in the UK and summarised above, have not yet been conducted in other countries. However, fairly small scale, preliminary studies have shown that feral male carp (*Cyprinus carpio*) captured near a major sewage treatment plant in the US have elevated plasma vitellogenin concentrations [25], and that vitellogenin was present in some male and immature brown trout (*Salmo trutta*) captured downstream of sewage treatment plants in Switzerland [26]. Very recently, a caged fish study in Sweden showed massive induction of vitellogenin in fish exposed to STW effluent [27]. All these studies suggest that the oestrogenic effects observed in the UK in fish exposed to effluent from STW [21] and in rivers receiving effluent [23, 24] will prove not to be unique to the UK, but rather to be a general phenomenon. However, it is likely that the oestrogenic potency of effluent will vary from effluent to effluent (and hence from river to river), depending on the 'strength' of the effluent, the size and efficiency of the sewage treatment plant, and the degree of dilution of the

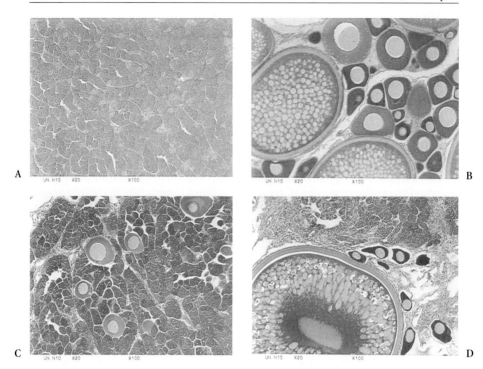

Fig. 1A-D. The histological apperance of the gonads of normal and intersex roach (Rutilis rutilis) caught in British rivers. **A** A normal male. The testis is full of lobules in which spermatogenis takes place. **B** A normal female. The overy is full of oocytes at different stages of growth. The large oocytes are undergoing vitellogenesis, whereas the smaller (primary) oocytes have not yet entered vitellogenesis. **C** A mildly intersex fish. Most of the gonad has the appearance of a testis: it is full of lobules in which spermatogenesis is taking place. Six primary oocytes are scattered amongst the testicular tissue. **D** A severely intersex fish. This picture shows one large, vitellogenic oocyte and a number of smaller, primary oocytes, apparently physically separated from an area of reasonably normal-looking testicular tissue. All pictures were taken at the same magnification. (× 100).

effluent in the receiving water, besides (probably) many other factors that will emerge as research proceeds.

The discovery that effluent from STWs is oestrogenic to fish raises the issue of whether wild fish, living in rivers receiving significant inputs of effluent, are being adversely affected. Until very recently, there was very little evidence to suggest that they are. If effluents were 'feminising' wild fish populations, then one might expect to find elevated vitellogenin concentrations, and intersexuality (specifically, ovarian tissue in the testes of males), possibly even all-female populations. In fact, reports of occasional, individual intersex fish, or of a small proportion of intersex fish amongst a large sample, have regularly appeared (e.g. [28, 29]). Sometimes these instances have even been linked to environmental conditions [28], although the specific cause has remained unknown. This situation has now changed, with the publication of an extensive field study of intersexuality in one species of native freshwater fish, the roach (*Rutilis rutilis*) [2]. This study set out to address the question of whether exposure to effluent from STWs caused oestrogenic responses in wild fish. Populations of roach were sampled both upstream and downstream of STWs on eight rivers and from five reference (control) sites throughout the British Isles; the rivers selected represented a range with regard to general water quality (from very good to poor). Histological examination of the gonads revealed that a high proportion of the 'males' were, in fact, intersex, as defined by the simultaneous presence of both male and female gonadal characteristics (Fig. 1). Intersex fish were found at all sites, although the incidence was much higher in rivers that received STW effluents than at the control sites; the incidence of intersexuality in 'male' fish ranged from 4% (at two control sites) to 100% in two populations of roach living downstream of major STWs in heavily impacted rivers. There was a highly significant positive correlation between the degree of intersexuality in the 'male' fish and their plasma vitellogenin concentrations [2], suggesting (but not proving) that both parameters were caused by the same factor (STW effluent). These results provide compelling evidence that populations of wild fish inhabiting many rivers in the UK are being exposed to oestrogenic chemicals, and that these are, in most cases, present at higher concentrations in stretches of river directly downstream from large STWs. However, the ecological significance of these results remains unclear presently.

Endocrine disruption has also been observed in marine fish. Following up on the report by Lye et al. [30] of reproductive problems in flounders (*Platichthys flexus*) exposed to effluents from a STW, Matthiessen and colleagues have conducted a very extensive investigation of endocrine disruption in flounder from estuaries and marine waters around the UK [31]. Vitellogenin concentrations were found to be significantly elevated, often markedly so, in flounder living in many estuaries, especially those that receive large amounts of industrial effluent. At two locations, some of the fish were hermaphrodite, with their testes containing large numbers of oocytes amongst apparently normal looking testicular tissue. The causative substances are unknown, but there was a clear relationship between vitellogenin induction and the volume of industrial (but not domestic) effluent discharged into the estuaries.

2.2.2
Amphibians

As far as I am aware, presently there is no strong evidence to suggest that wild populations of amphibians are experiencing endocrine disruption. However, a recent discovery may – and only may – turn out to be one such example. The first hint of a problem in frogs came in 1993, when a group of schoolchildren discovered frogs with deformed and missing limbs in a Minnesota farm pond, and posted their finding on the Internet! In the following years, similar deformities in many different species of amphibia have been reported from many different locations throughout the USA and Canada (e.g. [32]), despite the fact that they were (apparently) not present until very recently. The incidence of these limb abnormalities was as high as 60% (though usually considerably lower) amongst some frog populations.

No consensus has yet been reached regarding the cause of the deformities; possibilities include chemical pollutants, increased UV radiation (due to ozone depletion), and parasitic infestations. Many of the sites at which deformed frogs have been found are close to agricultural fields that are intensively sprayed with pesticides and herbicides at certain times of the year. Hence, it has been suggested that one or more of these chemicals might be disrupting normal frog development. It has long been known that retinoic acid, a metabolite of vitamin A, plays an important role in limb formation during metamorphosis of amphibians. It has been shown that exogenously applied retinoic acids (or antagonists) disrupt limb formation. A recent report [33] has now suggested that products of the degradation of S-methoprene, a very widely used insect growth regulator (i.e. a pesticide) dramatically interferes with normal amphibian development, inducing limb deformities similar to those observed in wild amphibians. Further, the metabolites probably cause these effects because they are structural similar to some known teratogenic retinoic acids. Although (as the authors state clearly) their research does not prove that the reported limb deformities in many wild amphibians are caused by degradation products of an insecticide, it certainly provides a plausible explanation. If this explanation is confirmed, it would be an example of endocrine disruption in amphibians.

2.2.3
Reptiles

One of the most influential studies concerned with endocrine disruption in aquatic wildlife is that of Guillette and co-workers on the alligators of Florida. The genesis of this ongoing study was the finding that one particular population of alligators (*Alligator mississippiensis*), living on Lake Apopka, showed a dramatic decline during the 1980s that continues to the present day [34]; population density on this lake is now only about one-tenth of what it was in the 1970s. Lake Apopka is a pollution "hot spot", due to the fact that it was contaminated by a major pesticide spill in 1980 (see [35] for details of exactly what pollutants were released). The lake is also located in a heavily agricultural area, and hence probably receives significant pollution via run-off.

Guillette et al. [3] studied juvenile alligators from Lake Apopka, and compared them with animals of similar age from Lake Woodruff, a control (unpolluted) lake. They found that juvenile female alligators from Lake Apopka had plasma oestradiol concentrations 50% higher than those in females from the control lake. The Apopka females also exhibited abnormal ovarian morphology, with large numbers of polyovular follicles and polynuclear oocytes. Males were also affected; juvenile male alligators from Lake Apopka had much lower plasma testosterone concentrations compared to the concentrations found in the control males. Further, males from Lake Apopka had poorly organised testes and abnormally small penises [3, 36]. It is thought that these effects were induced by pollutants at a very early stage of development of the alligators, probably in ovo; that is, they are so-called organisational effects. Subsequent research has shown that at least some of these effects persist for years [37], and that alligators in other lakes subjected to man's influence also appear to have been affected [37].

As will be discussed later, it has proved very difficult to determine exactly what caused these effects on the reproductive axis of alligators.

Very little evidence for (or against) endocrine disruption in other reptiles has been published. However, these are often large, and hence difficult, species to study, so perhaps this is not too surprising. A few papers (e.g. [38]) have reported preliminary data which have been interpreted to suggest that environmental contaminants might be adversely affecting some, but not all, reproductive indices in other species of reptiles, such as snapping turtles (*Chelydra serpentina*), but presently the available data are too sparse to reach any conclusions.

3
The Causative Chemicals

Probably the most difficult problem in this difficult area of research is to link exposure to specific chemicals to the effects observed in wild populations of organisms (such as those described above); that is, to identify the chemicals causing endocrine disruption in wildlife. The reason for this is that we know relatively little about what chemicals (either natural or man-made) are present in the aquatic environment, or about their concentrations (which will be variable, and depend on the specific situation). The problem is easily understood when it is realised that around 70,000 man-made chemicals are in everyday use. Many, if not most, of these will enter the aquatic environment, where they will degrade at varying rates, and varying extents, to produce many more chemicals. In addition there will be many, probably very many, natural chemicals present. This very large number of chemicals makes it likely that we will never have a complete picture of what chemicals are present, and at what concentrations. Even if all the chemicals in a particular aquatic environment could be identified and their concentrations determined, this may not help a great deal. In cases where direct exposure (e.g. water to organism) occurs, then knowledge of what is present in the water will be very helpful, but in cases where indirect exposure occurs (as might well occur in the case of top predators, which probably ingest contaminants in their food, and subsequently pass them to their offspring via the yolk in their eggs), knowledge of contaminant concentrations in the water and/or sediment might not be very informative, and may even mislead.

One of the most useful approaches adopted to date has been that of Toxicity Identification and Evaluation (TIE), which incorporates a chemical fractionation procedure with a method of detecting endocrine activity in the fractions. Essentially, a very complex effluent is split into fractions of decreasing complexity, with each fraction being analysed for oestrogenic activity (or whatever endocrine activity is of interest). Fractions identified as active are separated further, until they are simple enough to be analysed by a technique such as gas chromatography-mass spectrometry, leading to the identification of the active chemicals.

Such a TIE approach was used to identify the main oestrogenic chemicals in the effluents of seven UK STWs receiving primarily domestic influent. These were found to be the natural hormones oestradiol and oestrone, and the synthetic hormone ethinyl oestradiol [39], which presumably were excreted by (primarily) women. These steroids were detected as 'free' hormones, yet they are excreted as conjugates (primarily sulfates and glucuronides) which are biologically inactive. This suggested that de-conjugation must occur in the sewage system, a suggestion supported by the study of Panter et al. [40], who showed that conjugated steroids are very readily de-conjugated (to the active steroid) by minimal microbial activity.

To verify that the right chemicals had been identified, laboratory experiments were conducted in which fish were exposed to varying concentrations of these oestrogens, centred around the concentrations present in effluents. It was found that vitellogenin concentrations were elevated in a dose-dependent manner [41], thus demonstrating that environmentally-relevant concentrations of natural and synthetic oestrogens are alone sufficient to account for the elevated vitellogenin concentrations observed in both wild fish [2] and caged fish [24] living downstream of STW discharges in British rivers. Subsequent research from a number of other European countries has verified that natural and synthetic oestrogens are present in STW effluent at concentrations similar to those reported to be present in UK effluents [42–44].

Although this particular example demonstrated the power of the TIE approach, it was aided both by the fact that the chemicals of interest were known to be oestrogenic (because they led to elevated vitellogenin concentrations [22]), and that relatively straightforward assays for oestrogens are widely available. However, if it is unclear what type of endocrine activity is responsible for the effects observed, then the TIE approach becomes problematical, because it is not clear what bioassay should be used to guide the chemical fractionation. Ideally, the bioassay should be based upon the effects observed in the wild animal (perhaps, for example, vitellogenin induction), but this is rarely feasible; instead, a more simple (usually in vitro) assay is used. Such an approach can, however, mislead, because most relatively simple in vitro assays for endocrine activity do not represent factors such as bioaccumulation and metabolism, which occur in vivo and can be critical factors in determining the effects of chemicals. Furthermore, although many different in vitro assays for oestrogens (and mimics) are widely available, this is not true for other endocrine activities – for example, androgenic activity. Thus, further advances in the techniques used to measure different types of endocrine activity are needed before the TIE approach can be used widely. When the effects observed in wild animals are caused by many chemicals having different mechanisms of action (as may often be the case), then again the TIE approach is made much more difficult.

The only other realistic approach (besides TIE) to identifying the causative chemicals is to screen chemicals in different, usually in vitro, assays to identify those possessing appropriate endocrine activity. Rather than conduct a random screen (which would anyway be impractical, considering the very large number of chemicals present in the aquatic environment), the chemicals tested should be those likely to be present in the particular environment in which the affected organisms were found. For example, if pesticides are suspected of causing the adverse effects (as is the case with alligators on Lake Apopka), then it would obviously be appropriate to screen a selection of such chemicals. This approach has been quite widely used, sometimes because it is readily "doable", rather than because it is the most appropriate to the situation under investigation. Thus, for example, Jobling et al. [45] used such an approach to demonstrate that many common aquatic pollutants, including some phthalates, are weakly oestrogenic, findings that have since been corroborated by many other studies. However, despite phthalates being ubiquitous aquatic pollutants, there is presently no evidence showing that they cause endocrine disruption in wild fish (or any wildlife). This example demonstrates the difficulties associated with this approach; the ease of conducting many in vitro assays has led to many chemicals being tested in such assays, and to quite a few being shown to possess one, or more, types of endocrine activity. However, demonstrating that a particular chemical is present in the environment at concentrations which cause endocrine disruption to wildlife has proved infinitely more difficult. To date, this "shotgun" approach to identifying chemicals with endocrine activity has not led, to my knowledge, to a single example of the identity of a chemical causing endocrine disruption in wildlife actually being revealed.

These problems serve to highlight just some of the many difficulties that are encountered in trying to identify the chemicals responsible for endocrine disruption in aquatic organisms. They help explain why, with few exceptions (e.g. TBT and imposex in molluscs), the causative chemicals remain elusive. Yet, unless the chemical, or mixture of chemicals, is identified, appropriate laboratory studies (which can best address many aspects of endocrine disruption) cannot be undertaken, and remedial action, to reduce or eliminate the problem, cannot be considered.

Most of the issues associated with identifying the chemical, or chemicals, causing endocrine disruption in wildlife are summarised in Table 1. This considers just three chemicals (17β-oestradiol, ethinyl oestradiol and nonylphenol) which probably all play a role in causing the "feminisation" reported in freshwater fish in the UK. The factors that need to be considered are the potencies (efficacies) of these chemicals, their concentrations in the aquatic environment, their

Table 1. Factors to consider when attempting to assess the relative contributions of different oestrogenic chemicals to the "feminising" effects of effluent from sewage treatment works. Some of the information is taken from publications, but some is my own "best guess", based sometimes on relatively little information

Chemical	Concentration in effluent	Biocon-centration factor	Potency	Minimum effective concentration in vivo	Half-life
Nonylphenol	Low µg/l	500	Low	10 µg/l	Long (days → months)
Oestradiol	ng/l	Low?	High	10 ng/l	Short (hours?)
Ethinyl oestradiol	Low ng/l	500	Very high	1 ng/l	Intermediate (days?)

fate and behaviour in the aquatic environment, and their fate and behaviour in aquatic organisms (that is, their persistence, accumulation and metabolism within the organisms of interest). Note that, even with these three chemicals, which have been extensively studied in the last few years, quite a lot of the data required are not available, or supported by very little information. Nevertheless, Table 1 illustrates the difficulty in determining whether an extremely potent oestrogen, ethinyl oestradiol, but one which is present in the aquatic environment at extremely low concentrations (parts per trillion, or lower), is more, or less, important than nonylphenol, a rather weak oestrogen, but one which is widespread and often present at appreciable concentrations (parts per billion in solution and higher in sediments). Even this type of approach is very simplistic (and probably unrealistic), because aquatic wildlife are very rarely ever exposed to an individual chemical, but instead are nearly always exposed to highly complex, undefined, mixtures of chemicals simultaneously. Such an exposure scenario makes it extremely difficult to identify the chemical(s) causing the adverse effects; in many cases I suspect that it may never be possible to pin-down the causative chemicals (this has been achieved surprisingly rarely, despite more than two decades of research on endocrine disruption in wildlife). Further, the exposure regime is usually not static, but instead is constantly changing as the use of chemicals waxes and wanes, their rates of release into the environment change, and waste treatment processes (prior to release of effluent into the aquatic environment) change and improve. Thus, when effects are observed in wildlife, especially in adult animals, the exposure responsible for the effects may no longer persist by the time the (long-lived) effects are noticed, probably making it impossible ever to link cause with effect with certainty.

4
Laboratory Studies

To some extent because of the many difficulties associated with the study of endocrine disruption in wild populations of animals, a significant amount of laboratory-based research into endocrine disruption has been undertaken. Also driving such research has been the need to develop standardised test protocols that can be used to assess the (potential) endocrine-disrupting activities of chemicals to aid hazard and risk assessment. Results obtained from laboratory studies on the effects of EDCs on aquatic organisms (usually, to date, fish) are almost always much easier to interpret than those obtained from fieldwork studies, primarily because in the laboratory studies fish are usually exposed to only one (known) chemical, at defined concentrations, for fixed periods of time. For example, many studies have been conducted using alkylphenols, particularly nonylphenol, and these have shown that the chemical is indeed an oestrogen in vivo (as suggested by in vitro assays), leading not only to typical oestrogenic responses, such as elevated vitellogenin [46] and zona radiata protein [47] concentrations, but also in males to reduced testicular growth, impaired spermatogenesis, reduced secondary sexual characteristics, and formation of an oviduct [48–50], and in females reduced fecundity and egg quality (our unpublished results).

These effects are dose-dependent; for example, nonylphenol can cause intersexuality at "high" concentrations [51], but not at lower concentrations [52]. One challenge now is to try and relate the concentrations of the test chemicals that cause adverse effects in these laboratory studies to environmental concentrations of the same chemicals. Put another way, there is little, if any, doubt that many EDCs can cause adverse effects in laboratory studies, but whether they do so at the concentrations present in the environment is often less clear. It is good to see that in some recent studies this issue has been a major concern from design of the study through to its interpretation [48, 52].

Laboratory studies not only complement fieldwork; they can sometimes address, in a controlled manner, issues that are difficult, if not impossible, to tackle in any other way. For example, laboratory studies can address the question of when fish are most susceptible to EDCs. Such studies have shown that early life stages are particularly sensitive (e.g. (49]), and also that some effects are irreversible. They can also be used to study responses of fish to defined mixtures of chemicals (rather than to a single chemical). Such studies are extremely important because they much more realistically mimic the 'real world'. However, the design of such studies, including the analysis of the results obtained, is quite complicated (even uncertain), and to date no reliable studies have been published.

Laboratory studies have also been conducted on reptiles and, to a lesser extent to date, amphibians. Although many aquatic reptiles are very large (when adult), and hence unsuitable for laboratory studies, most species are oviparous (they lay eggs), and this presents an ideal situation in which the effects of defined chemical exposures on sex determination can be investigated. Many species of aquatic reptiles (e.g. most turtle species and all crocodilians) appear to lack sex chromosomes, and instead sex is determined by the incubation temperature of the egg. Thus, for example, in the red-eared slider turtle (a common species in the southern US), eggs incubated at 26 °C will produce male offspring, whereas eggs incubated at 31 °C produce females; intermediate temperatures produce mixed sex ratios. Because temperature can be manipulated experimentally so readily (in an incubator) to produce the desired sex, the effects of sex steroid hormones (androgens as well as oestrogens), their mimics, and chemicals which affect the synthesis of sex steroid hormones, can all be readily investigated. For example, applying oestrogens to the eggs of the red-eared slider turtle during incubation over-rides male-producing temperature effects, and produces females [53].

Using this very useful laboratory model, it was soon established that certain polychlorinated biphenyls (PCBs, which can have endocrine-disrupting effects) reverse sex in red-eared slider turtles [54]. Recent research has expanded these findings, and shown that environmentally-relevant doses of some xenobiotics (applied to turtle eggs at concentrations present in alligator eggs obtained from a contaminated lake), particularly chlorinated pesticides (e.g. nonachlor, arochlor, and metabolites of DDT) caused a significant degree of sex reversal [55]. Such results demonstrate the considerable promise of so-called egg assays, but much validation work still needs to be completed before the promise of these assays is fully realised. In particular, clear-dose-response relationships need to be established, and shown to be reproducible. The degree of variability between species in their responses to reference sex steroids, and to EDCs, needs to be determined; as the 'low' incubation temperature can produce males in some species, but females in others, it is likely that different chemicals will cause different effects (or degree of effect) to different species. Only when all of these studies have been completed should the interactive effects of chemicals (i.e. mixtures of chemicals) be investigated (I realise that this is the goal, of course, and hence there is an understandable temptation to "jump-the gun" before all the validation work has been completed).

Laboratory studies on endocrine disruption in amphibians are at an even earlier stage, and to date relatively little has been published. Recently, preliminary results showing that not only oestradiol, but also the xenoestrogens nonylphenol, octylphenol, bisphenol A, and even the extremely weak oestrogen butylhydroxyanisol, can "feminise" larval frogs (*Xenopus laevis*) [56] were published. These results, although undoubtedly preliminary, will hopefully encourage others to investigate the in vivo effects of EDCs on this important (and declining) group of vertebrates. One area in which amphibians should prove ideal experimental models is in the investigation of xenobiotics with thyroidal/antithyroidal activity, because metamorphosis (which is very easy to assess) is controlled by thyroid hormones.

One finding of considerable relevance that has come out of many of these laboratory-based studies is the exquisite sensitivity of some (possibly many) aquatic organisms to the chemicals of concern presently. For example, following on from the realisation that natural oestrogens are present in many effluents, many carefully-conducted studies have shown that fish are extremely sensitive to water-borne 17β-oestradiol; concentrations in the low tens of nanograms per litre range cause pronounced effects (e.g. [21, 52, 57]). Although less data are available on ethinyl oestradiol, it is likely to be even more potent, and concentrations below 1 ng/l are likely to cause significant, and probably adverse, effects [21, 27]. Similarly, nonylphenol is surprisingly potent as an oestrogen when administered to fish via the water; concentrations in the low microgram per litre range cause pronounced and adverse effects [46, 48]. The reasons behind such exquisite sensitivity to water-borne oestrogens and xenoestrogens probably include the continuous exposure (unlike the situation when a chemical is injected) for long periods of time (weeks to months), the ease with which the chemical enters the blood stream of the fish through the gills (though this is not proven), and the fact that the test chemical reaches most organs without passing first through the liver, where it would be metabolised (and probably de-toxified). Likewise, exogenous treatment of reptile eggs has shown that they are also exquisitely sensitive to applied chemicals; nanogram amounts of potent chemicals such as oestradiol, and low microgram, or even high nanogram, quantities of xenobiotics alter sex-determination [53, 55]. The preliminary experiments reported by Kloas et al. [56] also suggest that amphibian sex-determination is extremely sensitive to EDCs present in the water. As discussed previously, many species of molluscs are also adversely affected by very low (nanogram per litre) concentrations of TBT. Collectively, these results, although yet not always as firmly established as one would like, demonstrate that a variety of chemicals can cause effects at concentrations which might intuitively seem surprisingly low, and they strongly caution against any attempt to question the significance of endocrine disruption on the grounds that "the chemical is present at a very low concentration, surely too low to cause such effects".

Finally, when discussing laboratory studies, there is the pressing need to establish reproducible, standardised, and fully validated laboratory testing procedures that can be conducted by any competent laboratory wanting to assess the endocrine disrupting effects of a chemical. A start on developing such tests has recently been reported (e.g. [52, 58]). Factors to be investigated include the species of fish (or other organism) to be used in the tests, the age of the test organism (juveniles may be more sensitive to EDCs than adults), the length of exposure (generally the shorter the better) and the endpoints to be monitored. It is unlikely that a single species can be agreed upon by all countries; it is more likely that a small number of species, including the medaka (*Oryzias latipes*), the fathead minnow (*Pimepehales promelas*), the rainbow trout (*Oncorhyrichus mykiss*) and the zebrafish (*Danio rerio*) will be acceptable; each has advantages and disadvantages as a test species. It is possible that specific strains of species will form the basis of a testing protocol. For example, sex-linked colour variants of medaka exist, allowing easy external sexing of fish and, possibly, also providing a secondary sexual characteristic (body colour) that changes in response to hormones and hormone-mimicking chemicals. Another, related, approach would be to use single sex populations of fish, which can readily be produced in some species using techniques developed initially for use in aquaculture. Gimeno et al. [59] advocated the use of such fish, and used them very intelligently to demonstrate that a very weak oestrogenic chemical, 4-*tert*-pentylphenol, could feminize sexually undifferentiated juvenile male carp.

Agreement on a standardised suite of endpoints will also not be easy. In the case of oestrogenic (and possibly anti-oestrogenic) chemicals, there is little doubt that vitellogenin will serve as a very informative parameter, but it is less clear presently what parameters will serve as indicators of exposure to androgenic and anti-androgenic chemicals (and possibly other endocrine activities, if these are deemed important). It is also very likely that gonadal histology (as well as the size of the gonads) will be incorporated into any routine testing protocol. Other parameters that might well prove informative include frequency of egg laying (of females), number of eggs laid (i.e. fecundity), appearance and size (if quantifiable) of any secondary sexual characteristics (such as gonopodium length in some species of fish) and the plasma concentrations of some hormones (especially sex steroids and thyroid hormones). Standardised testing protocols, supported by the relevant regulatory bodies (e.g. OECD), will undoubtedly be developed, validated and then implemented, though the completion of these stages, prior to routine adoption of the tests, will take a number of years. Tests based upon other groups of organisms, such as amphibia, will probably take even longer to establish, primarily because our existing knowledge of their underlying reproductive endocrinology is poor, and hence baselines will need to be established before any disruption can be identified.

5
Mechanisms of Action

Surprisingly little is also known about the mechanisms of action of many (perhaps most) EDCs. Some are certainly oestrogenic, and much attention has been focused on these chemicals, primarily because some of the best-documented examples of endocrine disruption in wildlife (e.g. the 'feminisation' of male alligators in the US [3] and fish in the UK [2]) appear to be due to exposure to oestrogenic chemicals. Further, the wide availability of in vitro assays for oestrogens has meant that it is relatively easy to screen chemicals for oestrogenic activity. Such screening has demonstrated that a surprisingly diverse range of chemicals, such as some alkylphenols, phthalates, bi-phenolic chemicals and pesticides, all of which are well-documented aquatic pollutants, are oestrogenic, albeit usually only weakly so (e.g. [45, 60]). A relatively small proportion of these chemicals has also been shown to demonstrate oestrogenic activity in vivo; the well documented increase in the plasma vitellogenin concentration of fish in response to nonylphenol is probably the best example [46]. However, because a chemical demonstrates oestrogenic activity does not necessarily mean that any effects it causes in vivo are a consequence of the oestrogenic activity of the chemical. For example, p,p'-DDE is weakly oestrogenic in vitro, but shows anti-androgenic effects in vivo [61], leading to the suggestion that some of the 'oestrogenic' effects observed in wildlife are, in fact, caused by the anti-androgenic nature of some chemicals [61]. The general message that is slowly appearing is that many EDCs may well have multiple endocrine activities (i.e. a chemical could have oestrogenic and also anti-androgenic activities [62]), and hence presumably have multiple mechanisms of action in vivo, though this has yet to be demonstrated (and will be difficult).

Besides interacting directly (as agonists or antagonists) with steroid hormone receptors, and thus influencing the rates of transcription of the many genes regulated by steroid hormones, many EDCs probably also exert more indirect effects on reproduction. For example, it is likely that many of the enzymes involved in the biosynthesis of steroid hormones are themselves affected by EDCs. These effects could include altering the rates of expression of the genes coding for the enzymes (and hence affecting the levels of the enzymes), and acting as surrogate substrates for these enzymes (if chemicals with endocrine activity can mimic the structures of natural sex steroids, then they can presumably also act as substrates for enzymes that play roles in the biosynthesis of steroids). Although there are limited data available presently to support my contention that EDCs will have multiple mechanisms of action, especially in vivo, it has already been shown that nonylphenol affects a number of enzymes of the cytochrome P450-dependent monooxygenase system [63], which plays a central role in the oxidative metabolism or biotransformation of a wide range of foreign and endogenous compounds.

With so many potential mechanisms of action, it will not be an easy task to clarify exactly how a particular chemical, let alone a mixture of chemicals, causes a particular effect, or suite of effects; in the case of fish exposed to very complex, ill-defined, mixtures of chemicals, it may well never be possible to be certain exactly what chemicals cause the effects, or how they do so.

6
Conclusions

Our understanding of endocrine disruption in the aquatic environment has progressed considerably in the last few years, and is likely to improve further in the foreseeable future, as the intensive research activity underway presently in many countries comes to fruition. It is also likely that the ongoing research efforts will raise (possibly unexpected) new issues, which will in turn need exploring. I suspect that a great deal more research will be required (particularly in areas relatively uninvestigated presently, such as endocrine disruption in invertebrates) before it is possible to answer the many important questions that have arisen. Presently, the list of things that we do not know is much longer than the list of things we do know! The most important of these unanswered questions is whether or not endocrine disruption leads to population-level effects. In some cases it undoubtedly does; the example of TBT and its effects on molluscs should serve as a constant reminder that unexpected effects can be devastating. However, our understanding of the major factors (such as disease, habitat loss and over exploitation) controlling population numbers of most organisms (e.g. fish) is often very poor, and hence it will not be an easy task to put the importance of endocrine disruption into context with these other factors. Our aim should be to have enough understanding and knowledge to be able to prevent catastrophes, such as the complete elimination of some mollusc populations by TBT.

Endocrine disruption research is already having broad impact, and this is likely to increase. For example, the realisation that TBT causes imposex in many species of molluscs, leading ultimately to population-level declines, has led already to severe restrictions on the use of antifoulants containing TBT, and a worldwide ban on their use now looks, not before time, to be achievable. The rapidly increasing body of knowledge about the oestrogenic effects of nonylphenol (and related chemicals) has led already to a reassessment of the hazard posed by these chemicals to aquatic wildlife, and this is likely to lead to further restrictions in the permitted uses of the parent alkylphenol polyethoxylates from which nonylphenol derives. In other cases, where endocrine disruption is caused by natural chemicals (e.g. oestrogens excreted by humans and farm animals) or synthetic chemicals of profound importance, and where no obvious substitute exist (e.g. ethinyl oestradiol), the only feasible regulatory approach is to improve waste treatment processes, and hence increase the degradation of chemicals prior to their release (in effluents) to the aquatic environment. Thus, one of the outcomes of our rapidly increasing understanding of endocrine disruption will be further pressure to improve waste water treatment processes (this will have the added advantage of lowering the concentrations of other, non-endocrine active, chemicals entering the aquatic environment). This situation of regulatory action following conclusive evidence of adverse effects on wildlife caused by EDCs is reminiscent of the processes that followed the realisation many years ago that many chlorinated pesticides (e.g. PCBs and DDT) posed significant risks to wildlife, and in this regard endocrine disruption is no different from many other ecotoxicological issues.

Acknowledgement. I thank all my colleagues and students who have done all of the research which has come from my laboratory.

7
References

1. Bortone SA, Davis WP (1994) Bioscience 44: 165
2. Jobling S, Nolan M, Tyler CR, Brighty G, Sumpter JP (1998) Environ Sci Technol 32: 2498
3. Guillette LJ Jr, Gross TS, Masson GR, Matter JM, Percival HF, Woodward AR (1994) Environ Health Perspect 102: 680
4. Fry DM (1995) Environ Health Perspect 103(7): 165
5. Ashfield LA, Pottinger TG, Sumpter JP (1998) Environ Toxicol Chem 17: 679
6. Blaber SJM (1970) Proc Malacol Soc Lond 39: 231
7. Minchin D, Oehlmann J, Duggan CB, Stroben E, Keatinge M (1995) Mar Poll Bull 30: 633
8. Smith BS (1971) Proc Malacol Soc Lond 39: 377
9. Smith BS (1981) J Appl Toxicol 1: 15
10. Smith BS (1981) J Appl Toxicol 1: 22
11. Smith BS (1981) J Appl Toxicol 1: 141
12. Matthiessen P, Gibbs PE (1998) Environ Toxicol Chem 17: 37
13. Bryan GW, Gibbs PE, Hummerstone LG, Burt GR (1986) J Mar Biol Assoc UK 66: 611
14. Short JW, Rice SD, Brodersen CC, Stickle WB (1989) Mar Biol 102: 291
15. Baldwin WS, Graham SE, Shea D, Le Blanc GA (1997) Environ Toxicol Chem 16: 1905
16. Howell WM, Black DA, Bortone SA (1980) Copeia 1980: 676
17. Drysdale DT, Bortone SA (1989) Bull Environ Contam Toxicol 43: 611
18. Hegrenes S (1999) Copeia 1999: 491
19. Munkittrick KR, Van der Kraak GJ, McMaster ME, Portt CB, Van den Heuvel MR, Servos MR (1994) Environ Toxicol Chem 13: 1089
20. Karels AE, Soimasuo M, Lappivaara J, Leppanen H, Aaltonen T, Mellanen P, Oikari AOJ (1998) Ecotoxicology 7: 123
21. Purdom CE, Hardiman PA, Bye VJ, Eno NC, Tyler CR, Sumpter JP (1994) Chem Ecol 8: 275
22. Sumpter JP, Jobling S (1995) Environ Health Perspect 103(7): 173
23. Harries JE, Sheahan DA, Jobling S, Matthiessen P, Neall P, Routledge EJ, Rycroft R, Sumpter JP, Tylor T (1996) Environ Toxicol Chem 15: 1993
24. Harries JE, Sheahan DA, Jobling S, Matthiessen P, Neall P, Sumpter JP, Taylor T, Zaman N (1997) Environ Toxicol Chem 16: 534
25. Folmar LC, Denslow ND, Rao V, Chow M, Grain DA, Enblom J, Marcino J, Guillette LJ Jr (1996) Environ Health Perspect 104: 1096
26. Wahli T, Meier W, Segner H, Burkhardt-Holm P (1998) Histochem J 30: 753
27. Larsson DGJ, Adolfsson-Erica M, Parkkonen J, Pettersson M, Berg AH, Olsson PE, Forlin L (1999) Aquatic Toxicol 45: 91
28. Hutchinson P (1983) J Fish Biol 23: 241
29. Wiklund T, Lounasheimo L, Lom J, Byland G (1996) Dis Aquatic Org 26: 163
30. Lye CM, Frid CLJ, Gill ME, McCormick D (1997) Mar Poll Bull 34: 34
31. Allen Y, Scott AP, Matthiessen P, Haworth S, Thain JE, Feist S (1999) Environ Toxicol Chem 18: 1791
32. Ouellet M, Bonin J, Rodrigue J, Des Granges JL, Liar S (1997) J Wildl Dis 33: 95
33. La Clair JJ, Bantle JA, Dumont J (1998) Environ Sci Technol 32: 1453
34. Woodward AR, Percival HF, Jennings ML, Moore CT (1993) Fla Sci 56: 52
35. Semenza JC, Tolbert PE, Rubin CH, Guillette LJ Jr, Jackson RJ (1997) Environ Health Perspect 105: 1030
36. Guillette LJ Jr, Pickford DB, Grain DA, Rooney AA, Percival HF (1996) Gen Comp Endocrinol 101: 32
37. Grain DA, Guillette LJ Jr, Rooney AA, Pickford DB (1997) Environ Health Perspect 105: 528

38. De Solla SR, Bishop CA, Van der Kraak G, Brooks RJ (1998) Environ Health Perspect 106: 253
39. Desbrow C, Routledge EJ, Brighty GC, Sumpter JP, Waldock M (1998) Environ Sci Tech 32: 1549
40. Panter GH, Thompson RS, Beresford N, Sumpter JP (1999) Chemosphere 38: 3579
41. Routledge EJ, Sheahan DA, Desbrow C, Brighty GC, Waldock M, Sumpter JP (1998) Environ Sci Tech 32: 1559
42. Belfroid AC, Van der Horst A, Vethaak AD, Schafer AJ, Rijs GBJ, Wegner J, Cofino WP (1999) Sci Total Environ 225: 101
43. Stumpf M, Ternes TA, Iiaber K, Baumann W (1996) Vom Wasser 87: 251
44. Ternes TA, Stumpf M, Mueller J, Haberer K, Wilken R-D, Servos M (1999) Sci Total Environ 225: 81
45. Jobling S, Reynolds T, White R, Parker MG, Sumpter JP (1995) Environ Health Perspect 103: 582
46. Jobling S, Sheahan DA, Osborne JA, Matthiessen P, Sumpter JP (1995) Environ Toxicol Chem 15: 194
47. Aoukure A, Knudsen FR, Goksoyr A (1997) Environ Health Perspect 105: 418
48. Miles-Richardson SR, Pierens SL, Nichols KM, Kramer VJ, Snyder EM, Snyder SA, Render JA, Fitzgerald SD, Giesy JP (1999) Environ Res 80: S122
49. Gimeno S, Komen H, Venderbosch PWM, Bowmer T (1997) Environ Sci Technol 31: 2884
50. Christiansen T, Korsgaard B, Jespensen A (1998) J Exp Biol 201: 179
51. Gray MA, Metcalfe CD (1997) Environ Toxicol Chem 16: 1082
52. Nimood AC, Benson WH (1998) Aquat Toxicol 44: 141
53. Crews D, Wibbels T, Gutzke WHN (1989) Gen Comp Endocrinol 76: 159
54. Bergeron JM, Crews D, McLachlan JA (1994) Environ Health Perspect 102: 780
55. Willingham E, Crews D (1999) Gen Comp Endocrinol 113: 429
56. Kloas W, Lutz 1, Einspanier R (1999) Sci Total Environ 225: 59
57. Panter GH, Thompson RS, Sumpter JP (1998) Aquatic Toxicol 42: 243
58. Tyler CR, Van Aerle R, Hutchinson TH, Maddix S, Trip H (1999) Environ Toxicol Chem 18: 337
59. Gimeno S, Gerritsen A, Bowmer T, Komen H (1996) Nature 384: 221
60. Soto AM, Sonnenschein C, Chung KL, Fernandez MF, Olea N, Serrano FO (1995) Environ Health Perspect 103(7): 113
61. Kelce WR, Wilson EM (1997) J Mol Med 75: 198
62. Sohoni P, Sumpter JP (1998) J Endocrinol 158: 327
63. Arukure A, Forlin L, Goksoyr A (1997) Environ Toxicol Chem 16: 2576

Subject Index of Volume Part I and Part II

Normal page numbers refer to Part I, bold page numbers refer to Part II.

A

activator function (AF)
–, AF-1 5, 10, 17
–, AF-2 9–11, 16–17
activator protein 1 (AP-1) 18, 19
additive effects 43, 49–53
adenosine triphosphate **114**
adenylate cyclase **113**
adjuvant therapy **197**
AF, see activator function
African American 31, 34, 36–40, 42, 44, 46,
    47, 55, 59, 60, 141, 145, 147, 148
age 46, 47, 53, 61
agonist **116**
albumin 7, 140
alcohol use **31, 47**
alkylphenol polyethoxylates 75, 129,
    133–137, **287**
alkylphenols 75, 129, 132, 137–139, **286**
alligators 41, 132, **233, 278**
alpha-fetoprotein **86**
5-alpha reductase **84**
American alligator (*Alligator
    mississippiensis*) **259**
American mud snail (*Ilyanassa obsoleta*)
    273
aminoglutethimide **217**
amphibians **277, 283**
amsonic acid 79, **216, 220**
anabolic 32, 34, 47, 60,
androgen 131–132, 149
–, action 43
–, receptor 32, 34, 38, 40, 41–42, 43–45, 47,
    51, **52–55**, 54, **59–61**, 164
–, response element 46
androstanediol 35, 36, 39–42, 44, 45
androstenedione 7, 35, 38–41, 43, 45
androsterone 35, 36, 39–41
aneuploidy **193, 202, 203**
angiogenesis **79**
anogenital distance 47–48
antagonist **116**

antiandrogens 41, 46, 49, 51, 164, **220**
antiestrogenicity 15
antiestrogens 12, 14, 17–20
–, triphenylethylene-type 72
–, HO-PCBs 159, 161
antihormones 50–53
antioxidant 102, 114, 120
antithyroid 58
AP-1, see activator protein 1
apigenin 69
apolipoprotein-B **93**
apoptosis **79, 84**
aromatase 6, 45, 53, 56, 61, 86, 94, 112, **274**
arteriosclerosis **95**
Asian American 34, 36–38, 40
assays
–, for androgen receptor agonists and
    antagonists 54
–, cell proliferation assay 33–34
–, competitive binding assay 28–29
–, DNA binding and transcriptional activa-
    tion assay 55
–, Hershberger assay 42, 58
–, mammalian-based reporter gene assay
    31–33
–, N/C interaction assay 55
–, phage display assay 34
–, yeast-based steroid receptor assay
    30–31
astringin 69
atherosclerosis **91, 92**
Atlantic croaker (*Micropogonas undulatus*)
    228
Atlantic salmon 139
azoxymethane (AOM) **85**

B

BADGE, see BPA diglycidylether
BBP, see butylbenzylphthalate
bakelite 139, 149
behavior 131

beluga whale **224**
benchmark dose **121**
benign prostatic hyperplasia (BPH) **30, 42**
benzylbutylphthalate, see butylbenzylphthalate
beta-resorcylic acid lactone **98**
BHP, see benign prostatic hyperplasia
biochanin A **72, 73**
–, chemical structure **67, 104**
–, concentration in animal feed and
  silage **109**
–, exposure of domestic animals **108**
–, metabolism **104–106**
biologically-based models **121**
birds **132**
2,2-bis(p-chlorophenyl)-1,1,1-trichloro-
  ethane (p,p'-DDT) **41, 44, 46, 48, 77,**
  131–132, 134, 162, 216, 219, 231, 234, 251,
  253, 287
–, estrogenic action **260, 261**
2,2-bis(p-chlorophenyl)-1,1-dichloro-
  ethylene (p,p'-DDE) **41, 47–48, 58, 131,**
  156, 225, 231–233, 286
–, anti-androgenic action **261**
–, binding to androgen receptor **46, 164**
–, chemical structure **44, 77**
–, environmental levels **165**
–, inhibition of AR DNA binding **46**
–, inhibition of AR N/C interaction **57**
2,2-bis(p-hydroxyphenyl)-1,1,1-trichloro-
  ethane (HPTE) **41, 48–49, 52**
–, binding to androgen receptor **46**
–, chemical structure **44, 77**
–, inhibition of AR DNA binding **46**
–, inhibition of AR N/C interaction **57**
bisphenol A (BPA) **133, 152, 154, 183, 190,**
  202–204, 284
–, chemical structure **75**
–, derivatives **141–144**
–, diglycidyl ether (BADGE) **75, 141–142,**
  144
–, diglycerolmethacryate (Bis-GMA)
  141–143
–, dimethacrylate (BPA-DMA) **75, 142**
–, inhibition of AR N/C interaction **57**
–, triethyleneglycol dimethacryate
  (TEDGMA) **141–142**
–, in utero exposure **145–146, 148**
bleach kraft mill effluent (BKME) **258, 259**
–, and sex steroids in White suckers **259**
body mass index **32, 47**
bone **71, 89, 93–95, 142–143**
– mass **90, 95**
BOP, see N-nitroso-bis-(oxopropyl)amine
BPA, see bisphenol A

BPA diglycidylether (BADGE)
–, chemical structure **75**
–, occurrence **141–142**
–, pharmacokinetics in mice **144**
BPA-dimethacrylate (BPA-DMA)
–, chemical structure **75**
–, occurrence **142**
BPA-DMA, see BPA-dimethacrylate
BRCA1 gene **2, 30**
BRCA2 gene **2, 30**
breast **16**
breast cancer **2, 89, 164, 197**
–, risk factors **3**
breast cell lines **16**
breast development, premature **134**
breast milk **74, 95, 96**
brown trout (*Salmo trutta*) **275**
butylbenzylphthalate (BBP) **57, 78**
butylphenol **57**

C
C_{18} steroids **7**
C_{19} steroids **7**
cadmium **51**
cancer risk assessment **119**
Cape Cod, Massachusetts **135–136, 143**
cardiovascular system **94**
carbaryl **149**
carbohydrates **31**
carp (*Cyprinus carpio*) **138–139, 259, 275**
β-carotene **32**
casein **91, 94, 96**
catechin **69**
catechol estrogens **58, 191**
catechol-O-methyltransferase (COMT) **18,**
  191, 204
Caucasians **145**
CD-1 mice **196, 197**
CD-1 rats **202**
cell lines
–, CHO **57**
–, COS **54**
–, CV1 **55**
–, HepG2 **32–33, 158**
–, LNCaP **33**
–, MCF-7 **33–34, 48, 187, 189, 158**
–, MDA-MB-231 **158**
–, ZR-75 **187–188**
cell proliferation **16, 50, 53, 54, 56, 57, 60,**
  131–132, 139, 140–142, 147, 149
cell proliferation assay **33–34**
channel catfish (*Tetalurus punctatus*) **275**
chaperone **8**
chemical mixtures **135, 149**

childbirth 2
Chinese hamster ovary cells 228
Chinese hamster V79 cells 194, 199, 203
Chinese men 147
chlordecone (Kepone) 149, 216, 219, 253
–, chemical structure 77
–, estrogenic activity 163
cholesterol 90–93
2-(o-chlorophenyl)-2-(p-chlorophenyl)-1,1,1-trichloroethane (o,p'-DDT) 131–132, 231
–, chemical structure 77
–, estrogenic activity 163
–, inhibition of AR N/C interaction 57
CHO cells 57
clomiphene 72
clover 74, 76, 89, 98
– disease 218
co-activator 10–11, 16–17, 20, 51
cognitive function 95
Collaborative Perinatal Project 148
COMT, see catechol-O-methyltransferase
competitive binding assay 28–29
contaminants
–, and evolution 264
–, and IQ 264
–, and penis size 259
–, and sex ratios 265
–, and signal modification 266
–, mixtures 263
co-repressor 10, 20, 51
corpus luteum 90–91
corticosteroid-binding globulin 112, 140
corticosterone 259
COS cells 54, 229
coumestrol 73, 190, 194
CREST antibodies 196
cross-talk 114, 158
cryptorchidism 141, 142, 144, 145, 147–149, 160
coumestans 217
–, chemical structure 68
–, as a phytoestrogen class 103, 106
coumestrol
–, biological effects 115–117, 123–124
–, chemical structure 68, 104
–, concentration in plant foods 106
–, occurrence 104
CV1 cells 55, 228, 229
cyclic AMP 113
CYP, see cytochrome P450
cyproterone acetate
–, antiandrogenic activity 57
–, chemical structure 44, 73

cytochrome P450 (CYP) 16, 112, 197, 200, 203, 239, 286
–, CYP1A1 17
–, CYP1A2 17
–, CYP1B1 17, 191
–, CYP3A4 17
–, CYP3A5 17
–, CYP17 34, 37, 42, 44, 55
cytokine 91

D
daidzein 7, 72, 73, 194, 195
–, in breast milk 113
–, and cardiovascular disease 121
–, chemical structure 67, 104
–, concentration in animal feed and silage 109
–, exposure of domestic animals 108
–, occurrence and metabolism 104–106
danazol 217
DAS, see 4,4'-diaminostilbene-2,2'-disulfonic acid
DBCP, see dibromochloropropane
DBD, see DNA binding domain
o,p'-DDT, see 2-(o-chlorophenyl)- 2-(p-chlorophenyl)-1,1,1-trichloroethane
p,p'-DDE, see 2,2-bis(p-chlorophenyl)-1,1-dichloroethylene
p,p'-DDT, see 2,2-bis(p-chlorophenyl)-1,1,1-trichloroethane
degradation
–, aerobic 135–137
–, anaerobic 135, 137
dehydroepiandrosterone (DHEA) 6, 35, 36, 38–40, 45
– sulfate 91
demographic differences 164
DES, see diethylstilbestrol
O-desmethylangolensin 67
detergents 134, 148
DHEA, see dehydroepiandrosterone
DHT, see 5α-dihydrotestosterone
4,4'-diaminostilbene-2,2'-disulfonic acid (DAS) 149, 220
dibromochloropropane (DBCP) 79, 134, 149, 216, 219
dibutylphthalate 57–58
di-n-butyl phthalate 230, 235
dicarboximide fungicides 47
dicofol 234
dieldrin 231
–, chemical structure 77
–, activity in E-screen assay 164
dietary factors 31, 32, 55

dietary fiber 31
dietary habits 31, 47
di-(2-ethylhexyl) phthalate 78, **231**, **235**
diethyl phthalate **235**
diethylstilbestrol (DES) 1, 3, 11–12, 15–16,
     18, 49, 52, 56, 57, 59, **61**, 73, 133–135, 143,
     144, 150, 153, 159, 160, 173, 183, 190, 196,
     200, 201, 216, 217, 252, 256, 260
–, chemical structure 72, 103
–, DES lineage 181, 182
–, effects in females
–, –, cancer 174, 175
–, –, infertility 174, 175
–, –, leiomyoma 178
–, –, subfertility 174, 175
–, –, uterine tumors 178, 179
–, –, uterine carcinoma 178, 179
–, –, vaginal adenocarcinoma 176, 178
–, effects in males
–, –, cancer 174
–, –, infertility 174, 175
–, –, testes 174, 175
–, –, hypospadias 174
–, –, rete testes adenocarcinoma 176
–, –, multigenerational 182
–, –, tetrafluoro-DES 183
–, structural similarity with resveratrol
     106
–, in utero exposure 132, 146
dienestrol 72
dietary estrogens 111
5α-dihydrotestosterone (DHT) 33–41,
     43–45, **43, 45, 46, 49**, 53–55, 59, 154
–, chemical structures 71
–, transcriptional activation 52, 55
3,2′-dimethyl-4-aminobiphenyl (DMAB)
     50, 51, 57
7,12-dimethylbenzo[a]anthracene (DMBA)
     18, 81
diisononyl phthalate **236**
dimethyl phthalate **236**
dioctyl terephthalate **236**
DMAB, see 3,2′-dimethyl-4-aminobiphenyl
DMBA, see 7,12-dimethylbenzo[a]an-
     thracene
DNA adducts **54, 57, 58**, 191, 197–199, 202
DNA binding and transcriptional activation
     assay 55
DNA binding domain (DBD) 4–5, 9, 11
DNA damage **17**, 194
DNA repair 13
DNA strand breaks 17, **194**, 196
DNA synthesis 9
dog 29, 47
dogwhelks (*Nucella lapillus*) **273**

dopamine **124**
drinking water wells 134–135, 143

E
EACs, see endocrine active compounds
ectopic testis 47
EDCs, see endocrine disrupting chemicals
EDSTAC, see Endocrine Disruptor
     Screening and Testing Advisory
     Committee
EGF, see epidermal growth factor
eggshell thinning **232**
endocrine active compounds (EACs) 1, 12,
     14–15, 18–20, **115**
endocrine disruption 252
endocrine disrupting chemicals (EDCs)
     172, 184, 252, 253, 256, 263
Endocrine Disruptor Screening and Testing
     Advisory Committee (EDSTAC) 41,
     58–59
endometrial cancer **193**
endometrial hyperplasia 13
endometrium 132, 139
–, cancer 89
–, hyperplasia 89
endosulfan
–, activity in E-screen assay 164
–, chemical structure 77
enterodiol 73, **194**, 195
–, chemical structure 68
–, formation from plant lignans 103,
     105–106
–, production from plant food 105
enterolactone 73, **194**, 195
–, chemical structure 68
–, formation from plant lignans 103,
     105–106
–, production from plant food 105
Environmental Protection Agency (EPA)
     41, 49, 54, 58–59
epicatechin 69
epidermal growth factor (EGF) 3, 11–12,
     84
epididymis 45, 139, 146, **152, 154**
epigenetic 180, 182
epiphyseal closure 87, 90
epithelial cells 53, 55, **60**
epoxy resin 141, 143
equilenin 65, **190**, 194
equilin 65, **190**, 194
equine estrogens **193**
equol 72, 73
–, antioxidant effects 121
–, chemical structure 67

–, formation from daidzein 104, 106
erbB2/neu receptors **84**
ER, see estrogen receptor
ERE, see estrogen response elements
ERKO, see estrogen receptor knockout
E-screen 132, 140, 164, **253**
estradiol 1, 7, 18, 83, 117, 119
17β-estradiol **6, 7, 32, 33, 35, 36, 40, 41, 43,
    48, 49, 52, 53, 55–58, 60, 72, 73,** 145, 147,
    155, **158, 159,** 183, **189,** 228, 279, 281, 284
–, chemical structure 65, 103
–, 17α-ethinyl **189**
–, 2-fluoro **189**
–, 2-hydroxy 19, **32,** 58, 183, 191
–, 4-hydroxy 19, 58, 183, 191
–, hydroxysteroid dehydrogenase **6**
–, importance for development and health
    114
estriol 65, 83
estrogen 1, **2,** 3, 7, 9, 12, 17–20, **30–32,** 35,
    39, 44–47, 52, 53, 55–61, 112, 182, 183,
    184
–, estrous cyclicity 131, 146
–, estrus 146–147
–, excretion 84
–, gynecomastia 131
–, reference values 96
–, replacement therapy 95
–, resistance 86
–, stilbene-type 71
–, synthesis
–, – in adipose tissue 88
–, – in ovary 88
–, – in placenta 91
–, synthesis of transport proteins 92
–, target organs 85
–, ultrasensitive assay 96
–, uterine weight 142, 145, 147
–, vaginal cornification 147
–, vaginal opening 146
estrogen binding proteins
–, α-fetoprotein 140
–, albumin 7, 140
–, corticosteroid-binding globulin 140
–, sex hormone-binding globulin 84, 140
estrogen receptor (ER) 3, 42, **55–58, 61,**
    112, 139–140, 157, **181**
–, alpha (ERα) 85–86, 131, 161
–, beta (ERβ) 85–86, 131, 161
–, ERα 3, **11, 181**
–, ERβ 3, **15, 181**
–, knockout mice 85–86
estrogen receptor knockout (ERKO)
–, αERKO 7
–, βERKO 7

estrogen replacement therapy **193**
estrogen response element (ERE) **3,** 9–11,
    17–18, **113,** 140
–, consensus ERE (cERE) 9, 11, 17, 20
–, nonconsensus ERE (nERE) 11, 17, 20
estrogenicity 15, 155
estrone **32, 34, 35, 40,** 53, 65, 83, **189,** 279
–, 2-hydroxy 17, 58
–, 4-hydroxy 17, 58
– sulfatase **6**
estrus cycle model **123**
17α-ethinylestradiol 72
ethinyltestosterone 74
ethinyl estradiol 149, **218,** 280, 281, 284,
    287
ethisterone, see ethinyltestosterone
ethylene dibromide 134, 149
ethylene glycol ethers 149
etiocholanolone 37
European Committee for Food 142
European moles 223

F
familial aggregation **60**
familial disposition **46**
farming 31, **46, 59**
fat **31, 32, 47, 58, 59**
– head minnow (*Pimephales promelas*)
    228, 260, 285
female genital tract 131
–, endometrium 132, 139
–, ovary 131, 139
–, uterus 146–147, 149
–, vagina 132, 146–147, 149
feminization 132
fenarimol 79, **240**
Fenholloway river **220**
fetal adrenal gland 92
fetal development 45, 163
fetal positioning effects 131, 145–146
α-fetoprotein 140
Fischer 344 rats **202, 203**
fish
–, Atlantic salmon 139
–, carp 138–139
–, feminization 132
–, Japanese medaka 139
–, mosquitofish 57
–, rainbow trout 137–138
flavonoids
–, chemical structures 69
–, estrogenic activity 106
flaxseed **197**
–, biological effects 115–118, 122–124

flaxseed
–, as source of lignans 103, 105, 111
flounder (*Platichthys flexus*) 277
flutamide 73, **227**
fluorescence polarization 29
follicle-stimulating hormone (FSH) **88, 89,**
    112, 146, 149, **150**
follicular phase **88,** 90
Food Quality Protection Act 41
formononetin 72, 73
–, chemical structure 67, 104
–, concentration in plant foods 105 – 106
–, in livestock feed and silage 108 – 109
fos 79, 113
free radicals **16**
frogs 277
FSH, see follicle-stimulating hormone
fungal mycotoxins 216, 217
fungicide 46
Fusarium spp. toxins 101, 107, 109

G
galactorrhea 93
gene mutations **202**
genetic 180, 182
–, polymorphisms **6**
genistein 14, 18, **183,** 190, 194, 195
–, biological effects 115 – 121, 123 – 124
–, in breast milk 113
–, chemical structure 67, 104
–, concentration in animal feed and silage
    109
–, exposure of domestic animals 108
–, intake in Japan and USA 112
–, occurrence and metabolism 104 – 106
genital tract malformations 143
genotoxic 58, 61, 189
gerbils 155, 156, **222**
glycitein 67
GnRH, see gonadotropin releasing hormone
goiter **100**
gonadotropin (GtH) **89, 90, 98,** 253
–, GtH-II 258
gonadotropin releasing hormone (GnRH)
    87, 112, 253
groundwater 137
growth 87
growth factors **9**
guaiaretic acid 68
gynecomastia 88, 90

H
HDL, see high density lipoprotein
heat-shock protein (hsp) 8
heavy metals **259**
hepatocellular carcinoma **197**
hepatocytes 197, 198
HepG2 cells 32 – 33, 158
2,2', 3,4', 5,5', 6-heptachloro-4-biphenylol 163
hereditary prostate cancer **30**
hermaphrodite **275**
Hershberger assay 42, 58, **230**
hexestrol 72, **183**
high density lipoprotein (HDL) **90,** 92
HO-PCB3, see 2', 4', 5'-trichloro-4-bi-
    phenylol
HO-PCB4, see 2', 3', 4', 5'-tetrachloro-4-
    biphenylol
hormonal imprinting 58, 59
hormone replacement therapy (HRT) 8, 90,
    94
hot flashes **88,** 89
4-HPR, see N-(4-hydroxyphenyl)-retinamide
hprt gene locus 194, 196
HPTE, see 2,2-bis(p-hydroxyphenyl)-1,1,1-
    trichloroethane
HRT, see hormone replacement therapy
HSD, see hydroxysteroid dehydrogenases
HSD3B2 gene 38
hsp, see heat-shock protein
HT29 cell line **196**
human breast 8, 191
human lymphocytes **196**
human sexual differentiation **217**
hydrocele **141**
8-hydroxydeoxyguanosine **57**
16α-hydroxyestrone 65, **191**
17α-hydroxylase **154**
hydroxyflutamide 44, 47 – 48, 56 – 57
hydroxy-PCBs 155
4-hydroxytamoxifen (OHT) 15 – 18, 72
hydroxysteroid dehydrogenase (HSD) 8, 34,
    38, 55, 84, 189
hydroxysteroid sulfotransferases 199
hypercholesterolemic **92,** 93
hypospadias 41, 47, 54, 141, 144, 145, 148, 160
hypothalamic pulse generator 87
hypothalamus 253, 254, 258
hypothyroidism **100**

I
ICI164,384 72
ICI182,780 12, 15, 17 – 19, 72
IGF, see insulin-like growth factor
imposex **273, 287**

imprinting 87, **182**
inducible NO synthase **85, 91**
industrial accidents **148**
infant formula 101, 113
inhibin 146
insulin-like growth factor (IGF) 3, 11 – 12,
     18 – 19
interactions
–, additive 43, 49 – 53
–, synergistic 49, 53
intermediate metabolites **16**
intersex in wildlife 222, 277
intersexuality 274, 276, 283
in utero exposure
–, to BPA 145 – 146, 148
–, to DES 132, 146
–, fetal positioning effects 131, 145 – 146
inverted-U dose response 157, **159**
iodine **100**
ionizing radiation **31, 51**
ipriflavone **73, 94, 95,** 106
–, biological effects 122 – 124
–, chemical structure 107
iprodione 57
isoflavones 14 – 15, 101, **196, 260**
–, anticarcinogenic activity 118
–, chemical structures 67, 104
–, exposure 111 – 113
–, occurrence and metabolism 104 – 106
isoflavonoids **217**
isolariciresinol 68

J
Jackfish Bay **221**
Japanese medaka 139
Japanese quail **228**
jun **79,** 113

K
kaempferol 69
kelp bass (*Paralabrax clathratus*) **228**
kelthane 57
kepone **134**
ketoconazol 79, **239**
Ki67 **10**
killifish **221**
kraft mill effluent **216, 220, 274, 275**

L
Lake Apopka **214, 231, 233, 234, 278**
Lake Woodruff **234, 278**
large mouth bass (*Micropterus salmoides*)
     259

lariciresinol 68
larval frogs (*Xenopus laevis*) **284**
LBD, see ligand binding domain
LDL, see low density lipoproteins
Leydig cells 146, **153**
lemmings **223**
leukemia **196, 202**
LH, see luteinizing hormone
life-style **31, 47, 59**
ligand binding domain (LBD) 4 – 5, 9, 11,
     16
lignans 14 – 15, 101, **194, 260**
–, anticarcinogenic activity 118
–, chemical structures 68, 103
–, exposure 111 – 113
–, occurrence and metabolism 103 – 106
linuron **79, 229**
lipid peroxidation **57, 58, 191, 200**
lipoprotein **91, 93**
– a **90, 93**
liver
–, metabolism of steroids **259**
–, metabolism of steroids in alligator **260**
–, metabolism of steroids in fish **260**
–, sexual dimorphism **259**
LNCaP 33
lobules types **10**
Lobund Wistar **48, 51**
low density lipoproteins (LDL) **90 – 93**
low dose effects 145 – 146, 148 – 149
luteal phase **88,** 90
luteinizing hormone (LH) **88, 89**
17,20-lyase **37, 43, 44**

M
male genital tract 131
–, epididymis 45, 139, 146
–, feminization 132
–, Leydig cells 146
–, preputial gland 146
–, prostate 45, 47, 54, 132, 139, 145, 149
–, seminal vesicle 45, 47, 49, 139, 147
–, Sertoli cells 146
–, sperm count 47, 54, 131 – 132, 139, 146,
     156, 164
–, testis 131, 139, 146
mammalian androgens 71
mammalian-based reporter gene assay
     31 – 33
mammalian estrogens 65
mammalian lignans 68
mammalian progestins 71
mammary cancer **193**
mammary glands 131, 146 – 147, 149

MAPK, see mitogen-activated protein
    kinase
margin of exposure **121**
matairesinol 68, **73**, **194**, **195**
MCF-7 cells **12**, 33 – 34, 48, 187, 189, 158, **253**
MDA-MB-231 cells 158
MDA-MB-453 cells **230**
medaka (*Oryzias latipes*) **285**
medroxyprogesterone acetate (MPA)
–, chemical structure 44, 74
–, inhibition of AR N/C interaction 56 – 57
megestrol acetate 74
melengestrol acetate
–, chemical structure 74
–, as a growth promotor 73
memory 95
menarche 87
menopausal hot flushes 94
menopause **2**
menstrual cycle **10**
messenger RNA 113
methoxychlor 41, 46 – 47, **182**, **184**, **225**
–, chemical structure 44, 77
–, inhibition of AR N/C interaction 57
–, metabolite 44, 48
4′-methylcoumestrol 68
methyltestosterone 73, **217**
methyltrienolone (R1881) 44, 47, 54, 56
mibolerone 56
microflora 72, 74, 87, 95
micronuclei **194**, **196**
microtubules **194**, **202**, **203**
mifipristone (RU 486) 74
minks **216**
miroestrol **218**
mitogen-activated protein kinase (MAPK)
    11, 18
mitotane **216**, **219**
mitotic spindle **193**, **202**
mixtures 43, 49 – 53
models of action **119**
moldy corn syndrome **98**
mono-(2-ethylhexyl) phthalate **235**
mosquito fish (*Gambusia affinis*) 57, **221**, **274**
MPA, see medroxyprogesterone acetate
Müllerian duct **153**, **161**, **221**
multigenerational studies 58, 147
multiple chemical exposures 49
mutations 30, 43, 54
mycoestrogens 69 – 70, 107 – 111

N
NADPH-cytochrome c reductase **112**
NAF, see nipple aspirate fluid

naringenin 69
National Cholesterol Education Program
    (NCEP) **92**, **93**
NBL rat, see Noble rat
N/C interaction assay 55
NCEP, see National Cholesterol Education
    Program
neonatal estrogen exposure **49**, **59**
neoplastic cell transformation **202**
N-(4-hydroxyphenyl)-retinamide (4-HPR)
    32
nipple aspirate fluid (NAF) **99**
nitric oxide (NO) **91**
p-nitrophenol 149
N-methyl-N-nitrosourea (NMU) **50**, **81**
N-nitrosation **85**
N-nitroso-bis-(oxopropyl)amine (BOP) 50,
    51
NMU, see N-methyl-N-nitrosourea
Noble rat 48, **49**, 55 – 58
non-additive interactions 159
non-monotonic dose response **157**, **159**
nonoxynol 134, 138, 149
nonylphenol 57, 132 – 139, **182**, **274**,
    281 – 284, **286**, **287**
nonylphenol polyethoxylate 133
nordihydroguaiaretic acid 68
norethisterone 74
nulliparous females **10**
nutrition **31**

O
obesity **33**, **47**
occupational factors **31**
octylphenol 57, **151**, **153**, **260**, **284**
octylphenol polyethoxylate 133
Office of Economic Cooperation and
    Development (OECD) 41, 58
OHT, see 4-hydroxytamoxifen
oral contraceptives **144**
orchidopexy **142**
organochlorines 14
orphan receptors **262**
osteoblast **95**
osteoclast **94**
ovary **16**, 131, 139
ovarian follicle **88**
ovariectomy **90**
ovulation **99**
oxandrolone 56
oxidative DNA damage **191**, **194**, **200**
oxidative stress **18**, 53, 58

P
PAHs, see polycyclic aromatic hydrocarbons
paper mills 57
paper mill effluent 220
PCBs, see polychlorinated biphenyls
PCB169 238
PCDDs, see polychlorinated dibenzo-p-dioxins
PCDFs, see polychlorinated dibenzofurans
PCDF, see polychlorinated dibenzofurans
2,2′,3′,4′,6′-pentachloro-4-biphenylol 160
perchloroethylene 149
perinatal exposure 49, 50, 52, 58, 59, 61
permixon 216, 218
peroxidases 200
peroxidase activity 158
pesticides 1, 14, 46, 132–134, 155, **259, 286**
PgR, see progesterone receptor
phage display assay 34
phenol 253
phenol red
–, chemical structure 75
–, estrogenicity 75–76
–, impurity 75–76
phenolphthalein 205
phosphodiesterase 114
phthalates 14, **235, 281, 286**
–, activity in yeast-based assay 187–188, 193
–, chemical structures 78, 172–173, 190
–, effects on reproductive development 169, 183, 197
–, effects on vitellogenin 187, 191–192, 197
–, industrial consumption 171
–, in food packaging 177
–, metabolites 173, 178–183, 188, 193–195
–, monoesters 179–182, 184, 188, 194, 196
–, multigenerational studies 184
–, receptor binding assay 187–189, 195
–, solubility 171, 173, 175, 181, 185–186
–, structure-activity relationship 189
–, testicular toxicity 182
phosphorylation 8
photo-oxidation 144
physical activity 33
physiologically-based pharmacodynamic (PBPD) models 122
physiologically-based pharmacokinetic (PBPK) models 122
phytoantiandrogens 216
phytoestrogens 1, 9, 12, 15, 18, 67–69, 101, 103–107, **118, 194, 216, 217**
phytosterols 221

piceatannol 69
piceid 69
pinoresinol 68
pituitary **89, 90, 254, 258**
placenta 144
plasticisers 171, 173
platelet **91**
platelet-derived growth factor **91**
polar bear 223
polycarbonate plastics 139, 141, 143
polychlorinated biphenyls (PCBs) 15, 77, **133, 156, 162, 169, 204, 216, 219, 237, 259, 283, 287**
–, binding to thyroid hormone binding protein 260
–, chemical structure 77
–, hydroxylated metabolites 157
polychlorinated dibenzo-p-dioxins (PCDDs) 77, 156, **219, 237**
polychlorinated dibenzofurans (PCDFs) 77, 156, **219, 237**
polycyclic aromatic hydrocarbons (PAHs) **259**
polycystic ovarian syndrome **13**
polymorphism 34, 37–39, 42–45, 47, 54, 55
polyvinylchloride (PVC) 149, 173, 176–177, 179, 182
32P-postlabeling **198, 199, 200, 202**
postmenopausal 74, 87–90, 92–95, 99, 193
postmenopause 94
pratensein 67
predators 149
premature thelarche 87
premenopausal **82, 87, 88, 92, 96, 99**
premenopause 93
prenatal exposure **61**
8-prenylnaringenin 69
preputial gland 146
preputial separation 58, **227**
procymidone 47, **228**
–, as antiandrogen 52
–, chemical structure 44, 79
–, inhibition of AR N/C interaction 57
progesterone
–, chemical structures 71
–, receptor (PgR) **10, 132, 139, 142, 158**
prolactin 49, 57, 59, 124, 140, 148
prostate 45, 47, 54, 132, 139, 145, 149
– cancer **140, 141**
– specific antigen (PSA) **30, 84, 141**
prostatic fluid 74, **83, 84**
prostatitis **30, 54, 61**
protein kinase A **113**
protein tyrosin kinase (PTK) **79**
prunetin 67

PSA, see prostate specific antigen
pseudohermaphrodites **223**
pseudohermaphroditism **217**
psoralidin **68**
PTK, see protein tyrosin kinase
PVC, see polyvinylchloride

Q

quinones **191, 194, 200, 202**
quinone methide **191, 198**

R

rainbow trout (*Oncorhynchus mykiss*) **137,
    138, 262, 285**
raloxifene **17, 72**
RBA, see relative binding affinity
RBA-SMA, see relative binding affinity-
    serum-modified access
reactive metabolites **197, 200**
reactive oxygen species (ROS) **57, 58, 61,
    79, 191, 194, 200**
receptor **253**
–, estrogen **263**
–, membrane bound **261**
–, orphan **262, 263**
–, steroid **261**
–, steroidogenic factor-I (SF-1) **262**
–, steroid-xenobiotic (SXR) **262**
red-eared slider turtle **283**
5α-reductase **34 – 37, 39 – 47, 54, 59, 154,
    218, 239**
redox cycling **191, 194**
relative binding affinity (RBA) **29, 77, 160**
relative binding affinity-serum-modified
    access (RBA-SMA) **86**
reptiles **278, 283**
resveratrol **106**
–, chemical structure **69, 107**
–, health effects **119 – 121, 124**
retinoic acid **32, 262, 278**
retinoids **32, 157, 216**
risk assessment **204, 264**
rivers **132, 135 – 137, 144**
RNA polymerase **113**
roach (*Rutilis rutilis*) **276, 277**
ROS, see reactive oxygen species
RU 486, see mifipristone
rubber industry **31**

S

Safe Drinking Water Act **41**
salmon **139**

saw palmetto **218**
secoisolariciresinol **68, 73, 194, 195**
secoisolariciresinol diglycoside
–, as antioxidant **120 – 121**
–, biological effects **115 – 118**
–, chemical structure **103**
–, metabolism **105 – 106**
selective estrogen receptor modulator
    (SERM) **12, 14 – 20**
semen **74, 83**
– quality **135 – 139, 143, 148, 160**
seminal vesicles **45, 47, 49, 139, 148, 154**
semiquinones **191, 200**
SERM, see selective estrogen receptor
    modulator
Sertoli cell **135, 146, 150, 153**
serum **157**
– albumin **86**
Seventh Day Adventist **76, 83, 97**
Seveso **222**
sewage **132 – 133, 135 – 136, 138**
sex differentiation **221**
sex hormone binding globulin (SHBG) **32,
    33, 35 – 37, 40, 41, 84, 86, 90, 112, 140, 253,
    254, 260**
sex ratio **134, 149, 150, 221, 283**
sex reversal **150, 157, 216, 283**
sexual activity **33, 54**
sexual differentiation **43, 45, 48, 140, 146,
    151, 153, 156**
sexual factors **33**
SHBG, see sex hormone binding globulin
SHE, see Syrian hamster embryo
SHR, see steroid hormone receptor
signal modification **266**
skeletal ossification **145**
skeletal system **94**
skin atrophy **95**
β-sitosterol **57**
S-methoprene **278**
smoking **30**
snapping turtles (*Chelydra serpentina*) **279**
soybeans **104 – 105, 113, 118, 120**
sperm count **47, 54, 131 – 132, 139, 146, 149,
    156, 164**
sperm density **135, 138, 139**
sperm production **151, 152**
spermicides **134, 148**
spotted hyaena **223**
SRC-1 **10 – 11, 17**
SRD5A2 gene **37, 42**
steroid hormone receptor (SHR) **3, 4,
    10 – 12**
steroid hormones **253, 254, 264**
steroid receptor RNA activator (SRA) **10**

steroidal estrogens, see mammalian
    estrogens
steroidogenesis pathways 8
steroidogenic factor-I (SF-1) 262
steroid-xenobiotic receptor (SXR) 262
stress 259
–, and heavy metals 259, 262
–, and paper mill effluent 262
–, and pesticides 259
–, and polycyclic aromatic hydrocarbons
    (PAHs) 259
–, and polychlorinated biphenyls (PCBs)
    259, 262
–, and reproduction 262
stromal cells 52, 53, 59, 61
stromal-epithelial interaction 53
structure-activity relationships 159
sulfatase deficiency 86, 93
sulfotransferases 199
synergistic effects 49, 53
Syrian hamster kidney 189, 194, 200, 202
Syrian hamster embryo (SHE) fibroblasts
    193, 202, 203

T

taleranol, see β-zearalanol
tamoxifen 12, 14, 16, 18–20, 71, 72, 73, 78,
    86, 90, 94, 98, 182, 190, 197, 198, 228
TBT, see tributyltin
TCDD, see 2,3,7,8-tetrachlorodibenzo-
    p-dioxin
TeBG, see testosterone binding globulin
testicular atrophy 90
testicular cancer 135, 140–143, 145, 146, 149
testis 131, 139, 146
testosterone 30, 33–41, 43–46, 48–60, 71,
    87, 98, 112, 259
–, biotransformation 260
–, in juvenile alligators 259
–, in BKME-exposed fish 258
testosterone binding globulin (TeBG) 7
2',3',4',5'-tetrachloro-4-biphenylol
    (HO-PCB4) 157–159
3,3',4,4'-tetrachlorobiphenyl 163
3,3',4',5-tetrachloro-4-biphenylol 163
2,3,7,8-tetrachlorodibenzo-p-dioxin
    (TCDD) 77, 216, 222
TGF, see transforming growth factor
thoracic nipples 48
thromboxane 91
thyroid 100
– hormones 42, 151, 157, 262, 284
–, function 163
tier I testing 41, 58

tier II testing 41
topoisomerase 79
toxaphene 164
toxic equivalency factor 49, 51
transforming growth factor (TGF) beta 79
2',4',5'-trichloro-4-biphenylol (HO-PCB3)
    157–159
transthyretin 163
17β-trenbolone 73
tributyltin (TBT) 273, 284, 287
true hermaphrodites 223
tumor xenografts 12
type II hyperlipoproteinemia 92
tyrosine kinase 120

U

U-shaped dose response curves 214
uterine adenocarcinoma 196
uterine defense mechanisms 89
uterotrophic assay 147
uterus 146–147, 149, 191

V

V79 cells, see Chinese hamster V79 cells
vagina 132, 146–147, 149
variance 264
vas deferens 45
vasectomy 32, 33
vegetarians 72, 75
venereal disease 30, 46, 59
ventral prostate 151, 230
very low density lipoproteins (VLDL) 91, 92
vinclozolin 41, 46–47, 225
–, antiandrogenic metabolites 225, 261
–, as antiandrogen 52
–, chemical structure 44, 79
–, inhibition of AR N/C interaction 57
–, metabolites 44, 52
vitamin A 32
vitamin D 262
vitellogenin 135, 138–140, 216, 275, 277,
    280, 285
VLDL, see very low density lipoproteins

W

waste water treatment 134–137, 143–144
water
–, contamination 141, 143
–, drinking water wells 134–135, 143
– fleas 274
–, groundwater 137
–, rivers 132, 135–137, 144

water
–, waste water treatment 134–137, 43–144
white sucker fish (*Catastomus commersoni*)
   221, 258, 259
wildlife 132, 148–149
–, alligators, Lake Apopka 41, 132
–, predators 149
Wistar rats **197**
witch's milk 86
Wolffian duct 45, **154, 221**
wool mills 135, 138

X
xenoestrogens 20, **117,** 163
Xenopus laevis 140, 146

Y
yeast-based steroid receptor assay
   30–31

Z
α-zearalanol 70
β-zearalanol 70
zearalanone 70
α-zearalenol 70
β-zearalenol 70
zearalenone 70, **98, 218**
zebrafish (*Danio rerio*) **285**
zeranol, see α-zearalanol
ZR-75 cells 187–188

Printing (Computer to Film): Saladruck Berlin
Binding: Stürtz AG, Würzburg